Mathematical Modeling of the Healthy and Diseased Lung

Béla Suki • Jason H. T. Bates

Mathematical Modeling of the Healthy and Diseased Lung

Linking Structure, Biomechanics, and Mechanobiology

Springer

Béla Suki
Dept of Biomedical Engineering
Boston University
Boston, MA, USA

Jason H. T. Bates
Larner College of Medicine
University of Vermont
Burlington, VT, USA

ISBN 978-3-031-95321-7 ISBN 978-3-031-95322-4 (eBook)
https://doi.org/10.1007/978-3-031-95322-4

This Springer imprint is published by the registered company Springer Nature Switzerland AG
The registered company address is: Gewerbestrasse 11, 6330 Cham, Switzerland

Dedicated by Béla to his life-long companion and wife Erzsébet, with appreciation and love.

and

Dedicated by Jason to his mother Philippa and the memory of his father Richard, with thanks for everything.

Preface

Since the beginning of the twentieth century, civilization has increasingly relied on innovative technologies to advance the human condition. While this is obvious to the general public, it is perhaps less widely appreciated that the science behind modern technologies almost always involves mathematics. For example, building a bridge requires mathematical optimization of materials, time, and money under the constraints of maintaining the strength, stability, and longevity of the final construct. Similarly, the internal organs in the body are now routinely examined with the aid of magnetic resonance imaging or high-resolution computed tomography, both of which have revolutionized the practice of medicine. Behind the scenes, however, calculus and its offspring, Fourier analysis, are critical for creating the three-dimensional images that allow doctors to locate structural abnormalities and diagnose disease.

The widespread use of mathematics in science and engineering is manifest most profoundly in the creation of mathematical and computational models. A model represents a simplified version of reality that incorporates those elements of a system that are considered important to the purpose at hand. The model is then used as aid in understanding how the system works, or to predict its future behavior. In fact, creating models is a natural human activity. The brain creates internal models of reality whenever it constructs images of the world. In order to catch a ball thrown toward us, our brain makes a prediction of when and where the ball will be at a future point in time, and then it instructs the appropriate muscles to make the catch. The ability to create mental models of reality and use neuronal computing to predict future events is an evolutionary feat that has been critical to our survival as a species. With the development of symbolic representation (first language and then writing), humans acquired the ability to create models of reality outside the brain. The predictions of these models can only be quantitatively precise, however, when they involve mathematical calculations, the success of which is ultimately judged on the basis of their usefulness.

Many important advancements in biology can be traced to the use of quantitative modeling. This applies particularly to the field of physiology with its focus on how biological systems function and is no better exemplified than by the physiology of the lung, the primary topic of this book. When the first morphometric measurements cataloging the structure and dimensions of the pulmonary airway tree appeared in the literature, scientists were able to reduce the complexity of the structure by encapsulating it in simple recursive rules that define, on average, how the tree progresses from its root at the trachea to its distal branches corresponding to the terminal airways. Fluid dynamics equations based on the principles of physics then allowed the computational prediction of how pressure gradients drive airflow through each branch of the structure. These models have greatly increased our understanding of, for example, how breathing difficulties arise when the airways constrict in asthma. Similarly, microscopic images of alveolar structure have led to the formulation of spring network models of the lung parenchyma that have improved our understanding of how emphysema and pulmonary fibrosis progress. These models, and others like them, will be explained in this book as we describe a range of important developments in theoretical and computational modeling related to the lung in health and disease.

How was the idea for this book born? We (BS and JB) have enjoyed a very productive scientific collaboration spanning more than 30 years. We have, both individually and together, developed and published numerous theoretical and computational models related to lung structure and function. These activities accelerated when, seven years ago, we received NIH funding from the Multiscale Modeling Forum to develop computational approaches to the study of emphysema and fibrosis, allowing us to advance the field using combined agent-based/physics-based models. Our modeling ideas and methodologies have also reached the classroom; both of us have taught university courses on the quantitative aspects of the respiratory system using books we have written ("Structure and function of the extracellular matrix: A multiscale quantitative approach," B. Suki, Elsevier, Academic Press, 2021; "Lung Mechanics. An inverse modeling approach," J.H.T. Bates, Cambridge University Press, 2009). These experiences led us to the realization that the field needs a comprehensive treatment of the computational modeling of lung biomechanics and mechanobiology as well as their derangements in disease.

Our primary goal in this book is to present advanced material that will be useful for quantitative scientists and modelers undertaking research into the pathophysiology of lung disease. Accordingly, we will assume that the reader has a working knowledge of calculus, differential equations, and probability theory, the key mathematical areas that computational models are based on. The chapters are written with rigor and include derivations of key formulae from first principles whenever possible. Key physical principles and mechanisms are explained in a general context as well as being applied specifically to the lung. Our intent is not to tell readers what models they should be using, but rather to give them the tools to make their own models. As we point out in Chap. 13, the most useful models are those that eventually lead to their replacement by better ones.

At the same time, we would also like the book to be useful to trainees seeking to develop their biomedical modeling skills, so we have included a small number of problems at the end of each chapter suitable for a graduate level course in pulmonary and biomedical engineering. These problems are meant to stimulate thought and help students gain the kind of insight that can only be acquired by coding models oneself and exploring their behavior as if performing experiments in a virtual laboratory.

The material presented in this book is organized into three main parts.

Part I includes the application of modeling to the healthy lung. We start with a short introduction that warms the reader up to modeling as a discipline in general. Chapter 2 fills the reader in on basic lung anatomy and physiology, including descriptions of the airways and parenchyma along with their resident cells. Chapter 3 covers linear lumped parameter models and their nonlinear extensions. These are the most widely used types of lung model, but they can become very complicated when significant anatomic detail is included. Chapter 4 moves on to the more computationally intensive modeling of parenchymal tissue elasticity from both a continuum and a network perspective, including a discussion of the effects of gravity. Chapter 5, the final chapter of Part I, examines one of the most interesting but often neglected properties of lung tissues, viscoelasticity. Here, we discuss the physical basis of viscoelasticity as well as how to model it by invoking concepts from the realm of complexity theory, a topic that will accompany us throughout the rest of the book.

Part II focuses on the modeling of disease processes and their consequences, beginning in Chap. 6 with a general description of tissue homeostasis using "agents" that represent cells operating according to specified rules of behavior within a virtual landscape representing the parenchymal tissues. This is an exciting area that allows for the exploration of dynamics at the interface between analytic and computational methods. Chapter 7 through 11 then deal, respectively, with the application of computational modeling to the canonical lung pathologies of asthma, pulmonary fibrosis, emphysema, acute lung injury, and diseases related to aberrations in lung defense mechanisms. Part II concludes in Chap. 12 with how modeling can help us understand the changes taking place in the aging lung from a network perspective, including the consequences of gradually drifting homeostatic control of cells and tissues. This should be of interest to all readers because although aging is often not considered to be a disease, many of its consequences have features in common with diseases that can strike at any age.

Part III consists of a single relatively brief chapter providing an outlook on the future of lung modeling as well as modeling in biology and physiology in general. This chapter raises questions related to the usefulness and limits of modeling, and the possible impacts that machine learning might have on the modeling field. We don't profess to have the answers to these questions, but maybe our thoughts will stimulate others to find them.

Finally, while we have tried to be comprehensive and detailed in our presentation of lung modeling in this book, it is impossible to cover everything. For example, while the basic equations of fluid dynamics are invoked for the calculation of airway resistance, we have not delved into the application of computational fluid dynamics to gas flow or aerosol delivery in the airways, neither have we touched on the biophysics of the air-liquid interface. These important fields justly deserve an entire book on their own, as well as being perhaps more relevant for design of therapies as opposed to elucidation of pathophysiology. We have also not specifically covered a number of important lung diseases such as lung cancer, cystic fibrosis, and tuberculosis. While computational modeling has been applied in some fashion to these and other lung diseases, the pathologies we cover here represent those into which modeling has perhaps made the greatest inroads to date.

We would like to thank Merry Stuber, the editor at Springer, for her guidance during the course of the writing as well as for her initial insistence that pushed us over the threshold to actually start writing the book.

Boston, MA, USA — Béla Suki
Burlington, VT, USA — Jason H. T. Bates

Competing Interests BS is a co-founder of and consultant for Mechanobiologix, LLC. JHTB is a scientific advisor for and shareholder in Oscillavent, Inc. and Respiratory Sciences, Inc., and is a consultant for Chiesi USA, Inc. The authors have no competing interests to declare that are relevant to the contents of this book.

Contents

Abbreviations

Chapter 1

V	Gas volume
P	Gas pressure
C	Compliance
R_f	Fluid flow resistance
t	Time
σ	Stress
x	X coordinate of position
R	Ideal gas constant
T	Absolute temperature
n	Number of moles of gas

Chapter 2

d, d_n	Airway diameter
l, l_n	Airway length
n	Generation number
γ	Diameter scaling factor
λ	Length scaling factor
β	Tracheal scaling factor
N	Generation number of terminal bronchioles
O	Horsfield order
Δ	Horsfield order difference
Q	Flow
r	Flow partitioning coefficient
P_{tp}	Transpulmonary pressure
CT	Computed tomography
S_{alv}	Alveolar surface area
V_{alv}	Alveolar volume

a, b	Constants
OCT	Optical coherence tomography

Chapter 3

P	Pressure
P_{el}	Elastic recoil pressure
P_{res}	Resistive pressure
P_0	Baseline pressure at zero flow and pressure
V	Volume
$\dot{V}$	Flow
$\dot{V}_1$	Flow into compartment 1
$\ddot{V}$	Volume acceleration
R_{aw}	Airway resistance
μ	Gas viscosity
r	Radius
l	Length
Re	Reynolds number
ρ	Density
R_L	Lung resistance
R_{tis}	Tissue resistance
R_1, R_2	Pathway resistances in two-compartment model
R_c	Common pathway resistance in parallel two-compartment model
E_L	Lung elastance
I_L	Lung inertance
t	Time
K_1, K_2	Rohrer constants of nonlinear flow resistance
E_1, E_2	Linear and quadratic parameters for nonlinear elastance
A, B, K	Parameters of the Salazar-Knowles equation (Eq. 3.11)
a, b, d	Parameters of sigmoidal pressure-volume relationship (Eq. 3.12)
τ	Time shift
$A_1 - A_6$	Parameters of the integrated two-compartment model equation (Eq. 3.18)
δt	Time increment
A_{in}, A_{out}	Amplitude of system input and output
ϕ_{in}, ϕ_{out}	Phase of system input and output
f	Frequency
$A(f)$, $\phi(f)$	Amplitude and phase of a complex function of frequency
i	Square root of -1
$Z_L(f)$	Lung impedance
$R(f)$	Real part of impedance (resistance)
$X(f)$	Imaginary part of impedance (reactance)

Chapter 4

P	Pressure
V	Volume
a, b, c, d	Constants
P'	Normalized pressure
V'	Normalized volume
F	Force
ΔX	Spring extension
k	Spring constant
$l_{s1}, \ldots, l_{sM}$	Stop extensions
N	Number of spring-string units
Δl_s	Unit stop extension
l_s	Continuous stop extension
v_s	Continuous stop volume
x	Total extension
k_T	Total spring constant
n	Distribution of stop extensions or stop volumes
E_L	Lung elastance
C, D, K	Constants in P-V curve
χ	Constant, $1/KE_L$
$\boldsymbol{T}$	Stress vector
A	Area
σ, σ_{ij}	Cauchy stress tensor and its components
$\boldsymbol{n}$	Outward unit normal vector
x, y, z	Cartesian coordinates
ρ	Density
u_j	jth displacement component
t	Time
$\boldsymbol{u}$	Displacement vector
$\boldsymbol{X}$	Position vector in the undeformed state
$\boldsymbol{x}$	Position vector in the deformed state
ϵ, ϵ_{ij}	Linear strain tensor and its components
C_{ijkl}	Rank-4 elasticity tensor
Y	Young's modulus
ν	Poisson's ratio
δ_{ij}	Kronecker delta
ϵ_G	Green strain tensor
l_0, l	Stress-free and deformed lengths
λ, λ_i	Stretch ratio
F_d	Deformation gradient tensor
J	Jacobian
I	Identity matrix
C_G	Cauchy-Green deformation tensor
R	Pure rotation matrix

U	Pure stretch matrix
I_1, I_2, I_3	Invariants of C_G
e	Strain energy density
D_{mn}, D_{kmn}	Elasticity constants
S_I	First Piola-Kirchhoff stress tensor
S_{II}	Second Piola-Kirchhoff stress tensor
α_i, μ_i	Material constants in Ogden hyperelastic model
R, R_e, R_i	Alveolar radius in the undeformed state, and at the external and internal surfaces
Y_c, Y_e	Collagen and elastic fiber Young's moduli
Y_{ce}, Y_{ee}	Collagen and elastic fiber effective moduli
$\delta_{A,\ c}$, $\delta_{A,\ e}$	Collagen and elastic fiber area fraction
p	Probability distribution
w	Collagen fiber waviness
w_1, w_2	Minimum and maximum waviness
α, β	Parameters of the beta distribution
$B(\alpha, \beta)$	Beta function
S, S_e, S_i	Normalized wall thickness radius in the undeformed state, and at the external and internal surfaces
$\sigma_{\theta\theta}$	Total circumferential stress
$\sigma_{c,\ \theta\theta}$, $\sigma_{e,\ \theta\theta}$	Collagen and elastic fiber circumferential stress
P_{alv}, $P_{alv,\ f}$	Total and fiber-related alveolar pressure
h_0	Wall thickness in the undeformed state
V_0, V_{alv}	Alveolar volume in the undeformed and deformed states
γ	Surface tension
$P_{alv,\ s}$	Surface tension-related alveolar pressure
T_f	Fiber tension
r_i	Alveolar radius at the inner surface in the deformed state
P_{tp}	Transpulmonary pressure
P_{pl}	Pleural pressure
E	Total strain energy
Δl	Spring extension
$\Delta\theta$	Change in angle between adjacent septal walls
r_b	Bond-bending spring constant
FRC	Functional residual capacity
TLC	Total lung capacity
P_{ta}	Transalveolar pressure

Chapter 5

t	Time
$F(t)$, $\dot{F}(t)$	Force and its time derivative
$x(t)$, $\dot{x}(t)$	Length or displacement and its time derivative (velocity)
R	Resistance
E	Elastance

x_0	Initial length (value of x)
E_{stat}	Static elastance
E_{relax}	Viscoelastic elastance
$y(t)$, $\dot{y}(t)$	Second length or displacement and its time derivative (velocity)
y_0	Initial length (value of y)
i	Square root of -1
f	Frequency (Hertz)
ω	Angular frequency (radian.s^{-1})
$y(f)$	Fourier transform of $y(t)$
$F(f)$	Fourier transform of $F(t)$
$y(f)$, $\dot{y}(f)$	Fourier transform of $y(t)$ and the transform of the time derivative of $y(t)$
$Z(f)$	Mechanical impedance
R_{tis}	Tissue resistance
R_{aw}	Airway resistance
$V(t)$	Volume
$\dot{V}(t)$	Flow
$\ddot{V}(t)$	Volume acceleration
$P(t)$, $\dot{P}(t)$	Pressure and its time derivative
G	Tissue damping in the constant-phase model
H	Tissue elastance in the constant-phase model
α	Parameter of the constant-phase model that relates G and H
ω_0	Normalization factor for angular frequency
A, B	Constants
β	Power-law exponent of time
F_{stat}	Static force
q	Factor in exponential exponent
τ	Relaxation time-constant
$P(t)$	Pressure
$h(t)$	Impulse response function
u	Dummy variable of integration
L	Total length of model
n	Number of springs in series
ΔF	Change in force
Δt	Time increment
D	Length of dashpot barrel
n_0	Number of springs in series prior to application of force
a	Exponent of length
b	Exponent of velocity

Chapter 6

ABM	Agent-based model
N	Number of rows and columns of patches in a 2D tissue sheet
$M_{m,n}$	Mass of tissue on the mth, nth patch
δM_{catab}	Amount of tissue removed from a patch by a catabolic agent

δM_{anab}	Amount of tissue added to a patch by an anabolic agent
α, β	Constants determining δM_{catab} and δM_{anab}
ν_j	Probability of a single patch being occupied by an anabolic agent
δ_j	Change of tissue mass on a patch in a single time step
q	Probability that ν_j is zero
$\mathbb{E}\{\cdot\}$	Statistical expectation operator
M_j	Mass of tissue on a single patch at time step j
$\{M_j\}$	Temporal sequence of M_j values
M^*	Fixed point of $\{M_j\}$
σ^2	Variance over time of tissue match on a single patch
ECM	Extracellular matrix
γ	Activation factor of an anabolic agent
T	Half-life of agent lifetime
a'	Pro-fibrotic activity of an agent
k'	Mechanical stiffness of a tissue patch
t'	Time
$f(a', k')$	Time derivative of a'
$g(a', k')$	Time derivative of k'
W	Constant determined by initial conditions
A	Maximum possible value of a'
Q	Value of k' at which a' is half maximum
$h(k')$	Monotonic function of k'
D'	Collagen digestion rate
P'	Collagen production rate
q'	Threshold value for tissue stiffness
t	Dimensionless time
a	Dimensionless pro-fibrotic activity of an agent
k	Dimensionless mechanical stiffness of a tissue patch
D	Dimensionless collagen digestion rate
P	Dimensionless collagen production rate
q	Dimensionless threshold value for tissue stiffness
k^*, a^*	Fixed points for tissue stiffness and agent activity
J	Jacobian
τ	Trace of a matrix
Δ	Determinant of a matrix
H	Number of nodes in a Hopfield net
$w_{i,j}$	Link between nodes i and j in a Hopfield net
ψ_i	State of node i in a Hopfield net
$\{\psi_{i,j}\}$	Set of states of all nodes in a Hopfield net at time step j
$f\{\cdot\}$	Nodal activation function in a Hopfield net
$\{\psi_i\}_0$	Stable state configuration of a Hopfield net
M	Number of different stable states of a Hopfield net
μ	Mass of tissue per unit volume
V	Unit volume of tissue

ρ	Mass density of tissue
div	Divergence operator (acting on a vector)
tr	Trace operator (acting on a matrix)
$\boldsymbol{F}$	Growth deformation gradient
COPD	Chronic obstructive pulmonary disease

Chapter 7

ASM	Airway smooth muscle
Th1, Th2	T-helper cell 1 and 2
t	Time
$F(t)$, $\dot{F}(t)$	Force and rate of change of force
F_0	Isometric force
$L(t)$, $w(t)$	Two different length variables
L_{ref}, w_0	Baseline values of L and w
$\dot{L}(t)$, $\dot{w}(t)$	Rates of change of L and w
a, b	Constants in the Hill equation for muscle
E_p, E_{Mp}	Springs constants in lumped parameter model of muscle strip
α_p, α_{Mp}	Constants determining E_p and E_{Mp}
R_p	Dashpot resistance in lumped parameter model of muscle strip
$F_A(t)$	Active force generated by muscle
$x(t)$, $\dot{x}(t)$	A length variable and its time derivative
$y(t)$, $\dot{y}(t)$	A length variable and its time derivative
$z(t)$, $\dot{z}(t)$	A length variable and its time derivative
E_s, E_{Ms}	Springs constants in lumped parameter model of muscle strip
$F_0(t)$, $\dot{F}_0(t)$	Variable isometric force and its time derivative
F_{0max}	Maximum active isometric muscle force
k_{on}, k_{off}	Kinetic rate constants of cross-bridge binding and detachment
$n(x, t)$	Distribution of bound cross-bridges as function of distance along the muscle (x) and t
a, b	Upper and lower limits of x defining a section of $n(x, t)$
$N((a, b), t)$	Total number of bound cross-bridges with extensions between a and b
$G(x, t)$	Rate at which myosin cross-bridges attach to actin in Huxley model
$H(x, t)$	Rate at which myosin cross-bridges detach from actin in Huxley model
$v(t)$	Velocity of muscle shortening
$f(x)$	Force-length ($F - x$) relationship of a single cross-bridge
n_{max}	Maximum possible number of bound cross-bridges at length x
$g(x)$, $h(x)$	Rate constant functions for cross-bridge binding and unbinding in Huxley model
X	Limit of extension of a single cross-bridge in Huxley model
E	Stiffness (elastic constant) of a single cross-bridge
$N(t)$	Total number of bound cross-bridges at time t

v_{max} Maximum (unloaded) shortening velocity of muscle

F_{br} Mean cross-bridge force

N_{max} Maximum number of bound cross-bridges across muscle cross-section

α Constant

$l(t)$ Muscle length

l_{crit} Critical muscle length above which the contractile machinery becomes disrupted

k_{break} Rate constant of disruption of contractile machinery in muscle

k_{fix} Rate constant of repair of contractile machinery in muscle

$N_{max,0}$ Maximum value of N_{max} achieved when contractile machinery is fully intact

A, B, C Constants in model of the force-velocity relationship for muscle

M Constant determined by the initial conditions

F_i, F_e Internal and external loads opposing muscle contraction

A_{max} Maximum airway cross-sectional area

P Airway transmural pressure

a_n Normalized airway area

a_{n0} Value of a_n when $P = 0$

a_{n0}' Value of da_n/dP when $P = 0$

P_1, P_2 Parameters in the hyperbolic model of airway wall elasticity

b_1, b_2 Exponents in the hyperbolic model of airway wall elasticity

r Airway radius

r_{iso} Radius of the airway under conditions of isotropic parenchymal expansion

r_{isoTLC} Value of r_{iso} at total lung capacity

$r_{anisoTLC}$ Radius of the airway under conditions of anisotropic parenchymal expansion

P_{tp} Transpulmonary pressure

P_{tpTLC} Transpulmonary pressure at total lung capacity

TLC Total lung capacity

P_{tm} Transmural pressure across the airway wall

ΔP_{tm} Change in transmural pressure across the airway wall

μ_s Parenchymal shear modulus

F_{isoP} Passive airway hoop stress under conditions of isotropic parenchymal expansion

F_{isoA} Active airway hoop stress under conditions of isotropic parenchymal expansion

F_{anisoP} Passive airway hoop stress under conditions of anisotropic parenchymal expansion

F_{anisoA} Active airway hoop stress under conditions of anisotropic parenchymal expansion

F_P Passive hoop stress in the airway wall

F_A Active hoop stress in the airway wall

σ	Fraction of airway circumference that is inextensible
P_0	Transpulmonary pressure at which parenchyma is homogeneously expanded
R_{aw}	Airway resistance
Δ	Difference in Horsfield order between two daughter airways
Z_{aw}	Airway impedance
N	Number of airway orders in the airway tree (equal to the order of the trachea)
i	Square root of -1
Z_N	Impedance of entire airway tree
$Z_{j,tot}$	Impedance of subtree subtended by airway of order j
Z_L	Impedance of lung
Z_{tis}	Impedance of tissue
f	Frequency
l_j, r_j	Length and radius of airway of order j
ρ	Gas density
R_j	Resistance of airway or order j
E_j	Elastance of airway or order j
I_j	Inertance of airway or order j
G	Tissue damping in the constant-phase model
H	Tissue elastance in the constant-phase model
α	Parameter relating G and H in constant-phase model
C_P, C_V	Specific heats of air at constant pressure and volume
ψ	Ratio of specific heats
J_0, J_1	Zeroth- and first-order Bessel functions
R_T	Resistance of the airway wall tissue
c	Sound velocity in air
F_v, F_t, ϕ_t, r_v, D	Functions of transmission line model of an airway
C_g	Airway gas compliance
μ	Gas viscosity
κ	Thermal conductivity
R_N	Newtonian resistance of the lung
γ	Scaling factor between airways of successive order
l	Airway length
λ	Ratio of airway length to radius
β	Length factor for the trachea
r_N	Radius of the trachea
E_{gj}	Gas elastance of airway of order j
E_{tis}	Young's modulus of airway wall tissue
E_{wj}	Wall elastance of airway of order j
G_u, H_u	Tissue damping and elastance of single parenchymal unit
η	Ratio of tissue damping to elastance
n_t	Number of terminal airways
H_{rs}	Stiffness (elastance) of the respiratory system
H_{shunt}	Stiffness (elastance) of the proximal shunt compartment
Z_{shunt}	Impedance of the proximal shunt compartment

Chapter 8

IPF	Idiopathic pulmonary fibrosis
FVC	Forced vital capacity
TLC	Total lung capacity
FRC	Functional residual capacity
MEFV	Maximal expiratory flow-volume
FEV1	Forced expiratory volume in 1 s
X	Density of cells
D_X	Diffusion coefficient of X
r_p	Proliferation rate
r_a	Apoptosis rate
r_{ij}	Transition rate
t	Time
Δ	Laplacian
C_{kj}	Cytokine concentration
β	Hill exponent
a_{kj}	Sigmoid parameter
b_{kj}	Sigmoid parameter
e_{kjl}	Sigmoid parameter
f_{kjl}	Sigmoid parameter
γ	Hill exponent
E_{kjl}	Enzyme concentration
G	Concentration of gene
P	Concentration of protein
δ	Hill exponent
ν	Exponent
p	Percolation probability
p_c	Percolation threshold
S	Cluster size
Π	Distribution function
τ	Exponent
a	Activation
k, k_1, k_2, k_{th}	Network-related stiffnesses
Q_a, Q_k, Q_A	Proportionality constants
Y	Young's modulus
L_0	Initial length
A	Cross-sectional area
ε	Strain
c	Constant
w_1, w_2	Weighting factors
ρ_i	Agent density

Chapter 9

COPD	Chronic obstructive pulmonary disease
Y	Tissue Young's modulus
Y_0	Initial Young's modulus
σ	Tissue stress
λ	Stretch ratio
σ_0	Proportionality stress constant
α_1, α_2	Exponents for stress
G	Shear modulus
ν	Poisson's ratio
P_A	Alveolar pressure
σ_s	Traction stress in septal wall
FRC	Functional residual capacity
P	Probability
LAA	Low-attenuation area
D	Exponent in LAA distribution
K	Proportionality constant
S	Cluster size
X	Dummy cluster size
r_b	Bond-bending constant
C	Compliance
t_d	Doubling time
D_2	Equivalent diameter
γ	Stiffness decay factor
τ	Time constant
P_{on}	Probability of binding
P_{off}	Probability of unbinding
c	Concentration
P_r	Rupture probability
F_f	Fibril force
$F_{f,\ r}$	Reference fibril force
Y_f	Fibril stiffness
$Y_{f,\ n}$	Fibril stiffness at step n
P_d	Digestion probability
B	Proportionality constant
w_f	Fibril waviness
s	Parameter

Chapter 10

ARDS	Acute respiratory distress syndrome
PaO_2	Arterial partial pressure of oxygen
FiO_2	Fraction of oxygen in inspired gas
VILI	Ventilator-induced lung injury
PEEP	Positive end-expiratory pressure

$\dot{V}_A$	Alveolar minute ventilation
f	Breathing frequency (breaths per minute)
V_T	Tidal volume
V_D	Dead space volume
Ω	Lung injury cost function
t	Time
$P(t)$	Pressure
$\dot{V}(t)$	Flow
E_{tis}	Lung tissue elastance
$P_A(t)$	Alveolar pressure
$y(t)$	Lung tissue distension
A, B, K	Parameters of Salazar-Knowles equation of lung pressure-volume relationship
$z(t)$	Fraction of open lung
$E_{R/D}$	Stiffness of spring governing recruitment and derecruitment
α	Fraction of permanently recruited (i.e., non-collapsible) lung
γ	Fraction of potentially recruitable lung
$P_{1/2}$	Alveolar pressure at which half the lung is recruited
τ	Determines sigmoidal steepness
R_{aw}	Airway resistance
y_{inj}	Tissue distension threshold for injury
T	Duration of mechanical ventilation
$\Omega_{voltrau}$	Amount of volutrauma
$K_{voltrau}$	Constant of proportionality for volutrauma
$y'(t)$	Tissue distention variable for calculating amount of volutrauma
$z'(t)$	Recruitment variable for calculating amount of atelectrauma
$\Omega_{atelectrau}$	Amount of atelectrauma
$K_{atelectrau}$	Constant of proportionality for atelectrauma
Ω_{VILI}	Amount of ventilator-induced lung injury
ΔP	Difference between closing and opening pressures
E_{rs}	Respiratory system elastance
$R_{R/D}$	Resistance of dashpot governing rate of recruitment and derecruitment
APRV	Airway pressure release ventilation
$\mathrm{P_{Hi}}, \mathrm{P_{Low}}$	Pressures applied during inspiration and expiration in APRV
$\mathrm{T_{Hi}}, \mathrm{T_{Low}}$	Durations of inspiration and expiration in APRV
P_o, P_c	Critical opening and closing pressures
$x(t)$	Virtual trajectory variable associated with a lung unit
s_o, s_c	Constants of proportionality for speed of opening and closing
μ_c, μ_0	Mean closing and opening pressures
σ_c^2, σ_0^2	Variances of closing and opening pressures
$p(P_c), p(P_o)$	Probability distributions of closing and opening pressures
$p(s_c), p(s_o)$	Probability distributions of closing and opening speeds
S_c, S_c	Scale factors for closing and opening speed distributions

P_1, P_2	Two different pressure levels
$F(P)$	Steady-state fraction of open lung at pressure P
t_o	Opening time for a lung unit
CT	Computed tomography
λ	Number of injurious events acting on the blood-gas barrier per unit time
β	Fraction of injurious events causing existing holes to expand in area
A_0	Initial area of hold in the blood-gas barrier
$F[A_i(t) < A]$	Probability that a hole has an area less that A at time t
$\alpha(t)$	Time-dependent normalization factor
k	Magnitude of power-law exponent
2D	Two-dimensional
N	Number of cells in a monolayer
n	Number of focal adhesions between each cell and its neighbors
n_{max}	Maximum value of n
R	Resistance of a monolayer
R_{max}	Maximum resistance of a monolayer
R_{min}	Minimum resistance of a monolayer
r	Resistance of a single cell in a monolayer
p_{break}	Probability that bubble passage will break a cell attachment
n_{broken}	Number of broken attachments to a given cell
ρ	Constant
η	Critical fraction of broken attachments to a cell at which it starts to leak
p_{attach}	Probability that a broken attachment will be repaired
ζ	Constant
V_{Tcrit}	Critical tidal volume above which volutrauma starts to occur
D	Value of V_T at which $\rho = 0.5$
$\dot{V}_{in}$	Flow of liquid into the alveolar space
$\dot{V}_{out}$	Drainage flow of fluid from the alveolar space
$\dot{V}_{net}$	Net flow of material into the alveolar space
B	Constant
V_{excess}	Excess fluid volume in the alveolar space
$\dot{V}_{clear}$	Rate of fluid clearance from the alveolar space
k_{clear}	Clearance rate constant
$E_{rs,\ baseline}$	Elastance of the uninjured respiratory system
q	Constant
ϵ	Epithelial strain between end-expiration and end-inspiration
ϕ	Surfactant production rate
ϵ^*	Critical strain beyond which strain inhibits surfactant production

Chapter 11

M0, M1, M2	Three different macrophage phenotypes
t	Time

$E_h(t)$	Fraction of healthy epithelial cells
$E_d(t)$	Fraction of damaged epithelial cells
$E_e(t)$	Fraction of dead epithelial cells
$\kappa_{prolif}(t)$	Rate at which epithelial cells proliferate to produce new healthy cells
b_p	Baseline rate of production of new epithelial cells
k_{ep}	Rate constant for production of pro-inflammatory mediators
$p_{med}(t)$	Level of pro-inflammatory mediators
$\kappa_{repair}(t)$	Rate of epithelial repair
b_r	Repair rate constant
C_R	Initial concentration of mediator
C_S	Initial concentration of receptor
C_P	Concentration of product
k_{er}	Maximum repair rate due to mediators
x_{er}	Mediator concentration at half maximum rate
$\kappa_{damage}(t)$	Rate of damage of healthy epithelial cells
$N(t)$	Concentration of activated neutrophils
$M_1(t)$	Concentration of M1 macrophages
k_{mne}	Maximum rate of Hill-function reaction
x_{mne}	Hill-function reaction shape parameter
κ_{vent}	Rate of damage of healthy epithelial cells by mechanical ventilation
s_d	Rate constant of ventilatory damage of epithelial cells
$\kappa_{Nphag}(t)$	Rate of phagocytosis of damaged epithelial cells by neutrophils
$\kappa_{M1phag}(t)$	Rate of phagocytosis of damaged epithelial cells by M1 macrophages
k_{en}, k_{em1}	Rate constants
a	Concentration of anti-inflammatory mediators
a_∞	Constant governing rate of phagocytosis by M1 macrophages
$\kappa_{death}(t)$	Rate at which damaged epithelial cells die
b_d	Rate constant for epithelial cells death
$P(t)$	Concentration of pneumococcal bacteria
P_0	Bacterial proliferation rate
k_p	Initial growth rate constant
P_{max}	Bacterial carrying capacity of the tissue
$k_{p_{AM}}$	Govern the rates of bacterial killing by M_0 macrophages
k_{p_N}	Govern the rates of bacterial killing by neutrophils
$k_{p_{IM}}$	Govern the rates of bacterial killing by M_1 inflammatory macrophages
k_{CFUBp}	Rate at which pneumococci enter the blood stream
n	Concentration of pneumococcal bacteria at half-maximum rate
NOX-D19	An agent for treating *Streptococcus pneumoniae* infection
SARS-CoV-2	Virus responsible for the Covid-19 pandemic
ACE2	Angiotensin-converting enzyme 2
$V(x, y, t)$	Viral load on a surface

D_V	Diffusivity of virus
k	Rate constant
c	Rate constant
I	Concentration of infected T cells
$\boldsymbol{ij}$	xy
$T(x, y, t)$	Concentration of T cells on epithelial surface
D_T	T cell diffusivity
d_T	Rate constant of T cell death
k_T	Rate constant of T cell infection by virus
L_6, k_{L6T}	Concentration of the cytokine IL-6 and its rate constant
L_{10}, k_{L10T}	Concentration of the cytokine IL-10 and its rate constant
k_{M1T}	Rate constant fo1 M1 macrophages
M_2, k_{M2T}	Concentration of M2 macrophage and its rate constant
$\alpha_{attract}$	Strength by which uninfected T cells are attracted by infected T cells
$\alpha_{repulse}$	Strength by which uninfected T cells are repulsed by virus
T_{max}	Carrying capacity of tissue for T cells
k_{infect}	Interaction rate constant between uninfected T cells and virus particles
d_I	Rate constant of T cell death
k_{L6I}	Interaction rate constant between uninfected T cells and IL-6
D_{M1}	Diffusivity of M1 macrophages
k_{L6M1}	Rate constant of increases in M1 macrophages due to IL-6
k_{NM1}	Rate constant of decreases in M1 macrophages due to IL-10
M_{1max}	Carrying capacity of the tissue for M1 macrophages
D_{M2}	Diffusivity of M2 macrophages
D_{L6}	Diffusivity of IL-6
D_{L10}	Diffusivity of IL-10
ABM	Agent-based model

Chapter 12

Y	Tissue stiffness
Y_f	Fiber stiffness
c	Proportionality constant
α	Stiffness exponent
σ_f	Fiber stress
ϵ_f	Fiber strain
b_f	Fiber nonlinearity parameter
E_f	Elastic energy of fiber
A_f	Fiber cross-sectional area
l_f	Fiber length
θ_0, θ_f	Angles between adjacent fibers in the undeformed and deformed states

r_b	Bond-bending spring constant
V_f	Fiber volume
E_t	Tissue strain energy
e_t	Tissue strain energy density
v_f	Fiber volume fraction
Z_f	Nonlinear elastic constant
q_f	Bond-bending constant per unit volume
δ_f	Change in angle
n_f	Number of fibers
n_w	Number of septal walls
S	Scaling function
n_c, n_e	Number of collagen and elastic fibers
v_c, v_e	Collagen and elastic fiber volume fractions
Y_c, Y_e	Collagen and elastic fiber stiffnesses
φ_f	Fiber alignment parameter
φ_c, φ_e	Collagen and elastic fiber alignment parameters
a	Activity
k	Spring constant
q	Threshold for switching between digestion and deposition
D	Rate of tissue digestion
P	Rate of tissue production
x	Damage extent
C	Rate of spontaneous decay
F	Cross-talk parameter between damage and activity
p	Probability distribution
K	Node degree
γ	Exponent of degree distribution
t	Time
$X_1, X_2, \ldots, X_N$	State variables
$\boldsymbol{\psi}$	State vector
ϕ, ϕ_c	Vitality and its critical value
p_0	Probability of a state being 0
p_1	Probability of a state being 1
s	Fraction of the population that remains alive
m	Mortality rate

Part I
Modeling the Healthy Lung

Chapter 1
Introduction

Abstract Mathematical and computational modeling is critical to advancing our understanding of structure-function relationships in the lung. Systems of coupled ordinary differential equations are used to describe the behavior of interacting lung compartments and are readily solved either analytically or numerically, but they have limited spatial resolution. Systems of partial differential equations describe behavior that is continuous both in time and space. Agent-based models are computationally expensive but have the advantage of representing reality more or less as we perceive it, making this type of modeling more accessible to biomedical scientists without strong mathematics backgrounds. Forward models, used for simulation, are implemented by a priori selection of model structure and parameter values, and may be highly complex. In contrast, inverse models, which are used to infer what may be taking place within a system from observed relationships between inputs and outputs, must have few adjustable parameters. Multiscale models account for phenomena that arise at different levels of length and/or time scale and are crucial for understanding the emergent behavior that appears ubiquitously in biology. Importantly, a model must be validated in terms of its usefulness for the task at hand.

Keywords Differential equations · Agent-based model · Forward model · Simulation · Inverse model · Model fitting · Multi-scale model · Emergent behavior · Model validation

1.1 Why Model the Lung?

Like all organs in the body, the lung has evolved an intricate structure that befits its function. The function of the lung is primarily to allow the diffusive exchange of oxygen and carbon dioxide between the atmosphere and the body tissues. The structure of the lung is thus a critical determinant of how it functions. Likewise, lung function is fundamentally reflective of its structure. Understanding how the lung works amounts to understanding its *structure-function relationships*.

In a healthy individual, the structure-function relationships that exist in the lung are as Nature intended them to be. That is, both structure and function are "normal" in the sense that they are similar to those in other healthy individuals, the term

B. Suki, J. H. T. Bates, *Mathematical Modeling of the Healthy and Diseased Lung*,
https://doi.org/10.1007/978-3-031-95322-4_1

"similar" meaning within the limits of normal interindividual variability. In disease, however, lung function becomes abnormal. Furthermore, abnormal function can invariably be traced to an abnormality in structure. For example, asthma is characterized by a reduced caliber of the conducting airways, an aspect of structure, which results in an elevated resistance to airflow, an aspect of function. Similarly, pulmonary fibrosis is characterized by a stiffening of the lung parenchyma, an aspect of function, which is due to the aberrant deposition of collagen, reflecting a change in structure.

Describing asthma and pulmonary fibrosis in such simple terms, however, does not do justice to the fact that they are both highly complex chronic diseases that change progressively over time. Their particular derangements, and the courses they follow, are never the same in any two patients. They involve vast arrays of interacting cellular and molecular processes. They can have a range of different etiologies. In fact, the complexities inherent in diseases such as asthma and pulmonary fibrosis are still far from fully understood. And given the apparently limitless complexities of biology in general, it is hard to imagine how our understanding of even the normal lung, let alone lung disease, will ever be complete.

How, then, do we gain any level of understanding of the link between structure and function in the lung? The answer is *modeling*. A model is a hypothesis—a formal statement about how a *system* works, stated in a way that permits predictions of system behavior. When the model is *mathematical*, it allows these predictions to be quantitative. The system itself is viewed as a set of interacting components that have a collective identity that receives *inputs* from its environment, and that produces *outputs* that are observable from the environment.

1.2 Model Complexity

A model is a representation of reality that purports to include all the details about the system being modeled that are in some sense important, while leaving out those details that are not. If you want to find your way to a friend's house for dinner, you might use a GPS map as a model of the local terrain. The map does not include an image of every blade of grass you will pass along your route. It shows only the roads to be traveled and the intervening intersections, and even these are usually represented as simple one-dimensional lines and curves. Additional detail is unhelpful, and indeed might even be confusing by distracting you with things that are not directly related to finding your way. Hence, the structure of a GPS map needs to manifest a level of detail relevant to the job at hand, namely arriving in time for dinner, and nothing more.

Models of the lung are no exception. There can never be a single definitive lung model that, once constructed, will serve every purpose hence forth. A model of the lung must always be constructed, or chosen from the available possibilities, with a specific purpose in mind, and the level of detail in the model must be appropriate to that purpose. There is no rule against constructing a model of the lung that represents everything currently known. However, this would be a major undertaking because rather a lot is now known about how the lung works. It would therefore be incumbent upon the modeler to justify the large amount of effort required in including everything. Importantly, the simple fact that creating such a comprehensive model might be possible is not sufficient justification for actually doing it. A model must be appropriate for the need at hand, and such needs constantly evolve. Modelers will never be out of work!

Most models of the lung thus do not include everything that is known about the organ, just as a GPS map does not include every conceivable detail about a route to be traveled. So, when making a lung model, what details do we include and what do we ignore? This is perhaps the central question in modeling, and there is no universal answer. Experience is crucial, as is learning from the mistakes and successes of others. There is now a huge literature going back decades that is devoted to modeling the lung, so it is important not to reinvent the wheel. By the same token, however, one must resist the temptation to reproduce every detail in a published model simply because each one was justified as being important by its creators. Some of these details may not be important for your study, in which case they should not be included in your model.

It is equally important, of course, not to leave out details that are important because the resulting model will fail to be useful at best, and at worst it will be misleading. To paraphrase Albert Einstein, "*a model should be as simple as possible, but no simpler*". This deceptively obvious statement implies the existence of a "sweet spot" defining the optimum in model complexity, but exactly where that spot lies for any particular application may be far from obvious. Finding the sweet spot usually relies on an iterative process in which individual model features are added, removed or fine-tuned, each time checking that the model predictions are not at odds with experimental observation, until the most parsimonious candidate is identified.

1.3 Types of Models

Models of the lungs are typically sets of mathematical equations embodying the laws of physics. In other words, they are not usually just diagrams like a GPS map. This is not to say that diagrams are unimportant. A diagram of the branching tree structure of the pulmonary airways, for example, can provide a useful representation of the numerous bifurcations that lead from the trachea to the terminal airways, the progressive narrowing of the branches as one proceeds distally into the lung, and possibly even the degree of asymmetry in the two daughter airways that emanate from each parent. Diagrams such as this are often referred to as models, but the understanding is that they imply the existence of mathematical equations or computational methods for quantitatively predicting the function of the structures they depict. For example, by applying mathematical formulae to calculate the flow resistance of each airway branch and then combining the individual resistances using electric circuit theory, the flow resistance of the entire structure can be calculated. This calculation can then be compared to experimental measurements of airway resistance in a real lung, thereby providing an assessment of model accuracy.

The simplest types of lung models involve algebraic relationships between one or more independent variables and a dependent variable. Suppose, for example, that you want to know how much gas volume, V, will enter a lung when it is subjected to a given inflation pressure, P. A simple model of the relationship between P and V is

$$V = CP \tag{1.1}$$

where V is expressed relative to the gas volume remaining in the lung when the applied pressure is zero (relative to atmospheric pressure). The quantity C is known as the compliance of the lung and is a measure of how easily it is inflated. Clearly, for a given increase in P, the lung gains more volume when C is large compared to when it is small. C is a *parameter*—a quantity having a fixed value that determines exactly how the model behaves. V and P are *variables*—quantities that can take on ranges of values and that can be measured and/or controlled experimentally in the laboratory. V is referred to as the *dependent variable* because Eq. 1.1 is written in a way that implies V to be the outcome of an experiment in which the parameter C is given a fixed value while the *independent variable* P is under experimental control. There is nothing fundamental about this particular assignation, however. One could just as easily write Eq. 1.1 as

$$P = \frac{V}{C} \tag{1.2}$$

in which case P is the dependent variable and V is the independent variable.

The model described by Eq. 1.1 is *linear* because the dependent variable, V, depends linearly on the independent variable and the parameter. That is, a change in either C or P induces a directly proportional change in V. If Eq. 1.1 is modified to

$$V = C\sqrt{P} \tag{1.3}$$

it becomes *nonlinear* in the variable P, while still being linear in C. Of course, nothing in the real world is perfectly linear, so the linearity assumption can only be invoked if the behavior being modeled is linear to a useful degree of approximation. This tends to be the case when the independent variables are examined over sufficiently small operating ranges. For example, Eq. 1.2 pertains to the lung quite accurately when the excursions in pressure and volume are within those of normal quiet breathing. As the depth of breathing increases, however, a better description of reality is obtained if Eq. 1.2 is replaced by something like Eq. 1.3.

The linearity assumption in modeling is typically made in the interests of mathematical tractability; linear models are easier to manipulate and solve than are nonlinear models. On the other hand, nonlinearities invariably lead to much more

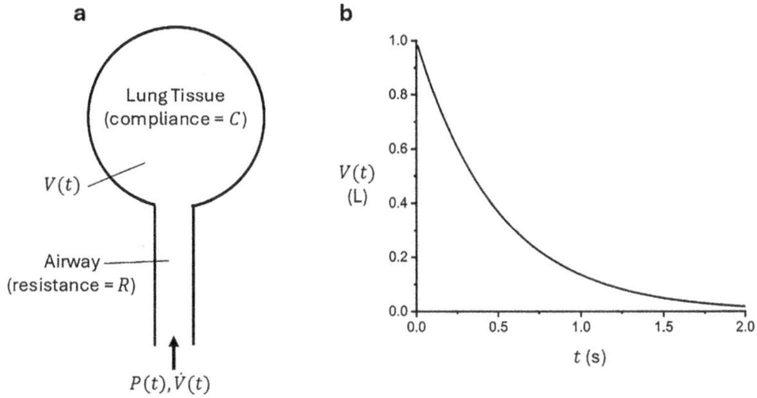

Fig. 1.1 (**a**) Single-compartment model of the lung consisting of an elastic balloon (compliance C) representing the lung tissue connected to a tube (resistance R) representing the conducting airways. The volume of air in the lung is $V(t)$. Pressure, $P(t)$, and flow, $\dot{V}(t)$, are measured at the airway opening. (**b**) Exponential decay of lung volume during a relaxed expiration, starting from an initial volume of 1 L

interesting model behavior. Nonlinear behavior is also typical of biologic systems themselves. Indeed, as we shall see later in this book, nonlinearities are often essential for biologic function, in which case they must be modeled explicitly.

Even though the variables V and P are usually functions of time, there is no explicit time dependence in Eq. 1.1. That is, Eq. 1.1 contains no information about how the state of the lung changes from one point in time to the next. It is thus a *static model*. Models describing behavior that depends explicitly on time are referred to as *dynamic models*. As an example, suppose the lung is modeled as a balloon, representing the elastic parenchymal tissues, connected to the end of a cylindrical tube representing the conducting airways (Fig. 1.1a). The behavior of this model during a relaxed expiration, when $P = 0$, is described by the equation

$$R\frac{dV(t)}{dt} + \frac{V(t)}{C} = 0 \tag{1.4}$$

where R represents the flow resistance of the tube and C represents the compliance of the balloon, and t is time. Equation 1.4 is referred to as an *ordinary differential equation* because it involves the time derivative of a single variable, $V(t)$, and thus it describes explicitly how volume changes at every instant. The complete time dependence of volume is given by the solution to Eq. 1.4, which is

$$V(t) = V_0 e^{-\frac{1}{RC}t} \tag{1.5}$$

where V_0 is the volume in the balloon at the beginning of the expiration (Fig. 1.1b).

The model in Fig. 1.1 is also what is known as a *lumped parameter model* because its *constitutive properties* (its resistance and compliance) are assigned to distinct and separate locations. That is, all the resistive properties are lumped together in the cylindrical tube, while all the compliant properties are lumped together in the balloon.

Nature is not usually this convenient, however. More often than not, properties such as resistance and compliance arise within exactly the same structures. Lung tissue, in particular, exhibits both resistive and elastic properties that can be ascribed to the fibers and fluids of which the tissue is comprised. Furthermore, these properties typically vary between different locations in the tissue. For example, when a strip of tissue is subjected to axial oscillations in length, its stress, σ, may be a function of both t and position, x, along the strip. We thus say that $\sigma = \sigma(x, t)$, and the expression relating stress to strain becomes a partial differential equation containing the partial derivatives $\partial\sigma(x, t)/\partial x$ and $\partial\sigma(x, t)/\partial t$. Partial differential equations are generally much harder to solve than ordinary differential equations, and when the boundary conditions are irregular, often can only be solved numerically using a computer.

In fact, most model equations cannot be solved in closed form. This greatly limited the scope of mathematical modeling before the advent of computers in the middle of the twentieth century despite the development of very sophisticated analytic methods prior to that time. Nowadays, it is possible to find the solution to virtually any model equation using numerical methods implemented on a computer. Such methods only find approximations to the actual solution, although in most cases the errors in these approximations are negligible. This leaves the only practical issues being whether the computer is fast enough to find the approximate solution in an acceptable amount of time, and whether it has enough memory to store the necessary parameters and variables in the process.

Computers also allow the simulation of models that are based on logic operations rather than mathematical equations. *Agent-based models* are a prime example. These models usually represent some environment of interest as a series of interconnected patches each of which is imbued with a set of properties, similar to a digital image being represented as a collection of pixels each having a specific location and set of properties such as intensity and color. The dynamics of an agent-based model result from the collective behavior of so-called *agents* that are placed in the environment and given rules by which they are to operate. An example might be a set of agents representing different cell types moving around in a two-dimensional landscape representing the airway epithelium. Agents representing inflammatory cells might, for example, be given instructions to inflict damage on every patch in the environment that they encounter, while anti-inflammatory agents would do the converse.

Agent-based models can easily become computationally expensive if the environment is large and there are many agents, each of which must parse through numerous rules at every time step. On the other hand, the rules the agents obey are usually expressible in terms of straightforward logic operations such as "IF you move to an environmental patch with less than a specified degree of integrity THEN undergo apoptosis". Such rules do not have to involve equations, which makes agent-based modeling more immediately accessible to biomedical scientists who may not have strong mathematical backgrounds but who do have a great deal of knowledge about biology. Thus, experimental scientists whose expertise lies in biological phenomenology are often able to crucially inform agent behavior in ways that computational scientists cannot. This makes agent-based modeling an undertaking in which the geeks and the lab rats are on an even footing, instead of the more usual modeling situation in which progress depends largely on the geeks.

Agent-based modeling also readily lends itself to being stochastic because the behaviors of agents can be probabilistic rather than purely deterministic. For example, decisions about where an agent moves to, or what it does when it gets there, can be made on the basis of the value of a pseudorandom number generated within the computer. This gives the resulting model the same kind of unpredictable behavior typical of nature. Of course, equation-based models can also be stochastic by having some of their variables adhere to probability distribution functions rather than deterministic equations, but the difficulties inherent in dealing with mathematics and statistics still apply.

1.4 Forward Versus Inverse Modeling

Models can be used, broadly speaking, in two distinct ways, known respectively as forward and inverse modeling. *Forward modeling*, also known as *simulation*, is used to predict how the system being modeled will behave under a specified set of circumstances. The parameters of the model are given values thought to be representative of the quantities they correspond to in the real system, and the model equations are driven by signals representing inputs received by the system from the environment. The result is a prediction of the outputs the system itself would produce under corresponding conditions. The accuracy of these predictions is, naturally, a reflection of how accurately and completely the model represents the real system. Prediction accuracy is never perfect, of course, so its quality must be based on a comparison between what the model predicts and how real systems behave. Since forward models are often most useful in circumstances where real behavior is not known, judgements about prediction accuracy must generally be inferred from those special cases in which the real behavior is known. If prediction accuracy is not up to the task at hand, the forward model must be appropriately modified, usually through the inclusion of more detail about the system. The extent to which a forward model can be upgraded is usually limited only by the modeler's knowledge of the system.

Inverse modeling, also called *system identification*, is used to infer the nature of a system on the basis of observed relationships between its inputs and outputs. The goal of inverse modeling is to identify a model that behaves as much like the real system as possible, while at the same time embodying the key mechanisms that are operative within the system. This requires two things. First, one has to decide which mechanisms need to be represented, which is equivalent to deciding on the structure of the model. Second, one has to evaluate the model parameters, a set of constants whose values govern the specific behavior of the model. *Model fitting* is the process of finding those values of the parameters that cause the behavior of the model to match the behavior of the system as closely as possible. In contrast to forward modeling, the complexity of an inverse model is severely constrained by the fact that only relatively simple models can be fit to a given data set in a unique manner. As the number of adjustable parameters in an inverse model become more numerous, there rapidly comes a point at which multiple different sets of parameter values provide essentially the same quality of fit to the data. Sometimes it is possible to identify the correct solution as the only one that makes physiological or physical sense, but frequently one faces an unresolvable ambiguity. It is thus extremely important to make as many simplifying assumptions as can be justified when constructing an inverse model in order to keep the number of free parameters to a minimum.

Forward and inverse models are often used in separate applications, but they can also function together in powerful ways, particularly when practical or ethical considerations make it difficult to investigate a real system in detail. In such cases, an inverse model can be fit to synthetic data generated by a complex forward model that takes the place of the real system. The advantage of this approach is that the mechanisms underlying the synthetic data are known precisely because the forward model is completely under the control of the investigator. Consequently, the performance of the inverse model can be evaluated exactly. Of course, the utility of this strategy rests on the extent to which the forward model constitutes an accurate representation of the real system. When the forward model is sufficiently accurate, however, using it as a "virtual laboratory" in this way can often lead to important insights that are not otherwise achievable.

1.5 Multiscale Modeling

There is currently a great deal of interest in the modeling community in what has come to be called *multiscale modeling*. This involves explicitly accounting for phenomena that arise at different levels of length and/or time scale. For example, it is clearly a gross oversimplification to assume that the entire lung can reasonably be represented as the balloon-on-a-pipe construct illustrated in Fig. 1.1a. One might argue that the balloon in this model should actually be composed of millions of tiny balloons, each one representing an individual alveolus. One could then construct a model incorporating the behavior of each of these tiny balloons. The advantage of such a model is that it allows exploration of the effects of topographic heterogeneities. For example, what happens if those tiny balloons are not all identical? How does overall lung behavior change if some of those balloons become filled with fluid, as could occur in pulmonary edema? A multiscale model in this case might address these questions by predicting the overall behavior of the lung from the ensemble behavior of all its myriad alveoli.

Multi-scale computational modeling is usually the only way to address questions like this effectively. This does not mean, however, that a multiscale model should strive to include every potentially relevant mechanism at every conceivable level of scale. Faithfully including every detail that is known about a complex system such as the lung can rapidly become unwieldy. Perhaps even more important, such a model is likely to be bewildering to interpret and thus unhelpful in the advancement of scientific comprehension. Taken to its logical extreme, a model that includes every detail of Nature is no different from Nature itself, raising the question as to what is gained by replacing one with the other. Scientific insight is typically characterized by the sense of enlightenment that arises when a set of seemingly unrelated empirical observations are suddenly appreciated as having a common explanation based on relatively simple underlying principles or mechanisms. Thus, when constructing a multiscale model, it is important to decide which mechanisms must be included, and which should be neglected as unimportant details. This is a non-trivial and often subjective process that relies on the knowledge and experience of the modeler.

The idea of multi-scale modeling is actually not new. It goes back at least to the middle of the nineteenth century with the development of statistical mechanics, starting with the kinetic theory of gases. A liter of gas at room temperature and ambient pressure contains an incomprehensible number of individual gas molecules. The ensemble behavior of all these molecules is what gives rise to the temperature and pressure of the gas, and how these variables change when the gas is compressed or expanded. It is possible, at least in principle and given a large enough computer, to model the random movements of each molecule in a liter of gas, along with their interactions with each other and the wall of the container, to calculate how pressure (P), temperature (T) and V co-vary. Doing so, however, would provide no insight whatsoever as to why these quantities are related the way they are. Contrast this with the effort of doing a little thinking about how the molecules behave on average, together with a few lines of algebra expressing the consequences of this average behavior, to arrive at the same perfect gas law, $PV = nRT$, where n is the number of moles of gas in the container and R is the ideal gas constant. The first option shows that the behavior arises. The second option shows how it arises. This is insight!

This brings us to the central concept in multi-scale modeling, namely emergent behavior. An emergent behavior is one that arises at a certain level of scale as a result of the ensemble behavior of system components at lower levels of scale. Emergence appears ubiquitously throughout the natural world, especially in biology. The lung is a case in point. The simple balloon-on-a-pipe model is certainly a gross oversimplification of an organ that is composed of thousands of airway branches and hundreds of millions of alveoli. From a mechanical perspective, however, the whole lung behaves remarkably like the model in Fig. 1.1a under a wide range of circumstances. How can this be? This is an interesting question, and its answer sheds light on the kind of insight that constitutes scientific advancement.

1.6 Model Validation

Finally, it is important to consider the issue of *model validation*. This is a foundational concept in the world of modeling and is widely considered to be an essential step in any respectable model development project. Model validation implies that modelers must somehow demonstrate that their model is "valid". But what does this mean? One thing it cannot mean is that the model is a correct representation of reality because, as we have already discussed, that is an impossibility. As one of the patron saints of modelers, George Box, so eloquently stated, "*All models are wrong, but some are useful*". Validity cannot be an expression of correctness.

So, what does validity mean in this context? In fact, "model validity" is an unfortunate term. A better one, per Box, would be "model usefulness". A model either is or is not useful for the specific purpose to which it is intended, and that eventually comes down to a judgement call. Examining the issue of usefulness—whether a model is up to the job or not—is crucially important and should be undertaken whenever models are used. But one can never use the term "valid" to mean that a model is correct and therefore applicable to all situations because while it may perform satisfactorily in one particular application, it will certainly fail in others. It is thus crucial to always acknowledge a model's limitations by noting those conditions under which the model fails or is incomplete in some important way.

Finally, having decided that a model has been sufficiently validated in the sense just described above, it is time to put it to use. This use depends on the problem being addressed, and in particular on whether the goal is inverse or forward modeling (Sect. 1.4). In the case of inverse modeling, it is critical to perform a sensitivity analysis to determine how much confidence can be placed in the estimates of the model parameters when they are evaluated by fitting the model to data. Sensitivity analysis is equally important for forward modeling so that an appropriate level of confidence can be placed in the model predictions. The rigorous pursuit of these steps can lead to a deeper and sometimes completely novel understanding of the system under study that cannot be obtained in any other way.

1.7 Outline of the Book

The overall goal of this book is to show how analytic and computational modeling has been used to advance our understanding of lung biomechanics and mechanobiology in health and disease. The technical aspects of modeling are dealt with as they arise. These aspects include the construction of lumped-parameter models governed by systems of differential equation (Chaps. 3, 4, 5, 7, 8, 10, and 11), the use of network models for predicting emergent behavior such as percolation (Chaps. 4, 8 and 9) and the existence of stable attractors (Chaps. 6 and 12), and the simulation of biological processes using agent-based models (Chaps. 6, 7, 8, 9, 10,

11, and 12). The book is organized in three parts. Part I (Chaps. 1 though 5) covers the modeling of function in the normal lung, including basic lung mechanics (Chap. 3) followed by detailed examinations of parenchymal elasticity (Chap. 4) and viscoelasticity (Chap. 5). The basics of pulmonary structure, physiology and pathophysiology are also covered (Chap. 2) in order to define the scientific problems that computational models have been designed to address. Part II of the book then deals with various important classes of lung disease including asthma (Chap. 7), pulmonary fibrosis (Chap. 8), chronic obstructive lung disease (Chap. 9), and acute lung injury (Chap. 10). Also covered are the more general questions of how the lung repairs itself after sustaining damage (Chap. 6) and how it defends itself against attack (Chap. 11). Part II concludes with an examination of something that affects every lung, namely aging (Chap. 12). Finally, Part III (Chap. 13) offers some brief predictions of where modeling of the lung, and indeed biological systems in general, may lead in the future in terms of revealing the mechanistic underpinnings of normal physiology and the various ways in which these mechanisms can fail in disease.

Further Reading[1]

General Mathematical and Computational Modeling

1. Wilson KG. Problems in physics with many scales of length. Scientific American, August 1979.
2. Prigogine I, Stengers I. Order out of chaos: man's new dialogue with nature. Flamingo; 1984.
3. DiStefano J III. Dynamic systems biology modeling and simulation. London: Academic Press; 2015.

Mathematical and Computational Modeling of the Lung

4. Suki B. Structure and function of extracellular matrix. A multiscale quantitative approach. London: Academic Press; 2022.
5. Bates JHT. Lung mechanics. An inverse modeling approach. Cambridge: Cambridge University Press; 2009.
6. Neelakantan S, Xin Y, Gaver DP, Cereda M, Rizi R, Smith BJ, Avazmohammadi R. Computational lung modelling in respiratory medicine. J R Soc Interface. 2022;19: 20220062. https://doi.org/10.1098/rsif.2022.0062.
7. Suki B, Bates JHT. Lung tissue mechanics as an emergent phenomenon. J Appl Physiol. 2011;110:1111–8. https://doi.org/10.1152/japplphysiol.01244.2010.
8. Suki B, Bates JHT, Frey U. Complexity and emergent phenomena. In: Fredberg JJ, editor. Handbook of physiology, vol. 1. American Physiological Society; 2011. p. 995–1029.

[1]The literature related to mathematical and computational modeling in general, and on pulmonary physiology in particular, is vast. Listed below are some selections that the reader might like to consult for further depth on the subject.

Chapter 2
Lung Struture and Function from a Modeling Perspective

Abstract The primary function of the lung is to supply the body with oxygen and eliminate carbon dioxide during breathing. To achieve this, the mammalian lung has evolved a complex bifurcating airway tree embedded in the parenchyma, comprised of alveoli, alveolar ducts and capillaries. The mechanical properties of the alveoli determine how easily the lung can be inflated. The overall stiffness of the parenchyma is an emergent property arising from the nature of its components and the way they are organized. The stiffness of individual alveoli is primarily determined by its load-bearing components, the collagen and elastic fibers. The other major contributor to stiffness is the surface tension at the interface between the air in the lungs and the fluid that lines the alveoli. The forces involved in the cyclic stretching of the tissue during breathing are sensed by resident lung cells that respond so as to maintain and repair the lung in the face of ongoing wear and tear as well as environmental exposures. Lung biology and mechanics are thus tightly coupled, with crucial implications for development, homeostasis, and response to injury and disease. The possibilities for investigating this rich functional and structural landscape through computational modeling are limitless.

Keywords Gas exchange · Airways · Alveoli · Extracellular matrix · Collagen fibers · Elastic fibers · Proteoglycans · Surface tension · Resident cells

2.1 Introduction

The primary function of the lung is to supply the body with oxygen while simultaneously eliminating carbon dioxide through the process of breathing. To achieve this critical function, the mammalian lung has evolved a complex structure in which a bifurcating airway tree transports gas between the environment and the alveoli where gas exchange takes place. The alveoli are extremely small and numerous, and so in aggregate have an enormous internal surface area. This feature is essential for allowing a sufficient number of gas molecules to diffuse to and from the capillary blood to sustain life. The main functional role of the airways themselves is to supply air to the alveoli as homogeneously as possible. This must be done within the constraints imposed by lung size, shape and the deformations caused by gravity.

B. Suki, J. H. T. Bates, *Mathematical Modeling of the Healthy and Diseased Lung*,
https://doi.org/10.1007/978-3-031-95322-4_2

Accordingly, the characteristic features of the airway tree are highly species dependent [1, 2]. For example, the number of airway generations and the degree of bifurcation asymmetry are different in humans compared to most animals [3]. The airways of the favorite species for laboratory experimentation, the mouse, are less space-filling, as measured by the fractal dimension (see below), than those of humans [4]. Average alveolar diameter is also species dependent, tending to increase with body weight [5].

The airways and alveoli are closely coupled mechanically because most of the airway tree is embedded within, and thus attached to, the alveolar parenchyma [6]. The parenchyma itself is comprised of the alveoli, alveolar ducts, the pulmonary capillaries, and the connective tissue that binds them all together. As the lung changes volume during breathing, the outward pull exerted by the alveolar walls on the embedded airways is modulated accordingly. This has major consequences for airway caliber and the extent to which it can change when the smooth muscle encircling the airway wall is induced to contract [7], as occurs during an attack of asthma. The mechanical stiffness (the ability to withstand deformation, see below and Chap. 4) of the alveoli themselves is a crucial determinant of how easily the lung can be inflated by the respiratory muscles during breathing. In a normal healthy individual, the mechanical stiffness of the parenchyma is easily overcome by the actions of the diaphragm and the intercostal muscles of the ribcage, but this may no longer be the case in diseases such as pulmonary fibrosis in which alveolar septal walls become markedly thickened and stiffened.

The overall stiffness of the alveolar parenchyma is an emergent property arising from the nature of the components of which it is comprised, and the way that these components are organized with respect to each other [8]. The mechanical properties of individual alveoli are determined in large part by the load-bearing components of the extracellular matrix (ECM), which include collagen and elastic fibers and their associated proteoglycan molecules [9]. The other major contributor to lung mechanics is the surface tension at the interface between the air in the lungs and the fluid that lines the airways and alveoli. Surface tension adds to lung elastic recoil because an input of energy is required to increase the surface area of the interface. Our survival is dependent on the surface tension in the lung being reduced below that of pure water by the secretion of pulmonary surfactant into the interface by specialized alveolar cells.

Depending on their contractile state, smooth muscle cells can also contribute in a major way to the stiffness of airway and blood vessel walls [7, 10]. In addition, to function properly throughout life, the ECM must be homeostatically maintained by various tissue-resident cells [11]. *Homeostasis*, the long-term maintenance of the physiological and biological properties of tissue, can be perturbed by external stimulants such as pollutants and cigarette smoke, which can affect both the airways and parenchyma as occurs in chronic obstructive pulmonary disease (COPD) [12]. Alterations in ECM structure can also be influenced genetically, as is suspected to be a contributing factor in pulmonary fibrosis [13]. Indeed, pathologic derangements in the mechanical properties of the airways and parenchyma are always accompanied by corresponding abnormalities in tissue structure, all of which invariably result in deterioration of the lung's primary function, namely gas exchange.

This chapter takes a multiscale approach to the basic anatomy and physiology of the lung, beginning with the airway and vasculature trees that converge on either side of the blood-gas barrier. This is followed by the structure and function of the acinus, a distal lung unit comprised of the very small respiratory bronchioles and their associated alveoli. The lowest level of scale is then represented by the structure and function of the ECM along with the various resident cells that maintain tissue homeostasis and respond to mechanical and chemical stimuli. In order to set the scene for later chapters, the consequences of various diseases for lung structure and function are mentioned as appropriate, along with the use of computational modeling to link structure and function.

2.2 The Airways

Quantitative assessment of lung anatomy dates back many decades [14, 15], so the complex multi-component organization of the organ is well recognized. A prime example of this organization is the airway tree, which consists of myriad branches of widely varying sizes and orientations all connected together in ways that might initially seem somewhat random. However, behind this apparent randomness lies a deterministic pattern [16] reflecting the primary function of the tree, which is to distribute inhaled air as uniformly as possible within the irregularly-shaped spatial domain of the lung so that gas exchange can take place efficiently [17]. The recursive organizational principles on which the structure of the airway tree is based [18] allow its seemingly complex branching pattern to be codified much more efficiently than if every branch required independent specification [19]. This no doubt reflects how the development of the airway tree is codified genetically. Indeed, it has recently been determined that up to 67% of regional variability in pulmonary blood flow is genetically determined [20], so it seems reasonable to assume that the airway tree, which runs parallel to the vascular tree, is similar. There are also systematic differences between the airway trees of different mouse strains [4], further supporting the role of genetics in determining structure.

The first efficient deterministic description of the branching pattern of the airway tree was formulated in 1962 [15]. This formulation is often referred to as the Weibel model, and its branches are characterized by generation numbers (n) beginning with the trachea ($n = 0$) and proceeding towards the periphery in a series of symmetric bifurcations, yielding 2^n identical branches at generation n. The diameter, d, of any daughter branch is given by that of its parent multiplied by a scaling factor $0 < \gamma < 1$, so if d_0 is the diameter of the trachea then the diameter of a branch from generation n is

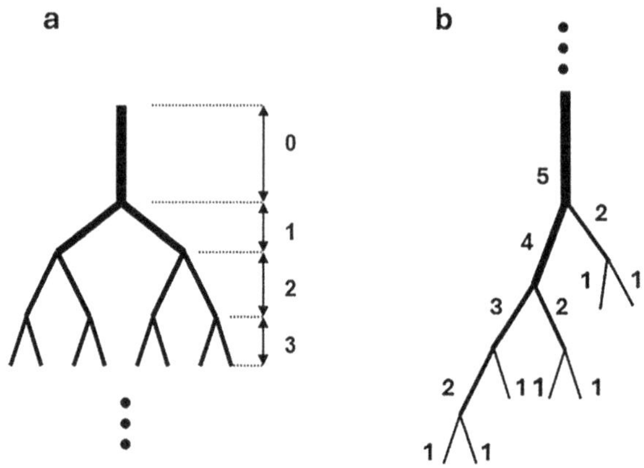

Fig. 2.1 (**a**) A symmetrically bifurcating Weibel airway tree with each branch given a generation number starting at the trachea with generation 0. (**b**) An asymmetrically branching Horsfield tree with each airway segment given an order number starting with the terminal bronchioles having a Horsfield order of 1. (Adapted with permission from Ref. [23])

$$d_n = d_0\gamma^n (n = 1, \ldots, N) \tag{2.1}$$

where N is the generation number of the terminal bronchioles, the smallest airways in the conducting zone of the airway tree. γ has a value of roughly 0.88 in the human lung [21]. In addition, if a fixed ratio between airway length, l, and d is assumed such that $l = \lambda d$, where λ has a value of roughly 3 [18, 22], then the length of any airway branch can be specified as

$$l_n = \lambda d_0\gamma^n (n = 1, \ldots, N). \tag{2.2}$$

The trachea departs from this scheme by being relatively longer than the higher generation branches, but if this extra length is specified by a factor β, the structure of the entire airway tree (Fig. 2.1a) can be completely specified in terms of only 5 parameters – d_0, γ, λ, β and N [24].

While the symmetrically branching airway tree model of Weibel is conceptually simple and computationally convenient, the pulmonary airway tree actually branches in an asymmetric fashion. This means that choosing either of the two daughter branches from any given parent results in different distances of travel to a terminal acinus, in principle resulting in regional heterogeneity of ventilation. A computationally efficient way to encode this asymmetry, introduced by Horsfield [25], is to assign each airway branch an order number (O) starting with the terminal bronchioles for which $O = 1$. At each bifurcation, the daughters have orders O_1 and O_2 that are typically different, while the parent has order $O_0 = [\max(O_1, O_2) + 1]$ (Fig. 2.1b). The degree of asymmetry emanating from any parent branch is determined by the difference $\Delta(O_0) = | O_1 - O_2|$. An asymmetrically branching Horsfield tree can thus be completely specified by tabulating airway branch diameter, length and Δ as functions of O. Such tables have been created for humans [25], dogs [26], and rodents [27]. The mechanical impedances of such trees can be calculated using computationally efficient recursive algorithms [24].

The Weibel and Horsfield airway trees are both fractals because, in both cases, the recursive application of scaling rules leads to self-similar structures. If the rules specified above are followed precisely, the resulting trees are entirely deterministic. Stochastic versions of these models, however, can be readily produced by randomly varying the dimensions of each branch about their nominal mean values during the recursive calculation process [28, 29]. Even more realistic airway tree models can be obtained by separately specifying each individual branch [30]. Such models, however, are computationally expensive and do not convey the insight inherent in an appreciation of the compact underlying design principles conferred by evolution, such as the scaling laws of Eqs. 2.1 and 2.2.

It has been suggested that various biological branching patterns might have evolved to conform to some kind of optimization strategy [18, 31, 32]. One such possibility for the airway tree is that gas flow is directed to the alveoli so as to minimize metabolism and hence entropy production in the respiratory muscles [33]. In general, the primary function of a tubular distribution network such as the pulmonary bronchial or vascular tree is to link a source (the airway opening or heart) to a spatially distributed receptacle (the alveoli or pulmonary capillary bed) so as to supply critical material (oxygen, carbon dioxide, nutrients) in a fashion that best satisfies the demand of the receptacle. This raises the question as to what kind of structure best satisfies demand in terms of the flow of material from the source to the receptacle. Assuming incompressibility, the flow Q_0 through a parent branch is the sum of the flows through the two daughters. That is,

$$Q_0 = Q_1 + Q_2 = rQ_0 + (1 - r)Q_0 \tag{2.3}$$

where $0 < r \leq 0.5$. Starting with the main feeder conduit (the trachea or pulmonary artery), sequential bifurcations of this type continue until Q_1 and Q_2 are of the order of Q_c, an amount small enough such that either air or blood is distributed with sufficient smoothness to its respective side of the blood-gas barrier. Starting with $Q_0 = 1$, the two parameters r and Q_c then fully determine the flow pattern throughout the tree. The tree is symmetric when $r = 0.5$, and becomes increasingly asymmetric as r decreases toward zero. Finally, to determine the structure of the tree, the flow through each branch must be linked to its diameter.

In 1927, Murray proposed that $Q \sim d^3$ represents an optimal relationship between function (Q) and structure (d), a statement now known as Murray's law [34]. This law allows diameters to be assigned to each branch of the tree according to Eq. 2.3. If flow partitioning is symmetric (i.e., $Q_1 = Q_2$), then substitution of Murray's law into Eq. 2.3 results in Eq. 2.1. By utilizing the empirical fact that branch diameter is approximately proportional to length as describe by Eq. 2.2, the full symmetric tree structure can then be specified. Functionally optimized asymmetric trees with realistic variability and complexity can be created with the additional specification of r as a random variable on the interval (0,0.5], [17]. Other details can be incorporated as appropriate, such as flow inertial effects that have been suggested to influence flow asymmetry [35]. This approach has been used to devise a 3D model of the airway tree [18] which, after a few adjustments to account for the irregular

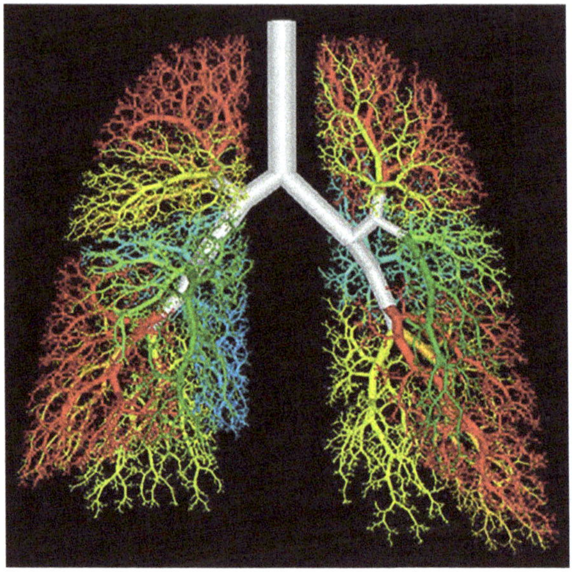

Fig. 2.2 A three-dimensional model of the human airway tree. (Reproduced with permission from Ref. [18])

shape of the chest cavity, produces a highly realistic structure (Fig. 2.2). Although this model is generic in terms of its structural rules, it can be made subject specific through the incorporation of an image-derived central airway geometry.

Another approach to the generation of a space-filling airway tree is to employ a tree growth algorithm [36]. Starting with a 3D CT image of the lung, a finite element mesh is generated and filled with a uniformly distributed set of points representing the pulmonary acini. Next, the proximal portion of the airway tree is generated by placing an initial 1D mesh along the centerlines of those airways large enough to be reliably identified in the CT image. The terminal ends of this truncated tree become the initial seeds for a subsequent stochastic algorithm in which the centroids of subsets of acinar points are utilized to recursively direct the growth of new branches until each acinus is connected to the tree by a single terminal segment. Both this approach and that based on Murray's law create *fractal* networks that are statistically self-similar with fractal dimensions (a measure of self-similar space-filling capacity) close to 3, indicating that these trees are space-filling and thus satisfy their primary function of bringing air uniformly to all corners of the lung [23]. Both these models have been used to explore relationships between structure and function in the lung [37, 38].

The personalization of these airway models to specific lungs has recently been facilitated by novel computational tools such as convolutional neural networks that can accurately and conveniently segment airways from 3D CT images [39]. In combination with such computational methods, these models have been used to predict inhaled aerosol deposition patterns [40], optimal inhalation strategies [41], and how the SARS-CoV-2 virus travels within [42] and spreads throughout [43] the pulmonary airways. However, in order to determine the true impact of inhaled substances, it is necessary to understand not only the airways but also the distal alveolar parenchyma into which the terminal airways empty.

2.3 The Parenchyma

The diameters of the respiratory bronchioles and alveolar ducts deviate from the simple scaling relationship represented in Eq. 2.1 because of the enormous increase in cross-sectional area that occurs in approaching the blood-gas barrier [15]. In this region of the lung, the primary mode of transport of O_2 and CO_2 is thought to be diffusion rather than convection. The alveolar ducts and their subtended alveoli together comprise a highly convoluted structure accounting for the majority of the lung's volume (Fig. 2.3). This structure is essentially a curved 2D sheet that is extremely thin and has a very large surface area. These two attributes together optimize the blood-gas barrier for the diffusive exchange of gases between the alveolar air and the pulmonary capillaries that run in the alveolar walls [21, 44]. Indeed, the functional requirements of gas exchange impose severe constraints on alveolar dimensions [44, 45]. In order to achieve the necessary surface area of more than 60 m^2 [45], the alveoli are very small (approximately 200 μm in diameter at total lung capacity, TLC) and numerous (approximately 480 million in

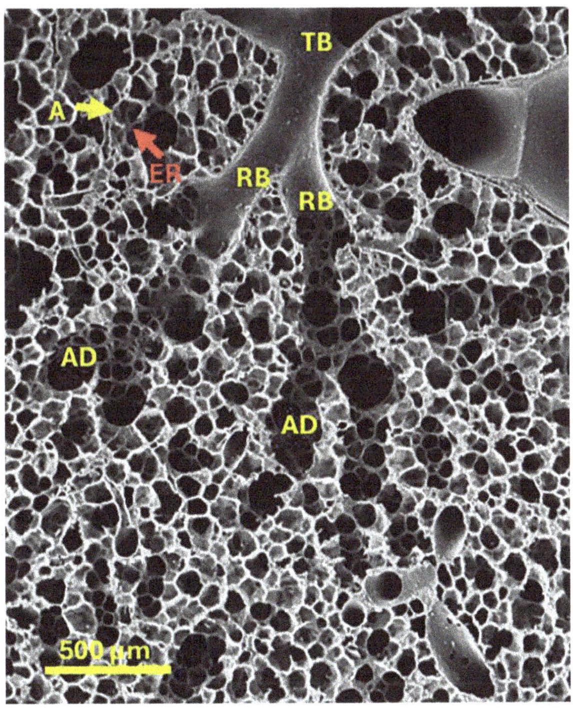

Fig. 2.3 A scanning electron microscopic image of the lung parenchyma. A terminal bronchiole (TB) at the top divides into respiratory bronchioles (RB) leading to alveolar ducts (AD) and subsequent alveoli opening from ADs and surrounding RBs and TBs in a space-filling manner. The yellow arrow marks a single alveolus and red arrow points to the entrance ring (ER) of the same alveolus. (Reproduced with permission from Ref. [44])

the human lung) [46]. In order to minimize diffusion distance, the thickness of the alveolar septal wall is less than 8 μm. The lung parenchyma is thus essentially a large collection of tiny thin-walled interconnected balloons ventilated via a duct system. The result is a fractal structure that has been fine-tuned by evolution to meet the gas exchange requirements of life [44, 47].

Gas exchange within the parenchyma is driven by the rhythmic inflation and deflation of the delicate structure shown in Fig. 2.3, which must therefore be compliant enough to allow easy extensibility during inspiration while maintaining stability against collapse during expiration. The former is permitted by highly flexible elastin fibers in the alveolar septal walls together with wavy collagen fibers that become straightened at high lung volumes to protect the tissue from rupture. Alveolar stability is threatened by the fact that the entire inner surface of the lung is lined with a layer of aqueous fluid, which can degrade alveolar ventilation in two distinct ways. First, if this surface tension equaled that of pure water, the pressures required for inspiration would be impossibly large, and the alveoli would buckle and collapse [48]. Second, increased surface tension predisposes to the formation of liquid plugs in small airways at low lung volume [49]. Such plugs completely occlude the airway lumen, thereby eliminating the downstream lung from participation in gas exchange. To deal with this, Nature has evolved a substance called pulmonary surfactant that is continually secreted by specialized cells, the alveolar type II epithelial cells. Surfactant lowers surface tension substantially in a lung volume dependent manner that both stabilizes the alveoli against collapse and greatly reduces the stiffness of the parenchyma so that it is easily inflated by the respiratory muscles.

Alveolar stability is also provided by the stress across the lung due to the difference between the pressure at the airway opening and the negative (with respect to atmospheric) pleural pressure surrounding the lungs in the thoracic cavity. This pressure difference is known as transpulmonary pressure (P_{tp}). The pleura transmits the stress arising from P_{tp} to the subpleural alveolar ducts and septal walls, which in turn pull on the more internal tissue structures primarily through the elastin and collagen fiber networks of the parenchyma [50]. P_{tp} thus provides a distending prestress to the alveolar walls at functional residual capacity (FRC) that protects the alveoli from collapse at end-expiration [6]. Pleural pressure becomes increasingly negative during inspiration due to contraction of the diaphragm and, to a lesser extent, expansion of the thoracic cavity which causes air from the environment to flow into the airways and ultimately the alveoli. The amount of air entering the respiratory zone of the lung, where gas exchange takes place, depends on how much the alveolar parenchyma inflates, which in turn is determined by the mechanical properties of both the ducts and the alveoli. Thus, even though oxygen travels mostly via diffusion within the acinus, the efficiency of gas exchange is critically dependent on alveolar mechanics.

This stress transmission pathway driven by P_{tp} is mediated to a large extent by the ECM [50], the mechanical properties of which are determined by its constituent elastin and collagen fibers and a space-filling "ground substance" composed mostly of proteoglycans in which the fibers are embedded [50]. P_{tp} is also influenced

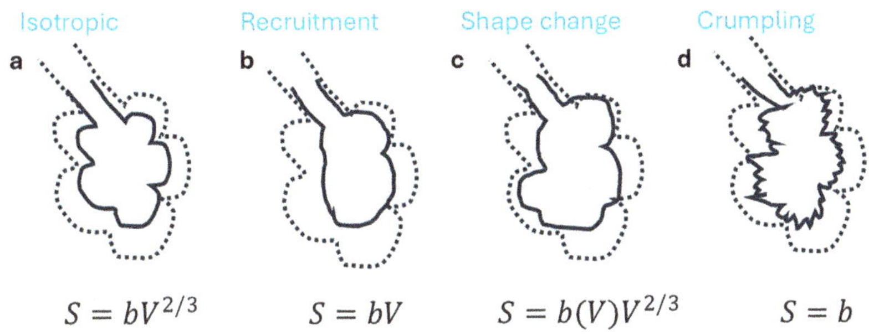

Fig. 2.4 Four types of alveolar expansion from a low (solid lines) inflation state to a high (dashed lines) inflation state (**a**) Alveoli expand in an isotropic fashion like a smooth balloon. (**b**) Some alveoli collapse at end-expiration and are then recruited again during inspiration. (**c**) Expansion is accompanied by shape change. (**d**) Toward end-expiration, septal walls fold and crumple like an accordion and then smoothly unfold during inspiration. The surface area (S_{alv})-volume (V_{alv}) relationships are shown for each expansion mechanism

substantially by the surface tension of the air-liquid interface [51–54] and to a lesser extent by the properties of adherent cells [55–57]. The static end-expiratory component of P_{tp} determines FRC along with a number of crucial physiologic processes that include regional ventilation and perfusion, contractile ability of airway smooth muscle cells, surface tension, and alveolar stability [58]. The dynamic component of P_{tp} due to breathing causes cyclic stress to be applied to the pulmonary cells that are attached to the fibers of the ECM. This affects surfactant production and secretion [59, 60] as well as other functions of lung cellular biology [61, 62].

Because total alveolar surface area (S_{alv}) is such an important determinant of both gas exchange efficiency and parenchymal mechanics, it is clearly important to understand the relationship between S_{alv} and alveolar volume (V_{alv}) during breathing. The apparently simple question of how an alveolus and alveolar duct expand during inflation has been the subject of ongoing discussion for decades [52, 63, 64]. Even now, exactly how alveolar shape changes with lung volume remains unresolved. In particular, it is not fully understood whether septal walls and basement membranes pleat and crumble [52, 65] or inflate and deflate smoothly [66]. In fact, four modes of alveolar expansion have been proposed [52]: (i) smooth balloon-like stretching, (ii) recruitment, (iii) expansion with a change in shape, and (iv) accordion-like unfolding, as illustrated schematically in Fig. 2.4. There are quantitative and qualitative differences between these modes. For example, recruitment implies that $S_{alv} = bV_{alv}$ whereas smooth expansion implies that $S_{alv} = bV_{alv}^{2/3}$, where b is a constant. A shape change occurs if b is not constant, whereas crumpling may be viewed as a reduction in the fractal dimension.

Early investigations into this question examined reconstructed images made from serial histologic sections [67] or electron micrographs [65] obtained from lungs fixed in various states of deflation from TLC. These studies showed that alveolar diameters decrease very little, from 102.7 ± 2.4 to 100.3 ± 3.3 μm, as the lung deflates from a P_{tp} of 30 to 10 cmH_2O [67]. In contrast, when P_{tp} decreases from 10 to 5

cmH_2O, the diameters drop significantly to 43.3 ± 2.3 μm. This is consistent with the existence of the plateau seen along the upper part of the pressure-volume relationship of the lung. It must be remembered, however, that these results are likely affected by the tissue shrinkage that is associated with fixation [68]. Direct micro-CT visualization of a single mouse alveolus fixed at a P_{tp} of 30 cmH_2O showed the diameter to be 59 ± 2 μm, a value greater than those consistently obtained from 2D images [69]. This is likely because micro-CT does not involve the histologic processing and serial sectioning that can produce additional tissue shrinkage.

Recent developments in imaging methodology now permit alveolar visualization without the need for tissue fixation. Optical coherence tomography (OCT) and intravital microscopy (IVM) allow 3D images of subpleural alveoli to be obtained rapidly during lung inflation and deflation [70]. Both methods can visualize the lung surface in situ through a window cut in the chest wall, but both have their respective disadvantages; OCT lacks the resolution of optical microscopy, while IVM may suffer from artifacts caused by having to attach the lung to the chest wall around the window perimeter [63]. The perimeters of subpleural alveoli have also been imaged via confocal fluorescent microscopy, and suggest that the structural components of a single alveolus stretch heterogeneously during inflation [71].

Near artifact-free dynamic imaging of alveoli can be achieved with the so-called "crystal ribcage", a transparent container in the shape of the thoracic cavity into which an isolated, ventilated and perfused lung is placed. The lung can then be imaged through the transparent chest wall from the scale of the whole organ down to that of single cells [72]. The crystal ribcage has been used to determine the quasi-static pressure-diameter relationships of hundreds of subpleural alveoli as well as the diameter changes that occur during the respiratory cycle at different ventilation frequencies. These investigations suggest that (i) pressure-diameter relationships are quite heterogeneous even under quasi-static conditions due to the heterogeneous distribution of sub-alveolar strains, (ii) increasing ventilation frequency distorts the diameter-time profile, and (iii) alveolar mechanics are significantly altered by diseases such as pulmonary fibrosis, emphysema, pneumonia and cancer.

Synchrotron-based x-ray CT is another imaging technology that allows in vivo visualizing of deep lung structures with high spatial and temporal resolution [73, 74]. Imaging both alveoli and alveolar ducts throughout the respiratory cycle shows that individual alveoli follow the relationship $S = bV^{2/3}$, with b being a function of lung volume, indicating that alveolar shape changes during inflation [73]. Early in inflation, alveoli expand more than ducts, but the ducts catch up at higher lung volumes where the parenchyma stretches more uniformly. Interestingly, no folding or unfolding of septal walls has been detected during breathing with this technique. Both the crystal ribcage and synchrotron-based x-ray imaging have reported large variations in alveolar septal wall linear strains, ranging from ~0.04 to above 0.4, in response to a change in P_{tp} of ~12 cmH_2O. This indicates substantial micro-level heterogeneity in the mechanical properties of the alveoli, implying that the lung parenchyma has evolved to function well even in the presence of massive alveolar heterogeneity.

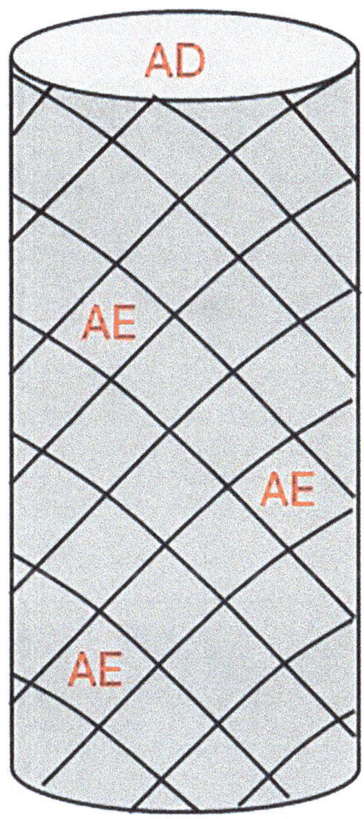

Fig. 2.5 Helical arrangement of intersecting line elements forming a regular lattice of alveolar entrance (AE) rings on the surface of the alveolar duct (AD)

The first quantitative model of parenchymal mechanics based on alveolar structure was a 2D hexagonal network of springs proposed by Mead in 1970 [6]. This model has subsequently been extended into 3 dimensions, along with the inclusion of more anatomic detail such as crisscrossing helical stress fibers along the alveolar duct that intersect to define a regular lattice of alveolar entrance rings [75] (Fig. 2.5). Despite the fact that this helical model mimics alveolar parenchymal mechanics rather well (see Chap. 4 for more details), it is the shapes of the alveoli to which the ducts connect that determine whether or not the model is an efficient packing structure for the parenchyma. Fung proposed various regular 3D geometric shapes (polyhedra) as candidate models for the alveolar duct and its surrounding alveoli [76]. The basic unit of the model is the space-filling truncated octahedron (tetrakaidecahedron) (Fig. 2.6a). The parenchyma is built up from a combination these units, with some internal faces removed to create the alveolar duct. The octahedral face corresponding to the alveolar entrance ring is also removed to open each alveolus to the duct (Fig. 2.6b, c). This model assumes that the alveoli are all identical in size and shape, that they connect uniformly to the alveolar duct, and that the alveolar entrance rings are reinforced and stabilized with additional structural ECM. Optimization of this polyhedral structure in terms of maximizing gas exchange and minimizing gas transit time to the blood-gas barrier has been proposed [77], while variations of the model have been used for analyzing the mechanical properties of the parenchyma in health and disease [78–83]. The

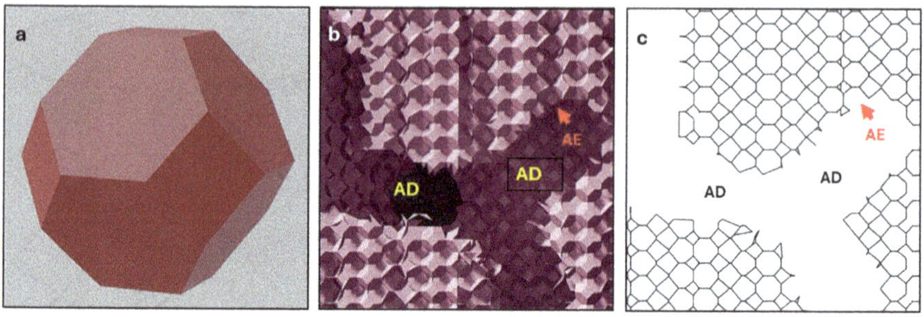

Fig. 2.6 (**a**) The basic unit of many computational models of the acinus is the truncated octahedron that consists of 8 hexagonal and 6 square faces. (**b**) Cut view of a large set of truncated octahedra assembled with selected faces removed to form bifurcating ducts (AD). An alveolar entrance (AE) ring is shown by the red arrow. (**c**) A 2D section of the structure in panel (**b**) showing the regular organization of hexagons and squares. (Images kindly provided by Joseph K. Hall)

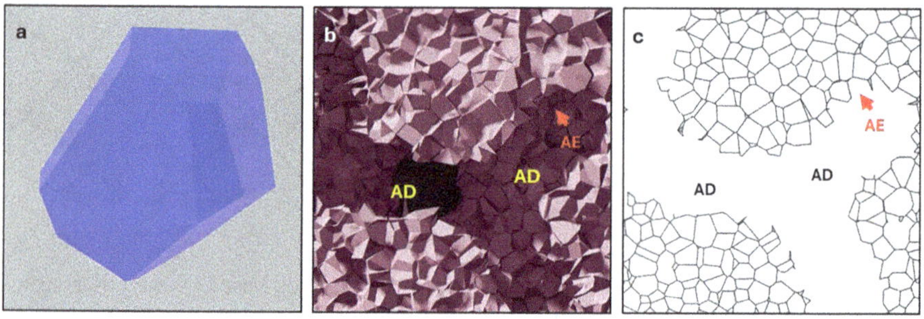

Fig. 2.7 (**a**) An example of a polyhedron, the basic unit of the amorphous space-filling network. (**b**) Cut view of a large set of polyhedra with selected faces removed to form a bifurcating duct (AD). An alveolar entrance (AE) ring is shown by the red arrow. (**c**) A 2D section of the structure in panel (**b**) showing the irregular organization of polyhedra. (Images kindly provided by Joseph K. Hall)

truncated octahedron model of the parenchyma has even been adapted to account for alveolar instability through the incorporation of hinges where septal walls join, allowing the alveoli to undergo collapse and re-expansion [84].

An issue with the truncated octahedron model, however, is that it gives rise to parenchymal structures that are unnaturally regular. There is significant natural variability in alveolar size and shape, as can be seen in Fig. 2.3. The apparent size heterogeneity that appears when the truncated octahedral model is sectioned (Fig. 2.6c) is merely a consequence of identical alveolar units (Fig. 2.6a) having been sliced through at different heights. The coefficient of variation (CV = standard deviation/mean) of alveolar volume in the model is actually zero, whereas micro-CT images of real parenchyma show the volume CV to be approximately 60% [69]. This natural variability is better recapitulated in an amorphous alveolar model created by surrounding a random set of points in 3D space with springs that form a Voronoi

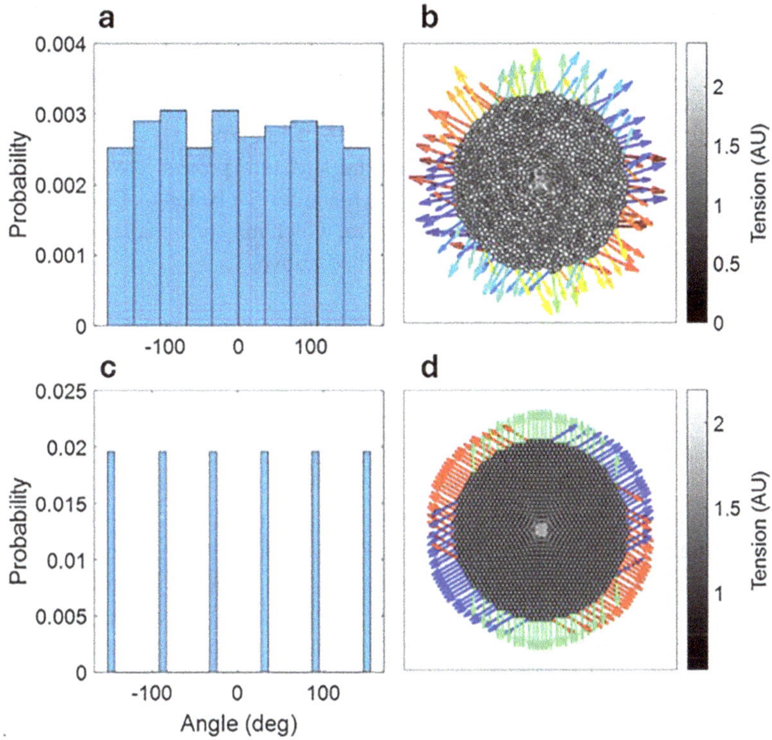

Fig. 2.8 (**a**) Distribution of reaction force angles in the amorphous network. (**b**) The amorphous network with a stiffened region in the center. (**c**) Probability distribution of reaction force angles in the regular hexagonal network. (**d**) The hexagonal network, with stiffened region in the center. In panels (**b** and **d**), the reaction forces at the boundary are shown as arrows. Color indicates the angle of the reaction force. (Reproduced with permission from Ref. [83])

network (Fig. 2.7) [83]. Alveolar ducts can be included in this model, as before, by eliminating internal faces, as shown in Fig. 2.7b.

An interesting consequence of random lattice irregularity is large-scale isotropy. In contrast to the 3D network of uniform truncated octahedra, which is not isotropic due to the limited rotational symmetry of its basic unit (Fig. 2.6a), the amorphous network exhibits isotropy because its alveolar units are all different and so on average have complete rotational symmetry [83], in agreement with experimental data [85]. This is illustrated in 2D in Fig. 2.8, which compares the periodic force vectors around the perimeter of a regular hexagonal network (Fig. 2.8a) to the continuously distributed vectors around an amorphous network (Fig. 2.8b). Anisotropy of force transmission means that mechanotransduction, defined as the conversion of mechanical information to chemical signaling leading to downstream biologic events [86, 87], is also not direction dependent. Direction-dependent mechanotransduction has not been observed in the lung, which supports the use of isotropic models.

Structural randomness may also be important for lung development. Too much regularity in the branching pattern of the airway tree could lead to amplification of small structural errors that occur during growth. The branching variations that are built into a heterogeneously branching fractal structure protect against such error propagation [19]. The same argument can be made for the benefits of random spatial variation versus regularity of alveolar structure, which may also explain the origin of the large alveolar strain heterogeneity described above.

2.4 The Extracellular Matrix

The principal structural framework for the lung is provided by the ECM that permeates the entire organ from the pleura, through the alveolar septa, alveolar ducts, terminal bronchioles, and conducting airway tree all the way to the trachea. The ECM holds all these structures together while exerting important biologic effects via its dynamic mechanical behavior [8]. The principal components of the ECM are the various types of collagen that confer tensile strength, the elastic fibers that permit distensibility, and the proteoglycans that contribute to tissue bulk. In addition to such mechanical attributes, these components have crucial biologic functions such as providing a framework for cell–cell and cell-substrate communication and maintaining an aqueous environment.

2.4.1 The Collagens

The primary structural protein of the ECM is collagen. Twenty-eight different types of collagen have been identified, including the fibril-forming types I, II and III, the regulatory fibril forming types V and XI, basement membrane type IV, the fibril-associated collagens with interrupted triple helices types IX, XII, XIV and XX, and other minor types. All collagen molecules have a helical part, called the collagen domain, which provides them with the ability to support structure. The helices are right-handed and consist of three α chains which are left-handed polypeptide coils. The major fibril forming types (I, II and III) consist of approximately 1000 amino acids [88]. The collagen domains are relatively stiff rod-like structures having lengths and diameters of approximately 300 and 1.5 nm, respectively, and can spontaneously assemble into fibrils [89]. Before being released from collagen-secreting cells, collagen exists in a precursor form known as procollagen in which the helical domains are decorated at both the carboxyl and amino ends with globular domains known as propeptides [89]. Once secreted, extracellular enzymes cleave the propeptides, which prepares the collagen molecules for fibril assembly first through lateral and then axial combination. The final fibril is a complex hierarchically packed set of molecules with some degree of fluid-like disorder [90, 91]. The fibrils themselves then undergo lateral and axial association to produce collagen fibers. Cross-linking of lysine and hydroxylysine residues within and between fibrils reinforces this cable-like structure to become the main loadbearing framework for many tissues. Further organization of fibers into higher-order structures such as the random networks that make up lung tissue and cartilage, or the quasi-parallel networks comprising tendon, is tissue and organ specific.

The fibers of the lung ECM are predominantly comprised of type I and type III collagen, and have diameters ranging from a few hundred nm to well above 1 μm [92]. The distribution of fiber thickness is skewed, with a heavy tail [92] that is similar to a power law, thus signifying complexity in fiber structure [93]. These fibers play a variety of roles in maintaining the structural and mechanical integrity of lung tissue. Axial fibers extend along the airway walls from the central airways down to the alveolar ducts. Peripheral fibers emanate into the lung from the highly collagenous visceral pleura. A network of fibers in the parenchymal septal walls links the other two networks together [94]. The sheet-like organization of the type IV collagen comprising the basement membrane allows epithelial and endothelial cells to adhere to alveolar/airway and vascular walls, respectively. The collagen content of the lung varies substantially during development [95, 96] and in diseases such as pulmonary fibrosis [97]. It can also be modulated *in vitro* through administration of proteases [98].

A key feature of the collagenous fiber system in the lung is that the individual collagen fibers are flaccid (have a wavy configuration under the microscope) at low lung volumes (Fig. 2.9a), implying they bear little stress. With increasing lung volume, however, the fibers progressively straighten (Fig. 2.9b) as they take on a load-bearing role [100] that protects the tissue from becoming overdistended. This

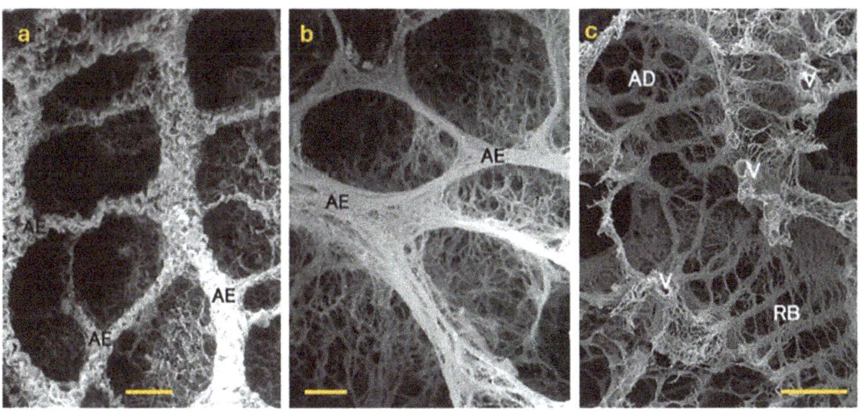

Fig. 2.9 Scanning electron microscopic images of the lung parenchyma. (**a**) Wavy organization of collagen fibers are seen at a low lung volume. (**b**) At a higher lung volume, collagen fibers appear straight. (**c**) Elastin fibers in the parenchyma are relatively straight at all lung volumes. *AE* alveolar entrance, *AD* alveolar duct, *RB* respiratory bronchiole, *V* vessel. Scale bars on panels (**a** and **b**) are 100 μm and on panel (**c**) is 200 μm. (Reproduced with permission from Ref. [99])

causes lung tissue to exhibit the phenomenon of strain stiffening, manifest as a nonlinear relationship between lung volume and inflation pressure [101, 102].

How stiff are collagen fibers? This is a question of great interest both from the physiological and the biological points of view. Collagen stiffness not only governs overall parenchymal stiffness and hence force transmission and ventilation at the macroscale [50], it also governs cellular behavior such as fibroblast migration [103] and phenotype [104]. The ability of a material to resist tensile strain (changes in length) is quantified by the Young's modulus defined as the ratio of a change in tensile stress (force divided by cross-sectional area perpendicular to the direction of force) to the corresponding change in tensile strain (change in length divided by starting length). The type I collagen molecule has a Young's modulus of between 3 and 9 GPa [89, 105]. The stress–strain behavior of lung collagen fibers is nonlinear for strains up to approximately 5% due to straightening of wavy fibrils within the fiber followed by a linear regime up to ~10% strain with Young's modulus values of 62+_40 MPa [148].Various models have been proposed to explain the mechanistic basis of the force-extension behavior of the collagen molecule, such as the worm-like chain model in which elasticity arises from a competition between continuous bending of the molecule and the entropic forces that resist unfolding of the molecule during stretching [106]. At the fibril and fiber levels, both analytic and computational approaches have been developed that take into account molecular elasticity, cross-linking, electrostatic interactions, bond formation and rupture as well as fibril waviness [8].

2.4.2 *The Elastic Fibers*

Elastic fibers are the second most important load-bearing constituent of the ECM after collagen. The core component of elastic fibers is elastin, a highly stretchable biopolymer. Elastin is synthesized by various lung cells including smooth muscle cells, endothelial cells and fibroblasts. It first emerges in a precursor form, known as tropoelastin, which is rich in hydrophobic amino acids [107]. This hydrophobicity together with a high crosslinking density creates a very stable and insoluble molecule that is able to form fibrils and fibers with a half-life of 70 years [108]. The surfaces of elastic fibers are decorated with microfibrils including fibrillins, fibulins, and microfibril-associated glyoproteins [107, 109, 110]. These microfibrils are essential for elastogenesis due to their ability to regulate the deposition of tropoelastin onto existing elastic fibers [111]. While the 3D molecular structure of elastic fibers is not fully understood [91, 112], they are known to be characterized by structural heterogeneity with diameter and length distributions having long tails similar to those observed for collagen fibers [92]. Elastic fibers are also mechanically connected to collagen fibers [113], possibly through microfibrils or proteoglycans [91, 114, 115].

Elastic fibers are highly compliant and exhibit linear stress–strain behavior up to at least 200% strain [116]. In addition, little energy is dissipated when elastic fibers are stretched, causing them to behave essentially like ideal linear springs [8], which is convenient for the construction of models of lung elastic networks [100]. Elastin is also highly resilient to rupture over a large range of strain. Evolution has exploited these superb attributes by incorporating elastic fibers into the ECM of tissues that need to undergo cyclic stretching throughout life, of which there is no better example than the lung which needs to recoil back to FRC after every inspiration.

How stiff are elastic fibers? They are 2 to 3 orders of magnitude less stiff than collagen fibers [116]. This is due to the entropic manner in which amorphous elastin stretches within the fiber as compared to the much more hierarchical and rigid organization of collagen fibers [8]. The Young's modulus of elastic tissue purified from the thoracic aorta is 0.4 MPa when measured in the circumferential direction at a strain of 0.4, and is 0.25 MPa in the longitudinal direction [117]. Stress–strain data for a single elastic fiber have also been reported [118], and yield a Young's modulus value of 0.74 MPa [8]. The Young's modulus of a microfibril can be as low as 0.2 MPa [119] and as high as 96 MPa [120], which is in the range of collagen fibril stiffness [116]. Microfibrils do not appear to contribute significantly to elastic fiber stiffness, however, as their elimination does not impact stiffness [121]. This is probably because of their low volume fraction within the fiber.

As with collagen, elastic fibers form a network that percolates throughout the lung tissue (Fig. 2.9c), allowing them to play a major role in lung elasticity. Indeed, elastin is thought to be the primary source of parenchymal stiffness at lung volumes in the range of normal breathing [122] while, as discussed above, collagen fibers become progressively more influential at higher lung volumes. Interestingly, separate digestion of elastin and collagen in lung tissue strips suggests that collagen also makes a significant contribution to stiffness at lower lung volumes [98], possibly as a result of the way that the collagen and elastic fiber networks are linked to each other mechanically [100].

2.4.3 The Proteoglycans

While both collagen and elastic fibers support the structure and biomechanical function of the lung, they need to be immersed in a microenvironment that is conducive to homeostatic maintenance and repair of the airway and parenchymal tissues. This conducive microenvironment is provided by a hydrated gel known as the ground substance, the composition and volume fraction of which are highly tissue specific [91] and can change markedly during development and in disease [123]. The primary constituents of the ground substance are the proteoglycans, which can bind to collagen, elastic fibers, cells and each other, gluing them all together to form a coherent living tissue. Proteoglycans are composed of a core protein to which sugar-like side chains are attached. These side chains are the glycosaminoglycans, long polymers of repeating disaccharides that are often sulfated and highly charged. This allows the glycosaminoglycans to sequester large numbers of water molecules, giving them the ability to modulate tissue bulk and mechanics [91]. There a numerous types of glycosaminoglycan including hyaluronic acid, dermatan sulfate, keratan sulfate, chondroitin sulfate and heparan sulfate, the latter two being the most abundant in the lung. Glycosaminoglycans exhibit substantial variability in composition, size and structure, and have molecular weights that span 3 orders of magnitude, corresponding to more than 10^4 monomers [124, 125]. Similar to collagen, glycosaminoglycans can form helical or randomly organized structures depending on local environmental factors such as ionic content and pH [125]. It is thus not surprising that electron micrographs show these structures to have extraordinary complexity [124, 126].

In addition to providing the tissue microenvironment with hydration and physical cohesion, proteoglycans have countless other biological functions related to the organization and maintenance of the ECM [127, 128]. Indeed, proteoglycans and glycosaminoglycans are essential for tissue development and regeneration [129]. Some proteoglycans serve as cellular transmembrane proteins that mediate various signaling pathways [130] including growth factor-regulated production and secretion of proteins such as enzymes involved in tissue remodeling and cell migration [131]. A key function of the proteoglycans decorin, biglycan, lumican and fibromodulin is to control the axial and lateral growth of collagen fibrils [91, 115, 132]. Proteoglycans have also been found to contribute to elastic fiber assembly [133].

Although less well understood, proteoglycans can play direct structural and biomechanical roles in the tissue [134–136]. For example, decorin, especially its core protein, has been shown to regulate the stiffness of collagen networks by influencing both individual fibril assembly and the formation of compact fibrillar networks [137]. Experimental and computational modeling studies have together shown that proteoglycans also exert a direct influence on the elastic behavior of the lung parenchyma by interfering with the ability of structural fibers to align in the direction of an applied stress (discussed in Chap. 4), especially at low lung volumes [135]. This may allow proteoglycans to help prevent the crumpling of septal walls as seen in Fig. 2.4d.

2.4.4 Lung Cells

The lung is home to approximately 50 different cell types that actively remodel the ECM during growth, maintain the airway and parenchymal tissues throughout life, and repair them after injury. These cells are affected by the prestress generated by P_{tp} and the force of gravity. This prestress is transmitted from the percolating fibers of the ECM to its adhering cells, which perceive force through the displacement of their focal adhesions that occurs when elastic fibers elongate, or wavy collagen fibers straighten. Cyclic variations in local tissue stress due to breathing are also transmitted by the ECM, triggering cellular mechanotransduction. For example, transforming growth factor beta (TGFβ), a cytokine that stimulates fibroblasts to produce and secrete ECM, is upregulated by mechanical stretch [138]. Tensile forces can stimulate the release of connective tissue growth factor and hence the production and secretion of ECM via a TGFβ-independent pathway [139]. Once the fiber network is remodeled via these processes, the nonlinear and viscoelastic properties of the tissue change both at the macro- and the microscales [140]. There are many mechanotransduction-related phenomena in the lung that operate in a similar fashion, such as stretch-induced surfactant release [141].

While it is clear that tissue mechanics can affect cellular behavior, the converse is also true. For example, when smooth muscle cells contract to narrow an airway, they stiffen the airway wall as well as cause a local distortion in the adjoining parenchyma. This induces local stresses in the ECM leading to distortion of the fiber network, which in turn alters the viscoelastic properties of the tissue (viscoelasticity is discussed in Chap. 5) [7, 55, 57, 142, 143]. The extent to which this alters the overall mechanical behavior of the parenchyma depends on the volume fraction of airways in the tissue [144]. Parenchymal mechanics are also affected, albeit to a much lesser extent, by contraction of the actin-myosin networks within epithelial cells [145], endothelial cells [146], and cells of the interstitium [55, 57]. Contractile cells such as fibroblasts and myofibroblasts also reside in alveolar ducts and entrance rings, as well as in blood vessel walls [147]. The contractile protein marker α-actin has even been found at the ends of alveolar septal walls, although not within the walls themselves [56]. The biologic or physiologic role of these contractile elements deep in the lung parenchyma is currently not known.

2.5 Conclusions

The lung is a complex system composed of a large number of disparate components that all work together to perform the crucial function of gas exchange. These components are arranged in a hierarchical multi-scale fashion that gives rise to interdependent collagen and elastic fiber networks glued together by proteoglycans. These structures in turn support hundreds of millions of alveoli served by a fractal branching airway system connecting them to the environment. This allows the tracheal cross-section of only a few square cm to transition to an area equivalent to that of a tennis court at the blood-gas barrier, a feature that makes mammalian life possible. The mechanical properties of the fiber networks together with the actions of pulmonary surfactant at the air-liquid interface allow the lung to be readily and repeatably distended in the act of breathing throughout life. The forces involved in this distension are sensed by resident lung cells that respond so as to maintain and repair the lung in the face of ongoing wear and tear as well as environmental exposures. Lung biology and mechanics are thus tightly and dynamically coupled, with crucial implications for development, homeostasis, and response to injury and disease. The possibilities for investigating this rich functional and structural landscape through computational modeling are limitless, as the following chapters will demonstrate.

2.6 Problems

1. Using Eqs. 2.1 and 2.2, calculate analytically how the total cross-sectional area of the symmetric airway tree at each generation increases with generation number n. Also calculate how the cross-sectional area depends on the physical distance (cumulative length) from the root of the tree. Code these equations up in your programming language of preference and, using $N = 24$, $d_0 = 1$ (normalized diameter), $\gamma = 0.8$ and $\lambda = 3$, plot the results. Try plotting these relationships on double logarithmic axes. What can you say about the increase of the total cross-sectional area in this model? Determine and plot the sensitivity of the cross-sectional area to γ as a function of generation number. Sensitivity can be defined as the percent change in cross-sectional area when γ is varied by a small percentage (e.g., ± 5%).
2. As a more advanced exercise, code up Eq. 2.3 in a sequential manner so that the entire tree can be efficiently stored in a linear array as follows. The generation numbers starting with 0 at the root allow us to track where the flow partitioning is in the tree. For example, consider the case of a symmetric bifurcation with $r = 0.5$ and $Q_c = 0.3$. Assuming $Q_0 = 1$ at generation $g = 0$, the first bifurcation will generate flows of $Q_1 = Q_2 = 0.5$ at generation $g = 1$. Both daughters give rise to two new daughters at $g = 2$ with identical flows of 0.25 in the 4 terminal segments with $g = 2$. The g array can be stored as $g = [0, 1, 2, 2, 1, 2, 2]$ which completely characterizes the connectivity of the tree. The end of the tree is found when two consecutive generation numbers are the same. Now consider an asymmetric tree with $Q_c = 0.25$. At the first bifurcation, assume $r = 0.4$. Moving to the left branch with flow of 0.4, assume that $r = 0.5$ which generates 2 daughter segments each with a flow of 0.2. As we move down the tree always to the left branch, this is stored in g as [0, 1, 2]. Since the local flow $< Q_c$, the tree terminates here. Moving back up one generation, we add the right branch: [0, 1, 2, 2]. Since the tree terminates here, we jump up to the first available unfinished bifurcation stemming from $g = 0$. The right daughter at $g = 1$ has a flow of 0.6 and is now added to the tree resulting $g = [0, 1, 2, 2, 1]$. Assume that branching occurs again with $r = 0.5$. Both daughters have flows of 0.3 and both further bifurcate into the next generation. Continuing this process, the generation number array is now $g = [0, 1, 2, 2, 1, 2, 3, 3, 2, 3, 3]$ and the corresponding flow array is $Q = [1, 0.4, 0.2, 0.2, 0.6, 0.3, 0.15, 0.15, 0.3, 0.15, 0.15]$.

   (a) To understand the bifurcation pattern and how the tree is stored, draw the tree structures corresponding to the above symmetric and asymmetric bifurcations. Write a code for an algorithm to create an arbitrary bifurcation pattern. Decrease Q_c to 0.1, fix $r = 0.3$, and generate the tree. Calculate and draw the tree manually and make sure your algorithm works correctly.
   (b) Create a large stochastic tree by varying r uniformly between 0.2 and 0.5 and assign $Q_c = 0.00006$. Show that the probability density distribution of terminal generation numbers is a Gaussian with a mean of approximately 16 and SD of 2. Show that the terminal flows follow a lognormal distribution.

Calculate the probability density distribution of all flows. Show that this distribution is approximately proportional to Q^{-2}. This is best seen on a double logarithmic graph (Hint: the first 1–2 bins would be omitted from the straight-line fit on the double log graphs). Calculate the diameter of each segment using a diameter exponent of $d_e = 2.8$. Scale the diameters such that the root, corresponding to the trachea, has a diameter of 2.5 cm. Calculate and characterize the distribution of diameters. Is this also a power law distribution?

(c) Generate 10 trees with identical parameters. How do the above generation number, flow and diameter distributions change from realization to realization? Change the uniform distribution of r to be between 0.2 and 0.4 and set $Q_c = 0.0008$. Which distributions are affected by this change in the model parameters? You might be interested to know that the latter is a model of the canine airway tree while the former is a model of the human airway tree.

3. Code up a structural network model of the parenchyma. For simplicity, use a 2D hexagonal network. Apply small random changes in the x and y positions of the internal nodes of the network. The amount of noise can be specified as a percentage of the side length of the hexagon.

   (a) Write an algorithm to compute the total internal 2D surface area (sum of the side lengths) of the network. How does the total area change when 1%, 5%, 10% or 20% noise is added to the node positions?
   (b) How is the total 2D volume (sum of the hexagonal areas), affected by the level of noise? Calculate the mean and SD of the areas and determine the coefficient of variation of the 2D volumes as a function of noise. Calculate the equivalent diameter of each enclosed area (diameter of a circle having the same area). What is the noise level that best mimics the normal lung structure if the coefficient of variation of the equivalent diameters of the normal lung is 25%?
   (c) Using the structure with the best noise level from above, carve out an alveolar duct bifurcation as in Fig. 2.6c. Determine how the total internal surface area decreases in this model compared to the one without the alveolar duct.

References

1. Plopper CG, Mariassy AT, Lollini LO. Structure as revealed by airway dissection. A comparison of mammalian lungs. Am Rev Respir Dis. 1983;128:S4–7.
2. Schlesinger RB, LA MF. Comparative morphometry of the upper bronchial tree in six mammalian species. Anat Rec. 1981;199:99–108.
3. Phalen RF, Oldham MJ. Tracheobronchial airway structure as revealed by casting techniques. Am Rev Respir Dis. 1983;128:S1–4.
4. Glenny RW, Krueger M, Bauer C, Beichel RR. The fractal geometry of bronchial trees differs by strain in mice. J Appl Physiol. *1985*;2020(128):362–7.

5. Lum H, Mitzner W. A species comparison of alveolar size and surface forces. J Appl Physiol. *1985*;1987(62):1865–71.
6. Mead J, Takishima T, Leith D. Stress distribution in lungs: a model of pulmonary elasticity. J Appl Physiol. 1970;28:596–608.
7. Bates JH, Lauzon AM. Parenchymal tethering, airway wall stiffness, and the dynamics of bronchoconstriction. J Appl Physiol. *1985*;2007(102):1912–20.
8. Suki B. Structure and function of the extracellular matrix: a multiscale quantitative approach. Academic; 2021.
9. Suki B, Stamenovic D, Hubmayr RD. Lung parenchymal mechanics. In: Fredberg JJ, Sieck GC, Gerthoffer WT, editors. Comprehensive physiology, the respiratory system, respiration mechanics: organ, cell, molecule, vol. 1. Wiley-Blackwell; 2011. p. 1317–51.
10. LaPrad AS, Lutchen KR, Suki B. A mechanical design principle for tissue structure and function in the airway tree. PLoS Comput Biol. 2013;9:e1003083.
11. Lloyd CM, Marsland BJ. Lung homeostasis: influence of age, microbes, and the immune system. Immunity. 2017;46:549–61.
12. Stolz D, Mkorombindo T, Schumann DM, Agusti A, Ash SY, Bafadhel M, Bai C, Chalmers JD, Criner GJ, Dharmage SC, et al. Towards the elimination of chronic obstructive pulmonary disease: a lancet commission. Lancet. 2022;400:921–72.
13. Knudsen L, Ruppert C, Ochs M. Tissue remodelling in pulmonary fibrosis. Cell Tissue Res. 2017;367:607–26.
14. Weibel ER, Gomez DM. A principle for counting tissue structures on random sections. J Appl Physiol. 1962;17:343–8.
15. Weibel ER, Gomez DM. Architecture of the human lung. Use of quantitative methods establishes fundamental relations between size and number of lung structures. Science. 1962;137:577–85.
16. Majumdar A, Alencar AM, Buldyrev SV, Hantos Z, Lutchen KR, Stanley HE, Suki B. Relating airway diameter distributions to regular branching asymmetry in the lung. Phys Rev Lett. 2005;95:168101.
17. Kitaoka H, Suki B. Branching design of the bronchial tree based on a diameter-flow relationship. J Appl Physiol. 1997;82:968–76.
18. Kitaoka H, Takaki R, Suki B. A three-dimensional model of the human airway tree. J Appl Physiol. 1999;87:2207–17.
19. West BJ. Physiology in fractal dimensions: error tolerance. Ann Biomed Eng. 1990;18:135–49.
20. Glenny R, Bernard S, Neradilek B, Polissar N. Quantifying the genetic influence on mammalian vascular tree structure. Proc Natl Acad Sci USA. 2007;104:6858–63.
21. Weibel ER. Morphometry of the human lung. New York: Academic; 1963.
22. Horsfield K, Cumming G. Morphology of the bronchial tree in man. J Appl Physiol. 1968;24: 373–83.
23. Suki B, Bates JH, Frey U. Complexity and emergent phenomena. In: Fredberg JJ, Sieck GC, Gerthoffer WT, editors. Comprehensive physiology, the respiratory system, respiration mechanics: organ, cell, molecule, vol. 1. Wiley-Blackwell; 2011. p. 995–1029.
24. Bates JHT. The role of airway shunt elastance on the compartmentalization of respiratory system impedance. J Eng Sci Med Diagn Ther. 2019;2:0110011–8.
25. Horsfield K, Dart G, Olson DE, Filley GF, Cumming G. Models of the human bronchial tree. J Appl Physiol. 1971;31:207–17.
26. Horsfield K, Kemp W, Phillips S. An asymmetrical model of the airways of the dog lung. J Appl Physiol. 1982;52:21–6.
27. Horsfield K. Axial pathways compared with complete data in morphological studies of the lung. Respir Physiol. 1984;55:317–24.
28. Wagers S, Lundblad LK, Ekman M, Irvin CG, Bates JH. The allergic mouse model of asthma: normal smooth muscle in an abnormal lung? J Appl Physiol (1985). 2004;96:2019–27.

29. Bates JHT, Kaminsky DA, Garrow OJ, Martin FK, Peters U, Tharp WG, Dixon AE. Lung de-recruitment in the allergic asthma of obesity: evidence from an anatomically based inverse model. J Appl Physiol (1985). 2023;134:356–64.
30. Gillis HL, Lutchen KR. How heterogeneous bronchoconstriction affects ventilation distribution in human lungs: a morphometric model. Ann Biomed Eng. 1999;27:14–22.
31. Glenny RW, Robertson HT. Fractal modeling of pulmonary blood flow heterogeneity. J Appl Physiol. 1991;70:1024–30.
32. Kamiya A, Togawa T. Optimal branching structure of the vascular tree. Bull Math Biophys. 1972;34:431–8.
33. Wilson TA. Design of the bronchial tree. Nature. 1967;213:668–9.
34. Murray CD. The physiological principle of minimum work: I. The vascular system and the cost of blood volume. Proc Natl Acad Sci USA. 1926;12:207–14.
35. Mauroy B, Filoche M, Andrade JS Jr, Sapoval B. Interplay between geometry and flow distribution in an airway tree. Phys Rev Lett. 2003;90:148101.
36. Howatson Tawhai M, Pullan AJ, Hunter PJ. Generation of an anatomically based three-dimensional model of the conducting airways. Ann Biomed Eng. 2000;28:793–802.
37. Tgavalekos NT, Venegas JG, Suki B, Lutchen KR. Relation between structure, function, and imaging in a three-dimensional model of the lung. Ann Biomed Eng. 2003;31:363–73.
38. Cl L, Tawhai MH, Mclennan G, Hoffman EA. computational fluid dynamics. IEEE Eng Med Biol Mag. 2009;28:25–33.
39. Garcia-Uceda A, Selvan R, Saghir Z, Tiddens H, de Bruijne M. Automatic airway segmentation from computed tomography using robust and efficient 3-D convolutional neural networks. Sci Rep. 2021;11:16001.
40. Miyawaki S, Hoffman EA, Lin CL. Numerical simulations of aerosol delivery to the human lung with an idealized laryngeal model, image-based airway model, and automatic meshing algorithm. Comput Fluids. 2017;148:1–9.
41. Sonnenberg AH, Herrmann J, Grinstaff MW, Suki B. A Markov chain model of particle deposition in the lung. Sci Rep. 2020;10:13573.
42. Islam MS, Larpruenrudee P, Paul AR, Paul G, Gemci T, Gu Y, Saha SC. SARS CoV-2 aerosol: how far it can travel to the lower airways? Phys Fluids. *1994*;2021(33):061903.
43. Moses ME, Hofmeyr S, Cannon JL, Andrews A, Gridley R, Hinga M, Leyba K, Pribisova A, Surjadidjaja V, Tasnim H, Forrest S. Spatially distributed infection increases viral load in a computational model of SARS-CoV-2 lung infection. PLoS Comput Biol. 2021;17:e1009735.
44. Weibel ER. What makes a good lung? Swiss Med Wkly. 2009;139:375–86.
45. Sapoval B, Filoche M, Weibel ER. Smaller is better--but not too small: a physical scale for the design of the mammalian pulmonary acinus. Proc Natl Acad Sci USA. 2002;99:10411–6.
46. Ochs M, Nyengaard JR, Jung A, Knudsen L, Voigt M, Wahlers T, Richter J, Gundersen HJ. The number of alveoli in the human lung. Am J Respir Crit Care Med. 2004;169:120–4.
47. Suki B, Bates JH. Lung tissue mechanics as an emergent phenomenon. J Appl Physiol (1985). 2011;110:1111–8.
48. Stamenovic D, Wilson TA. Parenchymal stability. J Appl Physiol. 1992;73:596–602.
49. Otis DR Jr, Johnson M, Pedley TJ, Kamm RD. Role of pulmonary surfactant in airway closure: a computational study. J Appl Physiol (1985). 1993;75:1323–33.
50. Suki B, Ito S, Stamenovic D, Lutchen KR, Ingenito EP. Biomechanics of the lung parenchyma: critical roles of collagen and mechanical forces. J Appl Physiol. 2005;98:1892–9.
51. Bachofen H, Hildebrandt J, Bachofen M. Pressure-volume curves of air- and liquid-filled excised lungs-surface tension in situ. J Appl Physiol. 1970;29:422–31.
52. Gil J, Bachofen H, Gehr P, Weibel ER. Alveolar volume-surface area relation in air- and saline-filled lungs fixed by vascular perfusion. J Appl Physiol. 1979;47:990–1001.
53. Smith JC, Stamenovic D. Surface forces in lungs. I. Alveolar surface tension-lung volume relationships. J Appl Physiol. 1986;60:1341–50.
54. Stamenovic D, Yager D. Elastic properties of air- and liquid-filled lung parenchyma. J Appl Physiol. 1988;65:2565–70.

55. Dolhnikoff M, Morin J, Ludwig MS. Human lung parenchyma responds to contractile stimulation. Am J Respir Crit Care Med. 1998;158:1607–12.
56. Oldmixon EH, Carlsson K, Kuhn C 3rd, Butler JP, Hoppin FG Jr. Alpha-actin: disposition, quantities, and estimated effects on lung recoil and compliance. J Appl Physiol. 2001;91:459–73.
57. Yuan H, Ingenito EP, Suki B. Dynamic properties of lung parenchyma: mechanical contributions of fiber network and interstitial cells. J Appl Physiol. 1997;83:1420–31. discussion 1418–1429
58. Stamenovic D. Micromechanical foundations of pulmonary elasticity. Physiol Rev. 1990;70: 1117–34.
59. Arold SP, Bartolak-Suki E, Suki B. Variable stretch pattern enhances surfactant secretion in alveolar type II cells in culture. Am J Physiol Lung Cell Mol Physiol. 2009;296:L574–81.
60. Bartolak-Suki E, Noble PB, Bou Jawde S, Pillow JJ, Suki B. Optimization of variable ventilation for physiology, immune response and surfactant enhancement in preterm lambs. Front Physiol. 2017;8:425.
61. Fredberg JJ, Kamm RD. Stress transmission in the lung: pathways from organ to molecule. Annu Rev Physiol. 2006;68:507–41.
62. Tschumperlin DJ, Boudreault F, Liu F. Recent advances and new opportunities in lung mechanobiology. J Biomech. 2010;43:99–107.
63. Roan E, Waters CM. What do we know about mechanical strain in lung alveoli? Am J Physiol Lung Cell Mol Physiol. 2011;301:L625–35.
64. Knudsen L, Hummel B, Wrede C, Zimmermann R, Perlman CE, Smith BJ. Acinar micromechanics in health and lung injury: what we have learned from quantitative morphology. Front Physiol. 2023;14:1142221.
65. Tschumperlin DJ, Margulies SS. Alveolar epithelial surface area-volume relationship in isolated rat lungs. J Appl Physiol. 1999;86:2026–33.
66. Dunnill MS. Effect of lung inflation on alveolar surface area in the dog. Nature. 1967;214: 1013–4.
67. Mercer RR, Laco JM, Crapo JD. Three-dimensional reconstruction of alveoli in the rat lung for pressure-volume relationships. J Appl Physiol (1985). 1987;62:1480–7.
68. Lum H, Mitzner W. Effects of 10% formalin fixation on fixed lung volume and lung tissue shrinkage. A comparison of eleven laboratory species. Am Rev Respir Dis. 1985;132:1078–83.
69. Parameswaran H, Bartolak-Suki E, Hamakawa H, Majumdar A, Allen PG, Suki B. Three-dimensional measurement of alveolar airspace volumes in normal and emphysematous lungs using micro-CT. J Appl Physiol. 2009;107:583–92.
70. Meissner S, Knels L, Krueger A, Koch T, Koch E. Simultaneous three-dimensional optical coherence tomography and intravital microscopy for imaging subpleural pulmonary alveoli in isolated rabbit lungs. J Biomed Opt. 2009;14:054020.
71. Perlman CE, Bhattacharya J. Alveolar expansion imaged by optical sectioning microscopy. J Appl Physiol (1985). 2007;103:1037–44.
72. Banerji R, Grifno GN, Shi L, Smolen D, LeBourdais R, Muhvich J, Eberman C, Hiller BE, Lee J, Regan K, et al. Crystal ribcage: a platform for probing real-time lung function at cellular resolution. Nat Methods. 2023;20:1790–801.
73. Kim MW, Yu SH, Yang U, Nukiwa R, Cho HJ, Kwon NS, Yong MJ, Kim NH, Lee SH, Lee JH, et al. Alveolar microdynamics during tidal ventilation in live animals imaged by SPring-8 synchrotron. Adv Sci (Weinh). 2024;11:e2306256.
74. Cercos-Pita JL, Fardin L, Leclerc H, Maury B, Perchiazzi G, Bravin A, Bayat S. Lung tissue biomechanics imaged with synchrotron phase contrast microtomography in live rats. Sci Rep. 2022;12:5056.
75. Wilson TA, Bachofen H. A model for mechanical structure of the alveolar duct. J Appl Physiol Respir Environ Exerc Physiol. 1982;52:1064–70.
76. Fung YC. A model of the lung structure and its validation. J Appl Physiol (1985). 1988;64: 2132–41.

77. Leeming A, Schroter R. A model morphology of the pulmonary acinus. Proc Inst Mech Eng H. 2008;222:429–37.
78. Denny E, Schroter RC. The mechanical behavior of a mammalian lung alveolar duct model. J Biomech Eng. 1995;117:254–61.
79. Denny E, Schroter RC. Viscoelastic behavior of a lung alveolar duct model. J Biomech Eng. 2000;122:143–51.
80. Parameswaran H, Majumdar A, Suki B. Linking microscopic spatial patterns of tissue destruction in emphysema to macroscopic decline in stiffness using a 3D computational model. PLoS Comput Biol. 2011;7:e1001125.
81. Fujioka H, Halpern D, Gaver DP 3rd. A model of surfactant-induced surface tension effects on the parenchymal tethering of pulmonary airways. J Biomech. 2013;46:319–28.
82. Burrowes KS, Tawhai MH, Hunter PJ. Modeling RBC and neutrophil distribution through an anatomically based pulmonary capillary network. Ann Biomed Eng. 2004;32:585–95.
83. Hall JK, Bates JHT, Casey DT, Bartolak-Suki E, Lutchen KR, Suki B. Predicting alveolar ventilation heterogeneity in pulmonary fibrosis using a non-uniform polyhedral spring network model. Front Netw Physiol. 2023;3:1124223.
84. Kitaoka H, Nieman GF, Fujino Y, Carney D, DiRocco J, Kawase I. A 4-dimensional model of the alveolar structure. J Physiol Sci. 2007;57:175–85.
85. Weed B, Patnaik S, Rougeau-Browning M, Brazile B, Liao J, Prabhu R, Williams L. Experimental evidence of mechanical isotropy in porcine lung parenchyma. Materials (Basel). 2015;8:2454–66. https://doi.org/10.3390/ma8052454. eCollection 8052015 May
86. Burgess JK, Gosens R. Mechanotransduction and the extracellular matrix: key drivers of lung pathologies and drug responsiveness. Biochem Pharmacol. 2024;228:116255.
87. Tschumperlin DJ. Matrix, mesenchyme, and mechanotransduction. Ann Am Thorac Soc. 2015;12(Suppl 1):S24–9.
88. Brown JC, Timpl R. The collagen superfamily. Int Arch Allergy Immunol. 1995;107:484–90.
89. Silver FH, Freeman JW, Seehra GP. Collagen self-assembly and the development of tendon mechanical properties. J Biomech. 2003;36:1529–53.
90. Holmes DF, Gilpin CJ, Baldock C, Ziese U, Koster AJ, Kadler KE. Corneal collagen fibril structure in three dimensions: structural insights into fibril assembly, mechanical properties, and tissue organization. Proc Natl Acad Sci USA. 2001;98:7307–12.
91. Hukins DWL. Connective tissue matrix. London: Macmillan; 1984.
92. Sobin SS, Fung YC, Tremer HM. Collagen and elastin fibers in human pulmonary alveolar walls. J Appl Physiol. 1988;64:1659–75.
93. Suki B. Fluctuations and power laws in pulmonary physiology. Am J Respir Crit Care Med. 2002;166:133–7.
94. Weibel E, Gil J. Bioengineering aspects of the lung. In: Lung biology in health and disease, vol. 3. New York: M. Dekker; 1977. p. 1–81.
95. Tanaka R, Al-Jamal R, Ludwig MS. Maturational changes in extracellular matrix and lung tissue mechanics. J Appl Physiol. 2001;91:2314–21.
96. Tanaka R, Ludwig MS. Changes in viscoelastic properties of rat lung parenchymal strips with maturation. J Appl Physiol. 1999;87:2081–9.
97. Dolhnikoff M, Mauad T, Ludwig MS. Extracellular matrix and oscillatory mechanics of rat lung parenchyma in bleomycin-induced fibrosis. Am J Respir Crit Care Med. 1999;160:1750–7.
98. Yuan H, Kononov S, Cavalcante FS, Lutchen KR, Ingenito EP, Suki B. Effects of collagenase and elastase on the mechanical properties of lung tissue strips. J Appl Physiol. 2000;89:3–14.
99. Toshima M, Ohtani Y, Ohtani O. Three-dimensional architecture of elastin and collagen fiber networks in the human and rat lung. Arch Histol Cytol. 2004;67:31–40.
100. Casey DT, Bou Jawde S, Herrmann J, Mori V, Mahoney JM, Suki B, Bates JHT. Percolation of collagen stress in a random network model of the alveolar wall. Sci Rep. 2021;11:16654.
101. Bou Jawde S, Karrobi K, Roblyer D, Vicario F, Herrmann J, Casey D, Lutchen KR, Stamenovic D, Bates JHT, Suki B. Inflation instability in the lung: an analytical model of a thick-walled alveolus with wavy fibres under large deformations. J R Soc Interface. 2021;18:20210594.

102. Maksym GN, Bates JH. A distributed nonlinear model of lung tissue elasticity. J Appl Physiol. 1997;82:32–41.
103. Lo CM, Wang HB, Dembo M, Wang YL. Cell movement is guided by the rigidity of the substrate. Biophys J. 2000;79:144–52.
104. Southern BD, Grove LM, Rahaman SO, Abraham S, Scheraga RG, Niese KA, Sun H, Herzog EL, Liu F, Tschumperlin DJ, et al. Matrix-driven myosin II mediates the pro-fibrotic fibroblast phenotype. J Biol Chem. 2016;291:6083–95.
105. Sasaki N, Odajima S. Stress-strain curve and young's modulus of a collagen molecule as determined by the X-ray diffraction technique. J Biomech. 1996;29:655–8.
106. Marko JF, Siggia ED. Stretching DNA. Macromolecules. 1995;28:8759–70.
107. Rosenbloom J, Abrams WR, Mecham R. Extracellular matrix 4: the elastic fiber. FASEB J. 1993;7:1208–18.
108. Powell JT, Vine N, Crossman M. On the accumulation of D-aspartate in elastin and other proteins of the ageing aorta. Atherosclerosis. 1992;97:201–8.
109. Daamen WF, Veerkamp JH, van Hest JC, van Kuppevelt TH. Elastin as a biomaterial for tissue engineering. Biomaterials. 2007;28:4378–98.
110. Reinhardt DP, Sasaki T, Dzamba BJ, Keene DR, Chu ML, Gohring W, Timpl R, Sakai LY. Fibrillin-1 and fibulin-2 interact and are colocalized in some tissues. J Biol Chem. 1996;271:19489–96.
111. Mithieux SM, Weiss AS. Elastin. Adv Protein Chem. 2005;70:437–61.
112. Kozel BA, Wachi H, Davis EC, Mecham RP. Domains in tropoelastin that mediate elastin deposition in vitro and in vivo. J Biol Chem. 2003;278:18491–8.
113. Brown RE, Butler JP, Rogers RA, Leith DE. Mechanical connections between elastin and collagen. Connect Tissue Res. 1994;30:295–308.
114. Kielty CM, Sherratt MJ, Shuttleworth CA. Elastic fibres. J Cell Sci. 2002;115:2817–28.
115. Raspanti M, Alessandrini A, Ottani V, Ruggeri A. Direct visualization of collagen-bound proteoglycans by tapping-mode atomic force microscopy. J Struct Biol. 1997;119:118–22.
116. Fung YC. Biomechanics: mechanical properties of living tissues. 2nd ed. New York: Springer; 1993.
117. Sherebrin MH, Song SH, Roach MR. Mechanical anisotropy of purified elastin from the thoracic aorta of dog and sheep. Can J Physiol Pharmacol. 1983;61:539–45.
118. Aaron BB, Gosline JM. Elastin as a random-network elastomer: a mechanical and optical analysis of single elastin fibers. Biopolymers. 1981;20:1247–60.
119. Thurmond F, Trotter J. Morphology and biomechanics of the microfibrillar network of sea cucumber dermis. J Exp Biol. 1996;199:1817–28.
120. Sherratt MJ, Baldock C, Haston JL, Holmes DF, Jones CJ, Shuttleworth CA, Wess TJ, Kielty CM. Fibrillin microfibrils are stiff reinforcing fibres in compliant tissues. J Mol Biol. 2003;332:183–93.
121. Koenders MM, Yang L, Wismans RG, van der Werf KO, Reinhardt DP, Daamen W, Bennink ML, Dijkstra PJ, van Kuppevelt TH, Feijen J. Microscale mechanical properties of single elastic fibers: the role of fibrillin-microfibrils. Biomaterials. 2009;30:2425–32.
122. Setnikar I. Origin and significance of the mechanical property of the lung. Arch Fisiol. 1955;55:349–74.
123. Juul SE, Kinsella MG, Wight TN, Hodson WA. Alterations in nonhuman primate (M. nemestrina) lung proteoglycans during normal development and acute hyaline membrane disease. Am J Respir Cell Mol Biol. 1993;8:299–310.
124. Buckwalter JA, Rosenberg LC. Electron microscopic studies of cartilage proteoglycans. Direct evidence for the variable length of the chondroitin sulfate-rich region of proteoglycan subunit core protein. J Biol Chem. 1982;257:9830–9.
125. Scott JE. Supramolecular organization of extracellular matrix glycosaminoglycans, in vitro and in the tissues. FASEB J. 1992;6:2639–45.
126. Fratzl P, Daxer A. Structural transformation of collagen fibrils in corneal stroma during drying. An x-ray scattering study. Biophys J. 1993;64:1210–4.

127. Carey DJ. Biological functions of proteoglycans: use of specific inhibitors of proteoglycan synthesis. Mol Cell Biochem. 1991;104:21–8.
128. Roseman S. Reflections on glycobiology. J Biol Chem. 2001;276:41527–42.
129. Wigen J, Elowsson-Rendin L, Karlsson L, Tykesson E, Westergren-Thorsson G. Glycosaminoglycans: a link between development and regeneration in the lung. Stem Cells Dev. 2019;28:823–32.
130. Carey DJ. Syndecans: multifunctional cell-surface co-receptors. Biochem J. 1997;327(Pt 1): 1–16.
131. Iozzo RV. Matrix proteoglycans: from molecular design to cellular function. Annu Rev Biochem. 1998;67:609–52.
132. Hulmes DJ, Wess TJ, Prockop DJ, Fratzl P. Radial packing, order, and disorder in collagen fibrils. Biophys J. 1995;68:1661–70.
133. Huang R, Merrilees MJ, Braun K, Beaumont B, Lemire J, Clowes AW, Hinek A, Wight TN. Inhibition of versican synthesis by antisense alters smooth muscle cell phenotype and induces elastic fiber formation in vitro and in neointima after vessel injury. Circ Res. 2006;98: 370–7.
134. Al Jamal R, Roughley PJ, Ludwig MS. Effect of glycosaminoglycan degradation on lung tissue viscoelasticity. Am J Physiol Lung Cell Mol Physiol. 2001;280:L306–15.
135. Cavalcante FS, Ito S, Brewer KK, Sakai H, Alencar AM, Almeida MP, Andrade JS Jr, Majumdar A, Ingenito EP, Suki B. Mechanical interactions between collagen and proteoglycans: implications for the stability of lung tissue. J Appl Physiol. 2005;98:672–9.
136. Fust A, LeBellego F, Iozzo RV, Roughley PJ, Ludwig MS. Alterations in lung mechanics in decorin-deficient mice. Am J Physiol Lung Cell Mol Physiol. 2005;288:L159–66.
137. Reese SP, Underwood CJ, Weiss JA. Effects of decorin proteoglycan on fibrillogenesis, ultrastructure, and mechanics of type I collagen gels. Matrix Biol. 2013;32:414–23.
138. Gutierrez JA, Perr HA. Mechanical stretch modulates TGF-beta1 and alpha1(I) collagen expression in fetal human intestinal smooth muscle cells. Am J Phys. 1999;277:G1074–80.
139. Schild C, Trueb B. Mechanical stress is required for high-level expression of connective tissue growth factor. Exp Cell Res. 2002;274:83–91.
140. Brewer KK, Sakai H, Alencar AM, Majumdar A, Arold SP, Lutchen KR, Ingenito EP, Suki B. Lung and alveolar wall elastic and hysteretic behavior in rats: effects of in vivo elastase treatment. J Appl Physiol. 2003;95:1926–36.
141. Wirtz HR, Dobbs LG. Calcium mobilization and exocytosis after one mechanical stretch of lung epithelial cells. Science. 1990;250:1266–9.
142. Fredberg JJ, Bunk D, Ingenito E, Shore SA. Tissue resistance and the contractile state of lung parenchyma. J Appl Physiol. 1993;74:1387–97.
143. Navajas D, Maksym GN, Bates JH. Dynamic viscoelastic nonlinearity of lung parenchymal tissue. J Appl Physiol. 1995;79:348–56.
144. Salerno FG, Dallaire M, Ludwig MS. Does the anatomic makeup of parenchymal lung strips affect oscillatory mechanics during induced constriction? J Appl Physiol. 1995;79:66–72.
145. Trepat X, Grabulosa M, Buscemi L, Rico F, Farre R, Navajas D. Thrombin and histamine induce stiffening of alveolar epithelial cells. J Appl Physiol. 2005;98:1567–74.
146. Kolozsvari B, Szijgyarto Z, Bai P, Gergely P, Verin A, Garcia JG, Bako E. Role of calcineurin in thrombin-mediated endothelial cell contraction. Cytometry A. 2009;75:405–11.
147. Leslie KO, Mitchell JJ, Woodcock-Mitchell JL, Low RB. Alpha smooth muscle actin expression in developing and adult human lung. Differentiation. 1990;44:143–9.
148. 148. Deng Y, Herrmann J, Wang Y, Nguyen M, Hall JK, Kim JH, Smith ML, Lutchen KR, Bartolak-Suki E, Suki B. Physiological mechanical forces accelerate the degradation of bovine lung collagen fibers by bacterial collagenase. Sci Rep. 2024;14(1):29008. Epub 20241122. PubMed PMID: 39578499; PMCID: PMC11584708.

Chapter 3
Lumped-Parameter Models of Lung Function

Abstract Lumped-parameter models of lung function assign the constitutive properties of the lung to distinct and separate structural entities. Such models are typically described by sets of coupled ordinary differential equations. The parameters of these equations correspond to specific physical attributes of the lung such as airway resistance or tissue elastance. The fundamental unit of a lumped-parameter model is the compartment, which corresponds to some space within which the time-varying quantity of interest is uniformly distributed. The parameters of a lumped parameter model can often be related to fundamental physical processes, such as viscous flow of gas through airways or the elastic recoil generated when the lungs are inflated. The simplest lumped parameter model of the lung consists of a single elastic compartment served by a single flow-resistive airway. Additional realism can be added to this model by making its constitutive properties nonlinear or by connecting it to additional compartments. If the lung behaves sufficiently like a linear dynamic system, its behavior can be encapsulated in the complex function of frequency known as mechanical input impedance, which itself can be described by lumped-parameter models.

Keywords Constitutive properties · Compartment · Nonlinearities · Degrees of freedom · Airway resistance · Lung elastance · Inertance · Linear dynamic system · Impedance

3.1 Introduction

The most readily solved mathematical models of dynamic lung function are those based on *ordinary differential equations* for which time is the only independent variable. The various *constitutive properties* of these models (those corresponding to specific physical attributes such as resistance to airflow or elastic stiffness) are assigned to distinct and separate structural entities. For example, the resistance of the conducting airway tree may be represented by a single parameter called *airway resistance* that is ascribed to the tree as a whole. In this case, no account is taken of the fact that flow resistance accrues continuously as one moves along the tree; these

B. Suki, J. H. T. Bates, *Mathematical Modeling of the Healthy and Diseased Lung*,
https://doi.org/10.1007/978-3-031-95322-4_3

properties are lumped together into a single quantity that neglects the distributed nature of reality. Accordingly, such models are called *lumped-parameter models*.

The simplest lumped-parameter model of the lung has only a single degree of freedom corresponding to the volume of air in the lung that accrues via transit through a single airway. The predictions of the linear version of this model can be calculated analytically. Further realism (and thus, complexity) can be conferred by making additions to the model in two distinct directions [1]. One is to make its constitutive properties nonlinear (e.g., have airway resistance depend on flow, or tissue elastance depend on volume). Solving such models usually requires numerical methods. Alternatively, additional mechanical degrees of freedom may be added to represent, for example, the compliance of the alveolar tissues versus the compliance of the conducting airway tree. If such models remain linear, they can also often be solved analytically. As the number of degrees of freedom increases, however, there comes a point where numerical methods are more convenient.

The fundamental unit of a lumped-parameter model is the *compartment*, which corresponds to some space within which the time-varying quantity of interest is "well mixed" at all times. For example, if the alveolar parenchyma of the lung is represented by a single compartment, the assumption is that pressure is always equal everywhere within this compartment. In other words, all alveoli behave identically and can thus be collectively represented as a single large alveolus. If this assumption is shown to be unreasonable, then it may be necessary to model the parenchyma as a collection of several distinct compartments. The pressures within each compartment are uniform, but they may differ between compartments. In fact, the concept of a compartment is ubiquitous in the modeling of biomedical systems in general, such as in the field of pharmacokinetics where it is common to consider the circulating blood to be a compartment following intravenous injection of a drug [2]. This is a good assumption so long as the rate at which drug is cleared from the circulation (e.g., by the liver or kidneys) is slow compared to the rate at which drug becomes uniformly mixed throughout the circulation.

Taken to extremes, lumped-parameter models reflecting any conceivable amount of anatomic realism can be constructed. Calculating their predictions, however, may involve substantial computational overhead. In this capacity, they function as forward models (Sect. 1.4). When lumped parameter models are sufficiently simple, with few compartments and adjustable parameters, they can serve as inverse models. In fact, most models of lung function that are fit to experimental data are of the lumped-parameter variety. Accordingly, this chapter focuses on inverse lumped-parameter models of lung function. We will begin with very simple models and then build in additional realism through increasing levels of complexity.

3.2 Lung Mechanics

Throughout life, the respiratory muscles must be able to change lung volume cyclically as we breathe. At rest, these changes amount to about half a liter roughly twelve times per minute, producing a *minute ventilation* in the order of 6 liters per minute, which can increase by several fold during exercise. The mechanical opposition to volume change in the normal lung is modest, being easily is overcome by the respiratory muscles without them fatiguing. In disease, however, the mechanical properties of the lung often become deranged in characteristic ways. Mathematical models of lung function attempt to represent these mechanical properties and their derangements in terms of the principles of physics.

The forces generated by the respiratory muscles during breathing are balanced by opposing forces generated within the lung. These opposing forces are largely of two types, *resistive* and *elastic*, although *inertive* forces may also be involved when flows change rapidly. Accordingly, a discussion of models of lung function begins with the physics of lung resistance, elastance and inertance.

3.2.1 Lung Resistance

Lung resistance refers to the energy cost associated with the flow of gas through the air passages of the lung, beginning with the airway opening (the mouth and/or nose in the case of a spontaneously breathing person, or the trachea in the case of a mechanically ventilated subject) and ending in the alveoli. This energy cost arises regardless of whether flow is positive (entering the lungs) or negative (leaving the lungs). In either case, the energy is dissipated in the form of heat.

The definition of lung resistance, R_L, is the derivative

$$R_L = \frac{dP_{res}}{d\dot{V}} \tag{3.1}$$

where P_{res} is the resistive pressure (the component of the pressure across the lungs that is driving the flow), and $\dot{V}$ is flow itself. (Note that placing a dot over a variable denotes its time derivative, flow being the time derivative of volume.) Understanding R_L amounts to understanding the physical processes it reflects.

3.2.2 Airway Resistance

About half of R_L in the normal lung during regular breathing is due to flow of air along the conducting airways of the pulmonary tree [3]. This component is called *airway resistance*, R_{aw}. The genesis of R_{aw} derives from the fact that even though the

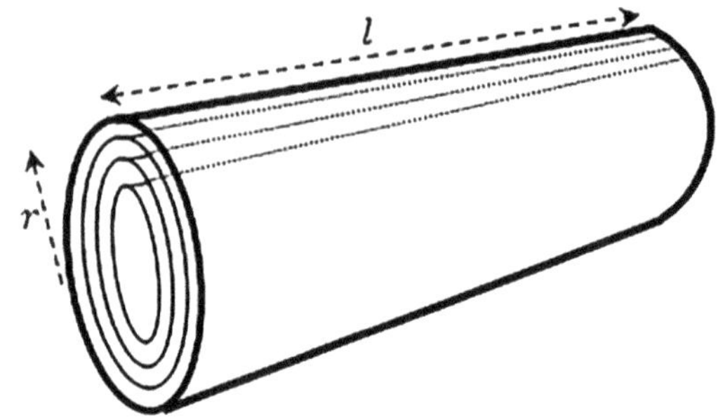

Fig. 3.1 A single cylindrical conduit of length l and radius r. Gas flows axially through the conduit in a series of concentric layers that generate friction as they slide past each other

bulk flow of air, and thus the average movement of each air molecule, is in the axial direction along the airway through which it travels, there is still a non-axial component to each molecule's path. The result is that adjacent molecules continually bump into each other, which causes them to exchange kinetic energy and momentum. As a result, some of the organized axial molecular motion gets converted into random molecular kinetic energy, causing the gas temperature to increase (although usually by an immeasurably small amount). A convenient way to visualize the bulk flow of gas along a conduit is as a nested series of annular layers traveling at slightly different axial velocities (Fig. 3.1). The relative velocities of adjacent layers are determined by a number of factors that can vary from situation to situation, but in all cases the adjacent layers rub against each other. This generates friction according to how different their velocities are. Even the outermost layer may move ever so slightly compared to the inner wall of the conduit, and thus rub against it. Gases that dissipate energy in this way are called *viscous*.

At the most basic level, R_{aw} can be assigned an empirical value, based on Eq. 3.1, from experimental measurements of pressure and flow. This, however, does not link R_{aw} to airway structure. The airway tree is a complicated structure with thousands of branching segments. However, it is frequently conceptualized as being equivalent to a single cylindrical conduit (Fig. 3.1) having a resistance equal to R_{aw}. The link between airway structure (conduit dimensions) and function (R_{aw}) can then be made on the basis of physical first principles provided certain conditions pertaining to the nature of the flow are satisfied. These conditions are:

- The airway has a uniform cross-section and smooth inner walls.
- Flow does not change with time or location along the airway.
- Air is incompressible and Newtonian.
- Flow is laminar, meaning it has smooth, well-defined streamlines (as depicted in Fig. 3.1).
- The flow streamlines are not affected by the sharp changes in diameter that typically occur at the airway entrance and exit, which is approximately true if the length of the conduit is much greater than its radius. (Note that in the actual airway tree, entrance effects occur at bifurcations between airway generations that usually have different diameters.)
- There is a no-slip condition such that axial fluid velocity is zero at the airway wall.

When these conditions are satisfied, flow along the airway proceeds in a way that minimizes the rate of energy dissipation, which leads to the *Poiseuille* formula for the resistance of a cylindrical conduit of length l and radius r when a gas of viscosity μ flows though it (Sect. 3.1.3 of [1]). R_{aw} is thus modeled as

$$R_{aw} = \frac{8\mu l\dot{V}}{\pi r^4}. \tag{3.2}$$

Equation 3.2 makes a precise link between structure and function. Of course, the airway tree is not actually a single cylindrical conduit, so this formula must be taken as a general indication of how r, l and R_{aw} are related. At a more accurate level, Eq. 3.2 can be applied to each individual branch of the airway tree and the individual branch resistance added in series or parallel, as the case may be, to arrive at a value of R_{aw} for the entire tree. The above assumptions still apply to each branch, however.

From a functional perspective, the most important aspect of the Eq. 3.2 is its inverse dependence on the fourth power of r. This means, for example, that narrowing a conduit by half increases its resistance by a factor of sixteen! It thus does not take much of a decrease in airway patency to have a major effect on the effort and sensation of breathing.

Like all models, the formula expressed in Eq. 3.2 is never satisfied perfectly in practice, but it is often a useful approximation to the truth when $\dot{V}$ is not too large. The definition of "large" in this case is provided by a dimensionless quantity known as the *Reynolds number* (*Re*) given by

$$Re = \frac{2\rho\dot{V}}{\pi\mu r} \tag{3.3}$$

where ρ is the density of air. When Re is significantly less than 2000 the flow is laminar, whereas when Re is significantly greater than 2000 the flow is *turbulent*, meaning that its streamlines are chaotic and wildly intermingled. In between these two extremes exists the so-called *transition region* where flow is neither purely laminar nor turbulent. However, Re is not the only factor that influences the flow regime. A sharp transition in diameter between the conduit and its upstream feeder can disturb laminar flow even when $Re << 2000$. Similarly, laminar flow becomes less likely when the inner walls of the airways are not smooth (such as the undulations due to cartilage), and when flow is unsteady, meaning it changes direction repeatedly so the streamlines are never allowed to reach steady state.

Flow through most of the pulmonary airways during quiet breathing is well approximated as being laminar, making Eq. 3.2 a useful formula for calculating the flow resistance of the individual airway branches. However, the laminar assumption is less satisfied in the larger central airways, particularly during exercise or coughing. There are two recourses to dealing with this in a model. By far the easiest is to increase the resistance of an airway branch above that predicted by Eq. 3.2 by a suitable empirical factor. Such factors are typically not much larger than unity and have been used to good effect in determining the overall resistance of the airway tree [4]. Also, R_{aw} exhibits a positive dependence on $\dot{V}$ (due to the onset of turbulence) and a negative dependence on lung volume (due to the increase in airway caliber), contrary to what Eq. 3.2 implies.

The alternative approach is to resort to solving the Navier-Stokes equations of fluid dynamics [4]. The advantage of this approach is that it can be applied in arbitrary geometries, not just cylindrical conduits, so it allows for the calculation of R_{aw} along structures that more closely resemble real airways. However, airway geometry varies widely between individuals and is never constant within any given individual due to the configurational changes wrought by breathing and occasional deep inspirations together with temporal variations in airway mucus. Exploiting the full fidelity of computational fluid dynamics would thus require calculating R_{aw} from high fidelity 3-dimensional images of a given lung throughout time. The computational cost of such an undertaking would be enormous. Computational fluid dynamics might be worth the effort when, say, calculating the airflow dynamics over an airplane wing where maximum accuracy can mean saving significant amounts of money and possibly even human lives. Such accuracy is simply not necessary for the diagnostic and therapeutic needs of clinical medicine. Equation 3.1 and its empirical derivatives are all that are required when it comes to estimating R_{aw} for the purposes of determining the functional effects of, for example, airway narrowing in asthma or airway widening in traction bronchiectasis, a manifestation of pulmonary fibrosis.

This is not to say, however, that computational fluid dynamics has no place in lung modeling. On the contrary, it is vital for determining the detailed fate of aerosols in the airways such as pertains to the inhalation of pollutants or nebulized medicines [5]. Computational fluid dynamics has also proven vital to understanding how the interfacial fluid flow that occurs at the air-liquid interface in the lung is affected by local surfactant concentrations [6].

3.2.3 Tissue Resistance

R_{aw} is not the only significant component of R_L in the lung. Another source of energy dissipation that can be just as important is the lung tissue, defined essentially as anything in the lung that is not air. Lung tissue is what is known as *viscoelastic*, meaning that it is capable of both dissipating and storing energy when stretched or distorted. The storage component is discussed in the next section. Here we focus on the mechanisms that give rise to energy dissipation in the tissue. This is encapsulated in the notion of "tissue resistance", R_{tis}. Like R_{aw}, the mechanisms giving rise to R_{tis} can ultimately be explained in terms of basic physics, although making the structure-function link is not as straightforward as for gas flow through airways such as, for example, encapsulated in Eq. 3.2.

The ultimate source of any form of resistance in the lungs is friction. In the case of R_{aw}, it is the friction between layers of gas sliding past each other (Fig. 3.1). In the case of R_{tis}, the source of the friction is less clear [7]. Nevertheless, there are certainly viscous fluid effects that occur as the constituent fibers, fluids and cells slide past each other when the tissue is stretched and distorted [8]. Also, since we define lung tissue as anything that is not air, some of the friction is generated at the air-liquid interface as the result of the breakage of the electrostatic bonds responsible for surface tension. This is demonstrated by the dramatic decrease seen in the hysteresis of the pressure-volume loop of the lung when it is inflated and deflated with saline instead of air [9]. Altogether, the situation is very complex and defies the kind of simplifications that, for example, justify the derivation of Eq. 3.2.

To complicate matters further, the viscoelastic character of lung tissue means that its resistance is a function of the way in which it is stretched. In particular, if sinusoidal length changes are imposed on lung tissue, R_{tis} is an inverse function of the frequency at which the length changes take place [10]. Consequently, R_{tis} is usually modeled empirically using constructs that behave appropriately but which do not purport to represent the specific structures involved in the generation of internal tissue friction. Furthermore, R_{tis} is nonlinear in the sense that it depends on both the magnitude and rate of stretch [10, 11]. We will return to R_{tis} and its possible underlying mechanisms in Chap. 5 when we discuss viscoelasticity in detail.

The combined effects of the airways and tissue on the dissipation of energy in the lung are typically expressed simply as

$$R_L = R_{aw} + R_{tis}. \tag{3.4}$$

3.2.4 Lung Elastance

When the lung is inflated, it develops an elastic recoil pressure, P_{el}, that opposes the inflating pressure. This occurs because the lung tissues, being viscoelastic, store energy in addition to dissipating it through friction as discussed above. This stored part of the total energy, that which is not dissipated, is returned during expiration in the form of kinetic energy of the recoiling tissues and the exhaled gas.

Empirically, the elastic character of lung tissue is encapsulated in a quantity called *lung elastance*, E_L, which is a measure of how rapidly P_{el} increases with increasing lung volume, V. That is,

$$E_L = \frac{dP_{el}}{dV}. \tag{3.5}$$

Linking E_L to the underlying physical processes responsible for it suffers from the same challenges of complexity faced by R_{tis} described above. These processes can ultimately be traced to forces at the atomic and molecular levels (see Chap. 4). Also, E_L exhibits a strong dependence on V, particularly as V approaches total lung capacity. This nonlinear behavior arises in large part because the primary load-bearing structure in lung tissue is an intertwined pair of collagen and elastin fiber networks. At low volumes, the load is carried most by the elastin fibers, but as volume increases, the much stiffer collagen fiber network assumes the load-bearing role to an increasing degree [12]. Accordingly, E_L is typically defined empirically accordingly to Eq. 3.5. (Note that E_L is the inverse of the quantity known as *lung compliance*, which is the more commonly used term in clinical medicine.)

3.2.5 Lung Inertance

Part of the pressure responsible for breathing is used to accelerate and decelerate entities in the lung that have mass, as dictated by Newton's law of motion. During normal breathing, rates of change of flow, also called volume acceleration, $\ddot{V}$, are small, so the effects on pressure due to mass can be neglected compared to the effects due to R_L and E_L. Mass starts to become important, however, when flow is oscillated in and out of the airway opening at increasing frequencies, beginning in the human lung at about 2 Hz. At the *resonant frequency*, the contributions of mass acceleration and tissue recoil to the total pressure across the lung become equal. This occurs in a normal human lung at about 10 Hz, although it can increase significantly in disease [13]. Frequencies in the range of the resonant frequency are relevant during application of the diagnostic methodology known as *Oscillometry* (also called the *Forced Oscillation Technique*) [1], as well as during high-frequency oscillatory mechanical ventilation.

The mass equivalent of R_L and E_L is known as *lung inertance*, I_L, and its contributing factors depend on the range of frequencies over which flow is being oscillated. For most applications of oscillometry, which typically extend to frequencies of 20–35 Hz, the mass that determines I_L is that of the air in the central airways [14]. If one models the conducting airways as a single cylindrical conduit, as in Fig. 3.1, then the mass of the gas that is to be accelerated is $\pi r^2 l\rho$. The force acting to accelerate this gas is the applied pressure acting over the cross-section of the conduit, $\pi r^2 P$. The acceleration of this mass is the acceleration of volume, $\ddot{V}$, divided by the cross-sectional area of the conduit (where placing two dots over a variable denotes the second derivative with respect to time). Expressing Newton's law in these terms gives

$$\pi r^2 P = \pi r^2 l\rho \frac{\ddot{V}}{\pi r^2} \tag{3.6}$$

so

$$P = \frac{l\rho}{\pi r^2}\ddot{V} = I_L \ddot{V}. \tag{3.7}$$

$I_L = \frac{l\rho}{\pi r^2}$ thus has the counterintuitive attribute of being inversely proportional to the cross-sectional area of the conduit.

As frequency approaches the point where one quarter of the wavelength of sound becomes comparable to the dimensions of the airway tree (around 150 Hz), it is no longer possible to think of a single value of inertance for the lungs. At these frequencies, the movement of gas in the airways must be viewed as a wave phenomenon rather than as a bulk flow [15]. Thus, for frequencies in the acoustic range, it is not possible to represent lung mechanics in terms of lumped-parameter models.

3.3 The Single-Compartment Model of the Lung

3.3.1 *The Linear Single-Compartment Model*

The simplest model of the lung that bears any reasonable resemblance to reality consists of a single elastic compartment (representing the inflatable lung tissues) connected to the outside world by a single flow-resistive conduit (representing the pulmonary airway tree). This model is frequently depicted as a balloon connected to a pipe. For the present purposes, the model will be represented as a pair of telescoping cannisters connected to each other by a spring, and to the outside world by a single conduit, because this representation is more readily extended to include additional physiological features such as the resistance of the tissues as illustrated in Fig. 3.2.

In its simplest instantiation, the flow resistance of the conduit corresponds to airway resistance characterized by the single parameter R_{aw} while the elastic stiffness of the lungs is represented by the spring constant E_L. Also connected between the two cannisters is a resistive element known as a dashpot representing the resistance, R_{tis}, of the tissues. The dashpot gives rise to a resistive pressure drop

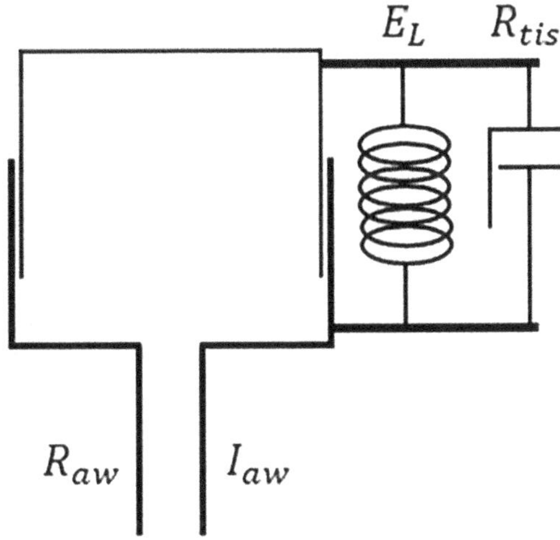

Fig. 3.2 Single-compartment model of the lung consisting of a pair of telescoping cannisters representing the alveolar tissues. The cannisters are connected by a spring representing tissue stiffness (E_L) and a dashpot representing tissue resistance (R_{tis}). A single cylindrical conduit having resistance R_{aw} represents the pulmonary conducting airway tree. I_{aw} represents the mass of the gas in the central airways

proportional to flow in the same way as airway resistance. The mass of the gas in the central airways also gives rise to an inertance, I_{aw}, that starts to become important as oscillation frequencies exceed about 2 Hz.

When a time-varying pressure, $P(t)$, is applied to the airway opening of this model, the resulting lung volume, $V(t)$, and its time-derivatives are related via the following equation:

$$P(t) = I_{aw}\ddot{V}(t) + R_L\dot{V}(t) + E_L V(t) + P_0 \tag{3.8}$$

where P_0 is the pressure needed to maintain the lung at $V = 0$, where V is defined relative to the elastic equilibrium volume that determines functional residual volume (FRC). The three coefficients in this model, I_{aw}, R_L, and P_0, can be conveniently estimated from experimental data by multiple linear regression (Sect. 3.2 of [1]).

Note that the model in Fig. 3.2 is being used to represent the lungs alone, so P corresponds to *transpulmonary pressure* (P_{tp}) defined as the difference between *airway opening pressure*, P_{ao}, and *pleural pressure, which is usually represented by its more easily measured surrogate, esophageal pressure*, P_{es}. In the case where $P = P_{ao}$, Eq. 3.8 would pertain to the entire respiratory system consisting of the lungs plus the chest wall. Conversely, if $P = P_{es}$, the equation would pertain to the chest wall alone. All of these instantiations of Eq. 3.8 are suited to being inverse models because they contain few parameters that can be estimated by fitting the equation to measurements of $P(t)$, $V(t)$, $\dot{V}(t)$, and $\ddot{V}(t)$.

3.3.2 Nonlinear Extensions of the Single-Compartment Model

Equation 3.8 describes the lung as a linear system because the dependent variable, $P(t)$, depends linearly on both the model parameters I_{aw}, R_{aw}, R_{tis} and P_0 and on the model variables $P(t)$, $V(t)$, $\dot{V}(t)$, and $\ddot{V}(t)$. Linearity is often a reasonable assumption when accounting for the mechanical behavior of the lung over limited ranges of flow and volume, but it fails toward the extremes of these variables as well as in certain disease situations. Of course, it is also not entirely correct even at low flows and volumes because nothing in the real world is ever perfectly linear. It is thus often desirable to extend Eq. 3.8 in a way that accounts for nonlinear behavior.

It is straightforward to make Eq. 3.8 account for nonlinearities in resistance and elastance by replacing the terms in R_L and/or E_L with terms that are nonlinear in $\dot{V}(t)$ and $V(t)$. The most straightforward way to do this is to use higher-order polynomials in $\dot{V}(t)$ and $V(t)$. For example, replacing the term $R_L\dot{V}(t)$ with the so-called Rohrer equation for flow resistance [16] gives

$$P(t) = K_1\dot{V}(t) + K_2\dot{V}(t)\left|\dot{V}(t)\right| + E_L V(t) + P_0 \tag{3.9}$$

where K_1 and K_2 are commonly referred to as the Rohrer coefficients. K_1 and K_2 do not correspond to any definable physical entity or process, so should be viewed simply as empirical constants that together define airflow resistance. Note that the variable for K_2 in the above equation is not the square of flow but rather the product of flow with its absolute value. This ensures that the sign of the second term on the right-hand side is always the same as that of the first term regardless of whether $\dot{V}(t)$ is positive (inspiration) or negative (expiration), as is necessary if both terms are to contribute to the pressure drop across the airways.

A simple way to account for the strain stiffening of the lung tissues is to give the elastic recoil pressure a quadratic dependence on lung volume thus:

$$\frac{ab}{cd}abP(t) = R_L\dot{V}(t) + E_1 V(t) + E_2 V^2(t) + P_0. \tag{3.10}$$

E_1 and E_2 are again empirical constants that do not have specific physical analogs. Nevertheless, they have proven useful as practical indicators of impending overinflation of the lung because when lung volume is within its normal tidal breathing range, the value of E_2 estimated from Eq. 3.10 is usually small enough to be safely neglected. As volume encroaches on the hyperinflated nonlinear portion of the lung's pressure-volume relationship, however, E_2 can start to become significant [17, 18].

Equations 3.9 and 3.10 can also be combined to yield an equation with 5 free parameters that is nonlinear in both resistance and elastance. Conveniently, the parameters of these equations can all still be estimated by multiple linear regression.

A special case of the nonlinear single-compartment model pertains to the quasi-static pressure-volume behavior of the lung, when $\dot{V}(t)$ is negligible (this is discussed in more detail in Chap. 4). The shape of this relationship from FRC to total lung capacity (TLC) is highly nonlinear, and is commonly modeled by the exponential relationship [19].

$$V = A + Be^{-KP}. \tag{3.11}$$

The parameters A, B and K are empirical, although they can be linked to the structural components of lung tissue to some extent. Below FRC, the relationship curves in the opposite sense, assuming a sigmoidal shape over the vital capacity range. An equation that has been used to model the entire relationship is [20].

$$V = a + \left[\frac{b}{1 + e^{-\left(\frac{P-d}{d}\right)}}\right] \tag{3.12}$$

where a, b, c and d again are empirical parameters. These parameters, however, cannot be estimated by multiple linear regression and so require some form of iterative model fitting method.

3.4 Two-Compartment Models of the Lung

The lung models considered so far are all *static* models in the sense that the dependent variable depends only on current values of the independent variables. For example, $P(t)$ is only influenced by $V(t)$ and not past values $V(t - \tau)$. When $P(t)$ is influenced by past values of any of the independent variables, the model is said to be *dynamic*. A dynamic model can still be linear, however, meaning that it is describing a *linear dynamic system*. A *system* in this sense is an entity that receives inputs and produces outputs, which for the lung usually means any flows that are administered at the airway opening and the corresponding airway opening pressures that are produced in the process. A dynamic system is *linear* if the magnitude of the output scales linearly with the magnitude of the input. Note that this does not mean the output waveform always has the same shape as the input waveform, but merely that scaling the input waveform by an arbitrary factor results in the output waveform being scaled by the same factor.

An important aspect of pulmonary physiology concerns the extent to which the lung is mechanically heterogeneous from one region to another. This is particularly important in disease, which is characterized by increased regional heterogeneities in lung function. Accounting for regional heterogeneity obviously cannot be achieved in a model with only a single alveolar compartment because the defining feature of a compartment is that all relevant variables are identical everywhere throughout. Accounting for heterogeneities thus requires a model of at least two compartments, but this introduces a new complication. How should these two compartments be arranged relative to each other? There are two canonical possibilities—either in series or in parallel. These two possibilities are illustrated in Fig. 3.3 using simplified versions of the model in Fig. 3.2.

Distinguishing between these two possibilities is important because of their different physiological interpretations. In the case of the series model (Fig. 3.3a), the distal compartment represents the alveolar parenchyma, while the proximal compartment represents the distensible upper airways (loosely defined as the cheeks and pharyngeal regions possibly together with the larger conducting airways). This model was introduced to the pulmonary physiology community by Mead in 1969 [21]. The parallel model (Fig. 3.3b), on the other hand, represents two distal alveolar compartments served by distinct airways, and was introduced by Otis et al. in 1956 [22]. The identities of these compartments are not obvious. One possibility might be right and left lungs, but they do not have to be anatomically distinct and so might correspond, for example, to intermingled alveoli having different emptying time-constants.

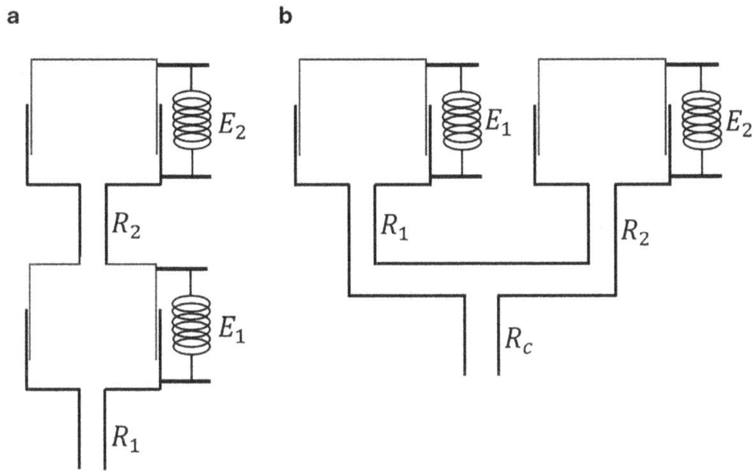

Fig. 3.3 (**a**) Series two-compartment model of the lung consisting of a proximal compartment representing the upper airways (resistance R_1, elastance E_1) connected to a distal compartment representing the alveolar parenchyma (resistance R_2, elastance E_2). (**b**) Parallel two-compartment model of the lung consisting of two alveolar compartments (resistances R_1, R_2 and elastances E_1, E_2, respectively) connected to a common central airway (resistance R_c)

The models illustrated in Fig. 3.3 depict lumped-parameter mechanical analogs of the lung because the components of these models have obvious correspondence to lung anatomy (specifically, airways and tissue). It is common, although not quite so intuitive, to express these models in terms of electrical circuits composed of resistors and capacitors (and inductors, when inertance is relevant). Converting between mechanical and electrical analogs is facilitated by considering how flow and pressure are apportioned between different branches of the mechanical model; the same apportionment pertains to current and voltage in the electrical model. For example, both models in Fig. 3.3 share the total flow between their two compartments, but the total pressure drop across the series model (Fig. 3.3a) is split between the two compartments whereas it is experienced equally by the two compartments in the parallel model (Fig. 3.3b). The corresponding electrical models are shown in Fig. 3.4 and demonstrate that both models are, in fact, parallel models in terms of the arrangement of their two capacitors. Thus, the apparent series and parallel nature of the mechanical models in Fig. 3.3a, b are actually only visual impressions.

The models in Figs. 3.3 and 3.4 each have two degrees of freedom, meaning that they have two separate repositories for either volume (the two alveolar compartments in Fig. 3.3) or electric charge (the two capacitors in Fig. 3.4). This means that the state of the models at any point in time is specified completely by the gas volumes in the two compartments in the case of the mechanical analogs, or the charges in the two capacitors in the case of the electrical analogs. Derivation of the

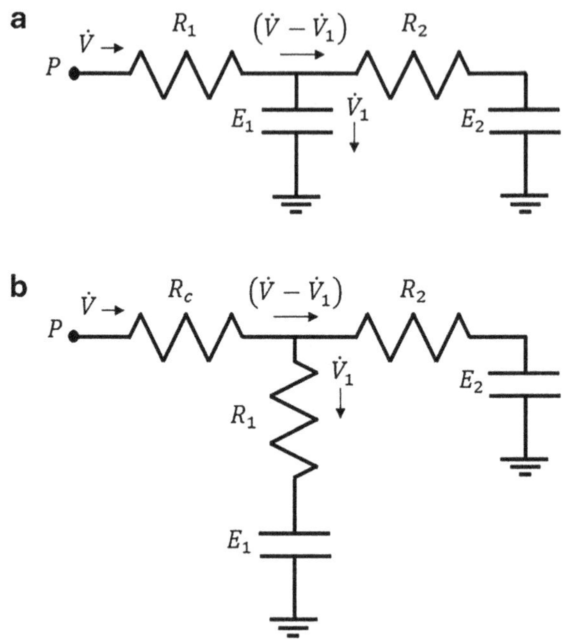

Fig. 3.4 (**a**) Electrical analog of the mechanical model shown in Fig. 3.3a. (**b**) Electrical analog of the mechanical model shown in Fig. 3.3b. Note that elastance (E_1 and E_2) is the inverse of capacitance, which is the usual parameter assigned to a capacitor in an electric circuit

equations of motion for these models begins with the separate differential equations for each of the two degrees of freedom.

In the case of the series mechanical model, inspection of its electrical analog (Fig. 3.4a) shows the pressure (voltage) drops across the two resistors are given in terms of the flows (currents) through the resistors by

$$P(t) - E_1 V_1(t) = R_1 \dot{V}(t) \tag{3.13}$$

and

$$E_1 V_1(t) - E_2[V(t) - V_1(t)] = R_2\left[\dot{V}(t) - \dot{V}_1(t)\right]. \tag{3.14}$$

These two first-order differential equations are coupled because $V(t)$ and $V_1(t)$ and their derivatives appear in both, but they can be reduced to a single second-order equation by eliminating $V_1(t)$ and its derivative (either using Laplace transforms or by differentiation and substitution in the time domain). The result is (Sect. 7.2 of [1])

$$\begin{aligned} R_2\dot{P}(t) + (E_1 + E_2)P(t) = R_1R_2\ddot{V}(t) + (E_1R_2 + E_2R_1 + E_1R_1)\dot{V}(t) \\ + E_1E_2V(t). \end{aligned} \tag{3.15}$$

The equation for the parallel mechanical model (Fig. 3.4b) is found in a corresponding way to be

$$\begin{aligned}(R_1+R_2)\dot{P}(t)+(E_1+E_2)P(t)&=[R_1R_2+R_c(R_1+R_2)]\ddot{V}(t)\\&+[E_1(R_2+R_c)+E_2(R_1+R_c)]\dot{V}(t)+E_1E_2V(t).\end{aligned} \tag{3.16}$$

It turns out there is another possible two-degree-of-freedom model of the lung that is more appropriate for the normal lung than the two models considered above. This third model consists of a single uniformly ventilated alveolus with walls composed of viscoelastic tissue [23]. We will leave considering of this model to Chap. 5, which covers tissue viscoelasticity, but it is also described by a second-order linear differential equation similar to Eqs. 3.15 and 3.16 (Sect. 7.2 of [1]).

Inspection of Eqs. 3.15 and 3.16 shows that the model parameters in each case appear in various combinations that are simple factors multiplying their respective variables. Equation 3.15, for example, can be rewritten

$$\begin{aligned}P(t)&=\frac{R_1R_2}{(E_1+E_2)}\ddot{V}(t)+\frac{E_1R_2+E_2R_1+E_1R_1}{(E_1+E_2)}\dot{V}(t)+\frac{E_1E_2}{(E_1+E_2)}V(t)-R_2\dot{P}(t)+P_0\\&=A_1\ddot{V}(t)+A_2\dot{V}(t)+A_3V(t)+A_4\dot{P}(t)+A_5\end{aligned} \tag{3.17}$$

where the offset pressure parameter P_0 has been added as in Eq. 3.8. This makes it obvious that the parameters A_i; $i = 1, ..4$ can be conveniently estimated from measurements of the independent variable $P(t)$ and the independent variables $\ddot{V}(t)$, $\dot{V}(t)$, $V(t)$, $\dot{P}(t)$ and 1 using multiple linear regression. (Note that although $\ddot{V}(t)$, $V(t)$ and $\dot{P}(t)$ are called "independent variables" here for regression purposes, in fact they are determined by $\dot{V}(t)$ and $P(t)$.) However, since the primary measurements from the physiology laboratory are usually $P(t)$ and $\dot{V}(t)$, this means that both these quantities must be numerically differentiated with respect to time to generate all the terms in Eq. 3.8. The problem with this is that differentiation is an inherently noise-amplifying process. It turns out that a more robust approach is to fit the time-integral of Eq. 3.17, which, after rearranging of terms, is

$$P(t)=A_1\dot{V}(t)+A_2V(t)+A_3\int_0^t V(t)dt+A_4\int_0^t P(t)dt+A_5t+A_6 \tag{3.18}$$

This results in the need to estimate a sixth parameter, A_6, because of the constant of integration, but this disadvantage is more than offset by the reduced noise in the independent variables [24]. Once the A_i have been evaluated, the parameters of the series mechanical model (Fig. 3.4a) can be determined via Eq. 3.17.

Similar equations can be developed for models with more than two compartments, but the number of parameters in such models rapidly becomes too numerous to be robustly supported by measured $P(t)$, $\dot{V}(t)$ and $V(t)$ signals. Accordingly, inverse models of lung mechanics have been limited to those no more complex than those exemplified by Eqs. 3.8, 3.9, 3.10 and 3.18.

3.5 Solving Large Systems of Model Equations

Inverse models of lung mechanics are thus, by necessity, limited to having only one or two compartments, possibly with the addition of nonlinear constitutive properties. On the other hand, there is no limitation on the number of compartments that a forward model can have when the parameters governing each of its compartments are already known (see Sect. 1.4). Deriving a single equation of motion for such a model rapidly becomes impractical as the number of compartments increases, as should be evident from the appearance of Eqs. 3.15 or 3.16 for models of only two compartments. However, deriving a single equation is not necessary for calculating the predictions of a forward model because this can be done numerically working from the system of coupled equations that describe the behavior of each individual compartment.

As an example, consider the two coupled first-order differential equations (Eqs. 3.13 and 3.14) that describe the series mechanical model illustrated in Fig. 3.4a. Beginning from specified initial conditions $V(0)$ and $V_1(0)$, and driven by a known $P(t)$ signal, these equations can be rewritten explicitly in terms of the time-derivatives of the model's two degrees of freedom at $t = 0$ to give

$$\dot{V}(0) = \frac{1}{R_1}P(0) - \frac{E_1}{R_1}V_1(0) \tag{3.19}$$

and

$$\dot{V}_1(0) = \dot{V}(0) - \frac{E_1}{R_2}V_1(0). \tag{3.20}$$

Since all quantities on the right-hand sides of these two equations are known, the values of the derivatives $\dot{V}(0)$ and $\dot{V}_1(0)$ are immediately determined.

Next, the values of V and V_1 are computed a short time δt later using a numerical approximation to the operation of integration with respect to time. There are a variety of such approximations in common use that have varying degrees of accuracy, but the simplest is the so-called *first-order Euler scheme* expressed for Eqs. 3.19 and 3.20 as

$$V(\delta t) = V(0) + \dot{V}(0)\delta t \tag{3.21}$$

and

$$V_1(\delta t) = V_1(0) + \dot{V}_1(0)\delta t. \tag{3.22}$$

The values for $V(\delta t)$ and $V_1(\delta t)$ determined in this manner are not perfectly accurate, of course, but if δt is small enough they can be extremely close to the correct answer.

All time arguments in Eqs. 3.21 and 3.22 are then incremented by δt to give $V(2\delta t)$ and $V_1(2\delta t)$, and so on over and over again until the desired time interval has elapsed. In this way, the model equations are boot-strapped from the initial conditions all the way through the period of simulation.

The above procedure can be generalized to a model comprised of an arbitrary number of compartments. The number of individual equations increases with the number of compartments, but the fundamental complexity of the task does not. Thus, the only thing required of a modeler in order to determine the predictions of a complex multi-compartment model is to derive the system of coupled differential equations.

3.6 Models of Mechanical Impedance

The lumped-parameter models discussed so far have all been derived in the time domain in the form of differential equations containing variables that are functions of time. This makes intuitive sense since we experience the passage of time naturally as we live our lives. Nevertheless, it often turns out to be more convenient to develop these models in terms of variables that are functions of frequency (f) so long as the lung behaves sufficiently like a linear dynamic system.

A linear dynamic system has the particular property that if it is excited by a sinusoidally varying input for long enough to allow all transients in its output to die out, the resulting steady-state output will also be sinusoidal. Furthermore, the frequency of the input will be the same as that of the input. The difference between the input and output thus only pertains to the amplitudes and phases of their respective sinusoids. Thus, if a sinusoidal flow

$$\dot{V}(t) = A_{in} \sin(\omega t + \phi_{in}) \tag{3.23}$$

is applied as the input, the pressure it will produce as the output is

$$P(t) = A_{out} \sin(\omega t + \phi_{out}). \tag{3.24}$$

Furthermore, the amplitude ratio $A = A_{out}/A_{oin}$ and the phase difference ($\phi = \phi_{out} - \phi_{in}$) depend only on f; they are independent of the amplitude and phase of the input. This means that a linear dynamic system is completely characterized by the functions $A(f)$ and $\phi(f)$ that specify how an input sinusoid is transformed to an output sinusoid for any value of f. These two functions may be written as the complex exponential function $A(f)e^{i\phi(f)}$, where $i = \sqrt{-1}$, which is often referred to as the system *transfer function* or *frequency response*.

When the input is a material of some kind moving through a physical system (electric current in the case of circuits, gas flow in the case of the lung) and the output is the force driving the movement of material (voltage or pressure), then the transfer function described above serves as general measure of the impediment to movement presenting by the system to the material in question. Accordingly, it is called *impedance*. The calculation of impedance is usually achieved by subjecting a system to a broad-band input (one consisting of the superposition of many different frequency components), measuring the resulting output, and then calculating the ratio of the Fourier transforms of the output to the input.

The linearity assumption is reasonably well met for the lung when modest variations in pressure, flow and volume are applied to the airway opening. When this is the case, lung impedance, $Z_L(f)$, can be calculated as

$$Z_L(f) = \frac{P(f)}{\dot{V}(f)} \tag{3.25}$$

where the argument f denotes the Fourier transform of the corresponding function of t.

As we have seen above, lumped-parameter models of lung mechanics are described by linear ordinary differential equations that relate $P(t)$ and its time-derivatives to $\dot{V}(t)$ to its derivatives and its integral, $V(t)$. These equations can be readily converted to the frequency domain using the fact that if $X(f)$ is the Fourier transform of $X(t)$, then the transform of the time-derivative of $X(t)$ is $i2\pi fX(f)$, and the transform of the time-integral of $X(t)$ is $-iX(f)/2\pi f$. This allows differential equations to be transformed into algebraic equations. For example, the Fourier transform of the equation for the single-compartment linear model (Eq. 3.8) is

$$P(f) = I_{aw}\ddot{V}(f) + R_L\dot{V}(f) + E_L V(f) = i2\pi f I_{aw}\dot{V}(f) + R_L\dot{V}(f) - i\frac{E_L}{2\pi f}\dot{V}(f) \tag{3.26}$$

which, through Eq. 3.25, leads to

$$Z_L(f) = R_L + i\left[2\pi f I_{aw} - \frac{E_L}{2\pi f}\right]. \tag{3.27}$$

The same process can be applied to the differential equations of the two-compartment models described above. For example, the Fourier transform of the equation for the series model (Eq. 3.15) is

$$\begin{aligned} i2\pi f R_2 P(f) + (E_1 + E_2)P(t) &= i2\pi f R_1 R_2 {}^{\cdot}V(t) \\ &+ (E_1R_2 + E_2R_1 + E_1R_1){}^{\cdot}V(t) - i\frac{E_1E_2}{2\pi f}{}^{\cdot}V(t). \end{aligned} \tag{3.28}$$

This can be rearranged to give a specific formula for $Z_L(f)$ as was done for the single-compartment model in Eq. 3.27. It should be evident at this point, however, that as the order of the differential equation increases, the algebra involved in deriving an explicit expression for $Z_L(f)$ rapidly becomes tortuous to the point of

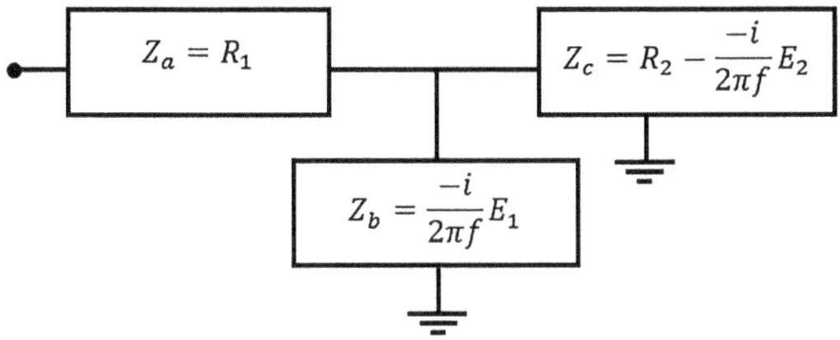

Fig. 3.5 Block structure t-network representation of the series model shown in Fig. 3.4a. The individual impedances Z_b and Z_c are added in parallel and the result added in series with Z_a. This process can be applied iteratively to a network of arbitrary size to calculate the impedance of the entire network

being impractical. Calculating the impedance of a complex structure thus needs to be done by computer starting from a block diagram of the sub-impedances that comprise the complete structure. The individual impedances of each structure must be expressible analytically, and then these individual impedances are added together.

For example, in the series model in Fig. 3.4a, the resistor R_1 is in series with the parallel combination of the capacitor E_1 and the series combination of R_2 and E_2. This can be expressed as the block structure shown in Fig. 3.5 in which the impedances $Z_a(f)$, $Z_b(f)$ and $Z_c(f)$ are arranged in a t-network. These three impedances are shown in Fig. 3.5 expressed in terms of the series model parameters. In general, each has a real part $R(f)$ and imaginary part $X(f)$ that combine together to give

$$Z_\alpha(f) = R_\alpha(f) + iX_\alpha(f); \alpha = a, b, c. \tag{3.29}$$

The parallel sum of $Z_b(f)$ and $Z_c(f)$ is

$$\begin{aligned} Z_\beta(f) &= \frac{Z_b(f)Z_c(f)}{Z_b(f) + Z_c(f)} \\ &= \frac{[R_b(f) + iX_b(f)][R_c(f) + iX_c(f)][R_b(f) + R_c(f) - i(X_b(f) + X_c(f))]}{[R_b(f) + R_c(f)]^2 + [X_b(f) + X_c(f)]^2} \\ &= R_\beta(f) + iX_\beta(f). \end{aligned} \tag{3.30}$$

Finally, the series sum of $Z_\beta(f)$ with $Z_a(f)$ is

$$Z_{total}(f) = Z_\beta(f) + Z_a(f) = R_\beta(f) + R_a(f) + i[X_\beta(f) + X_a(f)]. \tag{3.31}$$

The addition of the impedances expressed in Eqs. 3.29, 3.30, and 3.31 allows for the calculation of the impedance of a lung model that represents the branching structure of an entire airway tree once the separate impedances of the individual airway branches are known. In the case of a symmetrically branching tree in which all airways of a given generation are identical (Fig. 2.1a), one begins by adding the impedance of each terminal airway to that of its subtended acinar unit to produce a set of impedances are then combined in parallel pairs with the series impedance of their common parent airway. Next, these impedances are combined in parallel pairs with their parent airways, and so on all the way up to the trachea. A model of this nature has been used, for example, to determine the contributions of each airway generation to the overall resistance of the airway tree [4].

Of course, not all airways of a given generation are identical in a real lung. A straightforward way to mimic this in a model is to randomly assign characteristics to each airway branch according to specified probability distributions as its impedance is added into the total.

Another important characteristic of an actual airway tree is that it is not symmetric; the pathways to the acinus at any branchpoint in the tree generally traverse different numbers of branches. A convenient way to represent such asymmetry in a model is that due to Horsfield who identified the individual airway branches according to their *order number* as opposed to generation number as described in in Chap. 2.

A particularly convenient feature of the Horsfield airway tree structure is that its total impedance can be calculated recursively. The total impedance is the series sum of the tracheal impedance with the parallel sum of the two impedances subtended by the mainstem bronchi. But each of these subtended impedances is the serial sum of the impedance of the mainstem bronchus with its subtended impedance. Each of these next subtended impedances is itself the serial sum of the impedance of a lobar bronchus and the downstream impedance that it subtends. This calculation can be coded very efficiently in a recursive subroutine (one that can call itself).

Finally, by adding to the impedance of each terminal bronchiole an impedance representing its downstream acinar unit (which must take into account the mechanical properties of the parenchymal tissue), one obtains the impedance of the entire lung model [25–28]. Such models contain many degrees of freedom and a correspondingly large number of parameters. As such, they are much more faithful representations of the structure of a real lung than the one-compartment and two-compartments models introduced earlier in this chapter. Nevertheless, these anatomically based constructs are still lumped parameter models based on sets of coupled ordinary differential equations that are converted into their frequency-domain equivalents.

3.7 Conclusions

Lumped-parameter models based around the concept of the compartment serve as the mainstay for the quantitative description of lung function. When used for inverse modeling, such models must have simple structures and few adjustable parameters. As such, they can feature one or more (but not many) tissue compartments and interconnecting airway conduits having either linear or nonlinear constitutive properties. Forward models of lung function, in contrast, have no such limitations on degree of complexity, and thus may include a large amount of anatomic and physiologic detail such as the inclusion of the entire airway tree with all its myriad branches and parenchymal terminations. Lumped-parameter compartment models are described by systems of coupled ordinary differential equations that are usually readily solved on a computer in the time domain using standard numerical integration schemes. In the frequency domain, the mechanical input impedance of the lung can be calculated through the systematic summation of component impedances in series and parallel, as appropriate to the structure being simulated, but are based on the assumption that the lung behaves like a linear dynamic system.

3.8 Problems

1. Using Eqs. 3.13 and 3.14, simulate $V(t)$ and $\dot{V}(t)$ over 6 complete breaths where $P(t) = 15$ cmH_2O during each 3 s inspiration and $P(t) = 0$ during each 5 s expiration. Use the initial conditions that both $V(t)$ and $\dot{V}(t)$ are zero. Use the following model parameter values: $R_1 = 3$ cmH_2O.s.L^{-1}, $R_2 = 4$ cmH_2O.s.L^{-1}, $E_1 = 10$ cmH_2O. L^{-1}, $E_2 = 5$ cmH_2O. L^{-1}. Simulate the model at a sampling rate of 128 Hz.

   (a) Fit Eq. 3.18 to the $V(t)$, $\dot{V}(t)$ and $P(t)$ signals comprising the final simulated breath. Using Eq. 3.17, confirm that the regression parameters in Eq. 3.18 ($A_1 \ldots A_6$) correspond to the correct values of the model parameters.
   (b) Convert Eqs. 3.13 and 3.14 into the equations for a nonlinear two-compartment model of the lung in which the elastance of the distal alveolar compartment (E_2) depends linearly on the volume of air it contains. Re-simulate the model for 6 breaths and repeat a). How does the value of E_2 determined by fitting Eq. 3.18 to the simulated data relate to the elastance of the distal compartment in the model?

2. Repeat question 1, but this time drive the model with $\dot{V}(t)$ rather than $P(t)$. Use a 20-s $\dot{V}(t)$ signal composed of a sum of 5 sine waves of equal amplitude having frequencies of 0.5, 1, 2, 4 and 8 Hz. Isolate the last 2 s of the simulated $\dot{V}(t)$ and $P(t)$ signals. Calculate the fast Fourier transforms of the two signals and divide the transform of $P(t)$ by the transform of $\dot{V}(t)$ at each of the 5 frequencies in the $\dot{V}(t)$ signal.

   (a) Plot the real and imaginary parts of these 5 quotients against frequency. How do these plots compare to what you get if you calculate $Z_L(f)$ using Eq. 3.28?
   (b) How might you alter the parameters of the model to simulate asthma? Make appropriate adjustments to the parameter values and recompute $Z_L(f)$.
   (c) How might you alter the parameters of the model to simulate emphysema? Make appropriate adjustments to the parameter values and recompute $Z_L(f)$.

References

1. Bates JHT. Lung mechanics. An inverse modeling approach. Cambridge: Cambridge University Press; 2009.
2. Talevi A, Bellera CL. Compartmental pharmacokinetic models. In: Talevi A, Quiroga PA, editors. ADME processes in pharmaceutical sciences. Cham: Springer; 2024.
3. Ludwig MS, Shore SA, Fredberg JJ, Drazen JM. Differential responses of tissue viscance and collateral resistance to histamine and leukotriene C4. J Appl Physiol. 1985;1988(65):1424–9.
4. Pedley TJ, Schroter RC, Sudlow MF. The prediction of pressure drop and variation of resistance within the human bronchial airways. Respir Physiol. 1970;9:387–405.
5. Kuprat AP, Price O, Asgharian B, Singh RK, Colby S, Yugulis K, Corley RA, Darquenne C. Automated bidirectional coupling of multiscale models of aerosol dosimetry: validation with subject-specific deposition data. J Aerosol Sci. 2023;174
6. Smith BJ, Lukens S, Yamaguchi E, Gaver DP 3rd. Lagrangian transport properties of pulmonary interfacial flows. J Fluid Mech. 2011;705:234–57.
7. Fredberg JJ, Stamenovic D. On the imperfect elasticity of lung tissue. J Appl Physiol. 1985;1989(67):2408–19.
8. Mijailovich SM, Stamenovic D, Fredberg JJ. Toward a kinetic theory of connective tissue micromechanics. J Appl Physiol. 1985;1993(74):665–81.
9. Bachofen H, Hildebrandt J, Bachofen M. Pressure-volume curves of air- and liquid-filled excised lungs-surface tension in situ. J Appl Physiol. 1970;29:422–31.
10. Navajas D, Maksym GN, Bates JH. Dynamic viscoelastic nonlinearity of lung parenchymal tissue. J Appl Physiol. 1985;1995(79):348–56.
11. Bates JH, Maksym GN, Navajas D, Suki B. Lung tissue rheology and 1/f noise. Ann Biomed Eng. 1994;22:674–81.
12. Mead J. Mechanical properties of lungs. Physiol Rev. 1961;41:281–330.
13. Verbeken EK, Cauberghs M, Mertens I, Lauweryns JM, Van de Woestijne KP. Tissue and airway impedance of excised normal, senile, and emphysematous lungs. J Appl Physiol. 1985;1992(72):2343–53.
14. Bates JH, Irvin CG, Farre R, Hantos Z. Oscillation mechanics of the respiratory system. Compr Physiol. 2011;1:1233–72.
15. Jackson AC, Giurdanella CA, Dorkin HL. Density dependence of respiratory system impedances between 5 and 320 Hz in humans. J Appl Physiol. 1985;1989(67):2323–30.
16. Rohrer F. Der Stromungswiderstand der unregelmassigen Verzweigung des Bronchialsystems auf den Atmungsverlauf in verschiedenen Lungenbezirken. Pfluegers Arch. 1915;162:225–99.
17. Edibam C, Rutten AJ, Collins DV, Bersten AD. Effect of inspiratory flow pattern and inspiratory to expiratory ratio on nonlinear elastic behavior in patients with acute lung injury. Am J Respir Crit Care Med. 2003;167:702–7.
18. Bates JHT, Kaczka DW, Kollisch-Singule M, Nieman GF, Gaver DP. Atelectrauma can be avoided if expiration is sufficiently brief: evidence from inverse modeling and oscillometry during airway pressure release ventilation. Crit Care. 2024;28:329.
19. Salazar E, Knowles JH. An analysis of pressure-volume characteristics of the lungs. J Appl Physiol. 1964;19:97–104.
20. Venegas JG, Harris RS, Simon BA. A comprehensive equation for the pulmonary pressure-volume curve. J Appl Physiol. 1985;1998(84):389–95.
21. Mead J. Contribution of compliance of airways to frequency-dependent behavior of lungs. J Appl Physiol. 1969;26:670–3.
22. Otis AB, McKerrow CB, Bartlett RA, Mead J, McIlroy MB, Selver-Stone NJ, Radford EP Jr. Mechanical factors in distribution of pulmonary ventilation. J Appl Physiol. 1956;8:427–43.
23. Mount LE. The ventilation flow-resiostance anbd compliance of rate lungs. J Physiol. 1955;127: 157–67.
24. Sato J, Davey BL, Shardonofsky F, Bates JH. Low-frequency respiratory system resistance in the normal dog during mechanical ventilation. J Appl Physiol. 1985;1991(70):1536–43.

25. Wagers S, Lundblad LK, Ekman M, Irvin CG, Bates JH. The allergic mouse model of asthma: normal smooth muscle in an abnormal lung? J Appl Physiol. 1985;2004(96):2019–27.
26. Bates JHT, Kaminsky DA, Garrow OJ, Martin FK, Peters U, Tharp WG, Dixon AE. Lung de-recruitment in the allergic asthma of obesity: evidence from an anatomically based inverse model. J Appl Physiol. 1985;2023(134):356–64.
27. Lutchen KR, Greenstein JL, Suki B. How inhomogeneities and airway walls affect frequency dependence and separation of airway and tissue properties. J Appl Physiol. 1985;1996(80): 1696–707.
28. Herrmann J, Tawhai MH, Kaczka DW. Regional gas transport in the heterogeneous lung during oscillatory ventilation. J Appl Physiol. 1985;2016(121):1306–18.

Chapter 4
Parenchymal Tissue Elasticity

Abstract Upon stretching the lung during inspiration, the parenchymal tissue stores elastic energy in the form of a recoil force that helps lung volume return to functional residual capacity during expiration. This recoil force arises largely as a result of the stretching of collagen and elastic fiber networks together with the expansion of the surface film at the air-liquid interface. Linking the elastic properties of the parenchyma to the lung's multi-scale structures requires complex models based on the anatomic realities of the extracellular matrix. For example, unfolding and recruiting wavy collagen fibers greatly stiffens the tissue as it inflates toward total lung capacity. However, modeling predicts an instability during inflation leading to a runaway phenomenon at high transpulmonary pressures. At low pressures, the network properties of the parenchyma are also influenced by proteoglycans. Lung elasticity in all its manifestations is thus impossible to address in a quantitative fashion without the use of rather sophisticated mathematical and computational models. Predictions provided by such computational models can provide important insights into the nature of fundamental physiologic and biologic processes in the presence of gravity as well as how these processes become deranged in disease.

Keywords Pressure-volume curve · Stiffness · Spring constant · Stress-strain relation · Wavy collagen · Elastic fiber · Proteoglycans · Alveolar network · Gravity

4.1 Introduction

The alveolar parenchymal tissue of the lung is elastic, meaning that it has a resting configuration in which it bears no stress, and that it stores potential energy in the form of an elastic recoil force when stretched. The parenchymal recoil force that is generated during inspiration is an important contributor to the return of lung volume to functional residual capacity (FRC) during expiration. (The other contributors to expiration are the elastic recoil of the chest wall tissues and, during exercise, the active contraction of the expiratory muscles.) The expiratory forces generated by parenchymal elastic recoil are thus essential for breathing. At the same time, however, these forces must be readily overcome by the diaphragm and intercostal

B. Suki, J. H. T. Bates, *Mathematical Modeling of the Healthy and Diseased Lung*,
https://doi.org/10.1007/978-3-031-95322-4_4

muscles so that inspiration can be repeated roughly 12 times per minute throughout life without causing fatigue.

Parenchymal elastic recoil during inspiration arises largely as a result of distension forces generated within the elastic fiber network of the tissues together with the forces due to surface tension at the air-liquid interface. The link between these various forces and the elastic recoil pressure of the lung as a whole is mediated by the complexities of parenchymal structure, which must also be compatible with the stringent demands of gas exchange (see Chap. 2). Certain lung diseases can cause derangments in both the structure and function of the parenchymal tissue, with adverse consequences for both lung mechanics and gas exchange. Understanding the details of how parenchymal tissue elasticity arises in health, and how it becomes aberrant in disease, is thus essential to the understanding of pulmonary pathophysiology and to the development of effective treatments.

At a phenomelogical level, parenchymal tissue elasticity can be accurately represented by relatively simple lumped parameter models (see Chap. 3). Such models, however, do not link elastic function to the multi-scale structures that are ultimately responsible for it. Making this link requires more complex models based on the anatomic realities of parenchymal tissue. Structurally-based models are also necessary to accurately represent how parenchymal mechanics mediate physical interactions between the lungs and the chest wall to produce forces impacting the resident pulmonary cells. These forces can lead to stretch-induced and stiffness-related cellular mechanotransduction, with biologic consequences for the lung that have the potential to be pathogenic [1].

Fortunately, the mechanical behavior of the parenchyma can now be studied in detail over a wide range of length scales using a variety of sophisticated functional measurement and imaging techniques (see Chap. 2). Data provided by these techniques can inform computational models in numerous critical ways, and even allow them to be personalized to individual lungs. Computational modeling can also help fill in knowledge gaps that are beyond the capabilities of existing measurement methods. Accordingly, the present chapter examines parenchymal elastic modeling at a number of different levels of scale beginning with the bulk pressure-volume behavior of the lung. This is followed by an introduction to the world of nonlinear solid mechanics, which provides the tools needed to describe the nonlinear constitutive properties of parenchymal tissue. Also included are lower-level features including inter-alveolar interactions and the networks of wavy collagen and elastic fibers that support parenchymal stress in the presence of gravity.

4.2 Pressure-Volume Curves of the Lung Parenchyma: Phenomenological Description

The quasi-static mechanical behavior of the lung from a macroscopic perspective is best illustrated by its pressure-volume (P-V) curve, obtained by increasing and decreasing lung volume over the vital capacity range either continuously or in a stepwise manner while airway pressure relative to pleural pressure is recorded. Since airflow is negligible or zero under these circumstances, pressure is essentially equal to alveolar pressure, so the resulting P-V relationship reflects the elastic recoil of the parenchymal tissues. The first systematic analysis of the P-V curve in relation to surface tension was performed nearly a century ago [2]. Many studies have since reported P-V curves under various conditions and in different disease states [3]. An early and often used mathematical description of the P-V curve assumes an exponential relationship between P and V (Eq. 3.11) [4], but hyperbolic [5] and sigmoidal [6–8] functions have also been proposed. The parameters of these empirical equations can be shown to change in pathologic situations such as acute lung injury [9].

An interesting comprehensive P-V model [7] is given by

$$V = a + \frac{b}{1 + e^{-\frac{P-c}{d}}} \tag{4.1}$$

where a, b, c and d are adjustable parameters. An example of the shape of this P-V model is illustrated in Fig. 4.1. The parameter a defines the lower asymptote of V, which corresponds to residual volume (RV). The parameter b corresponds to the vital capacity of the lung, which is equal to the difference between RV and the upper asymptote of the curve that defines total lung capacity (TLC). The parameter c is the

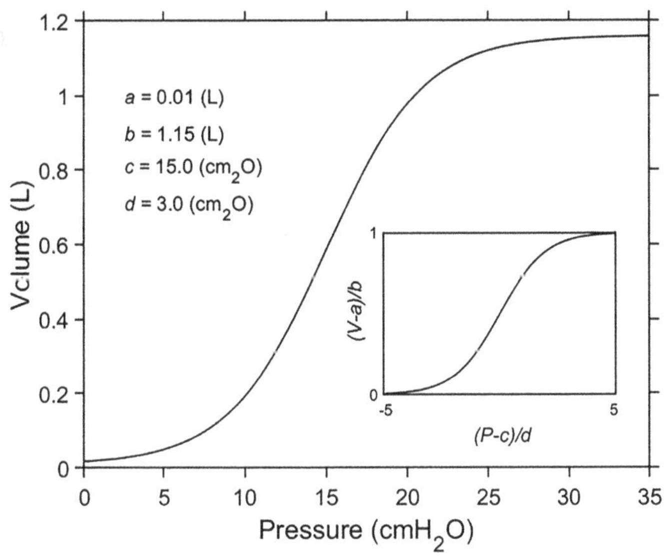

Fig. 4.1 Inflation pressure-volume (P-V) curve based on Eq. 4.1 with parameters approximately corresponding to those obtained from fitting the model to the P-V data in a patient with acute respiratory distress syndrome. The inset shows the normalize P-V curve corresponding to Eq. 4.2

value of P at the inflection point of the curve. Lung compliance, defined as the slope of the P-V relationship, is thus at its maximum when $P = c$. Lastly, the parameter d is related to the range of P over which most of the change in V occurs.

A non-dimensionalized scale-free version of Eq. 4.1 is obtained by introducing the variables $V' = \frac{V-a}{b}$ and $P' = \frac{P-c}{d}$, giving

$$V' = \frac{1}{1 + e^{-P'}}. \tag{4.2}$$

The intriguing fact about this transformation (see inset in Fig. 4.1) is that it defines a master relationship onto which individual P-V curves collapse regardless of whether they are obtained during inflation, deflation, in various disease conditions or in different species [7]. Furthermore, Eq. 4.2 does not contain any adjustable parameters, suggesting it harbors something fundamental about the nature of parenchymal elasticity. What might this be? One possibility is suggested by the fact that Eq. 4.2 is an analytic approximation to the integral of a normal distribution (which can only be precisely evaluated numerically). The sigmoidal shape of the P-V curve is thought to be significantly determined by the progressive recruitment of closed lung units as pressure increases. The specific shape of Eq. 4.2. might thus suggest that the associated reopening pressures are normally distributed [7]. This would only be true, however, if the closed lung units were arranged in parallel. In reality, these closed units include small airways that are part of a tree structure, so many of them are arranged in series. This causes the opening of closed airways during inspiration to occur in a series of avalanches that propagate to varying extents throughout the airway tree [10, 11], each individual opening event being driven by complex fluid dynamical processes [12].

In any case, the shape of the P-V curve is not determined only by opening events because the alveolar parenchyma exhibits the phenomenon known as strain stiffening, meaning that it becomes intrinsically stiffer the more it is stretched. This almost certainly accounts for much of the downward concavity in the relationship seen in Fig. 4.1 over the upper half of the lung volume range. Indeed, the point of inflection of the P′-V′ curve (where the second derivative of Eq. 4.2 is zero) occurs at $P' = 0$. The slope of the curve at this point is 0.25, which thus represents a universal maximum for lung compliance. On the original P-V curve, maximum compliance occurs at $P = c$ with a value of $b/4d$. One interpretation of the maximum compliance is that it corresponds to the point where few airways or alveoli remain collapsed, but strain stiffening has yet to exert a significant impact on tissue stiffness. The likely mechanism of strain stiffening, as illustrated in Fig. 2.9, is progressive straightening of initially wavy collagen fibers as V increases, causing these fibers to take on an increasing load-bearing role [13]. Since collagen fibers are orders of magnitude stiffer than elastic fibers [14], it is generally assumed that elastin determines lung compliance at low lung volumes where the collagen fibers are wavy and thus bearing little load [15].

4.3 An Isotropic 1D Recruitment Model of the P-V Curve

The first attempt to link progressive collagen recruitment to the nonlinear elastic behavior of lung parenchyma in a mathematical model is illustrated in Fig. 4.2 [16]. This model consists of a linear series of N units each comprised of a collagen fiber and an elastic fiber arranged in parallel. The elastic fibers are all identical, each behaving as an ideal linear spring with spring constant k. The collagen fibers associated with each elastic fiber behave like strings that bear no stress when flaccid (wavy) and cannot be extended when straight. These strings have different lengths, so they reach their straightened lengths at different points during extension of the model. When a unit is exposed to a force F, it stretches with an extension ΔX according to the spring law, $F = k\Delta X$, until its associated wavy string becomes taut at $\Delta X = l_s$. This corresponds to the *stop extension* of the unit.

Assume that the units have a range of discrete stop extensions denoted l_{s1}, l_{s2}, …, l_{sM}, and that there are n_m springs with stop extension l_m so that

$$N = \sum_{m}^{M} n_m. \tag{4.3}$$

Assume these stop extensions are uniformly spaced such that $l_{sj} = j\Delta l_s$, where Δl_s is the spacing between two consecutive stop extensions. To compute the force-extension behavior of this system, denote by F_1 the force at which the first group of units reach their stop extensions. Thus, $F_1 = kl_{s1} = k\Delta l_s$. Similarly, denote by F_2 the force at which the second group becomes taut, which occurs when $F_2 = kl_{s2} = k(2\text{-}\Delta l_s)$, and so on. In other words,

$$F_j = kj\Delta l_s. \tag{4.4}$$

The corresponding extensions of the entire system are expressed as follows. When $F < F_1$, all N units are stretched by the same amount, including those in the first group. When $F = F_1$, the total extension is $\Delta X_1 = N\Delta l_s$. For $F > F_1$, the $n_1$units in the first group are unable to stretch beyond Δl_s, so they collectively contribute a fixed amount $n_1\Delta l_s$ to the total extension. The remaining $N - n_1$ units continue to stretch uniformly until $F = F_2$, beyond which the second group of units contribute a fixed

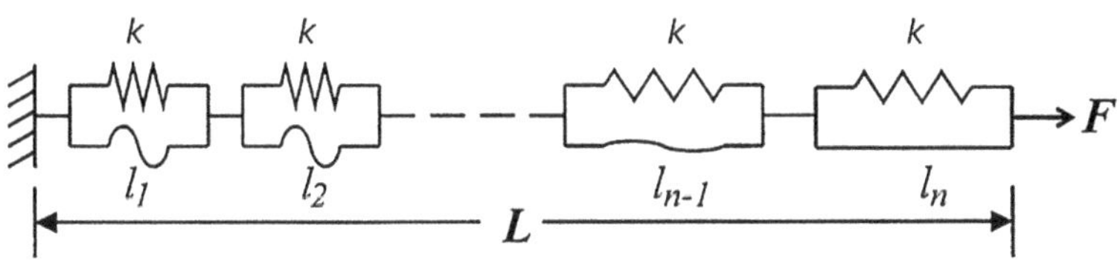

Fig. 4.2 A series model of parallel spring-string units. Springs are identical with stiffness k, whereas strings each have an associated stop length l at which further extension is not permitted. L length, F force. (Adapted with permission from Ref. [16])

amount $2n_2\Delta l_s$ to the total extension. The additional total extension between F_1 and F_2 is thus

$$\Delta X_2 = (N - n_1)\Delta l_s. \tag{4.5}$$

Likewise, the third group of units reaches its stop extension when F reaches F_3, at which point

$$\Delta X_3 = (N - n_1 - n_2)\Delta l_s \tag{4.6}$$

and so on. In general,

$$\Delta X_j = \left(N - \sum_{m=1}^{j-1} n_m\right)\Delta l_s. \tag{4.7}$$

The total extension of the model is the sum of the successive increments:

$$\begin{aligned} x_j &= \sum_{i=1}^{j} \Delta X_i = \sum_{i=1}^{j}\left(N - \sum_{m=1}^{i-1} n_m\right)\Delta l_s \\ &= \left(jN - \sum_{i=1}^{j}\sum_{m=1}^{i-1} n_m\right)\Delta l_s \end{aligned} \tag{4.8}$$

which gives

$$\Delta l_s = \frac{x_j}{jN - \sum_{i=1}^{j}\sum_{m=1}^{i-1} n_m}. \tag{4.9}$$

From Eq. 4.4,

$$F(x_j) = \sum_{s=1}^{j} \frac{k}{jN - \sum_{i=1}^{j}\sum_{m=1}^{i-1} n_m} x_j = \left[\frac{jk}{jN - \sum_{i=1}^{j}\sum_{m=1}^{i-1} n_m}\right] x_j = k_T(N,j)x_j \tag{4.10}$$

where $k_T(N,j)$ is a nonlinear stiffness function given by

$$k_T(N,j) = \frac{jk}{jN - \sum_{i=1}^{j}\sum_{m=1}^{i-1} n_m}. \tag{4.11}$$

As j increases and progressively more groups of units reach their stop extensions, the denominator in Eq. 4.11 becomes progressively smaller causing k_T to increase accordingly. Eventually, when all units are fully stretched, the term $\sum_{i=1}^{j}\sum_{m=1}^{i-1} n_m$ in the denominator becomes $\sum_{i=1}^{j} N = jN$ so the denominator itself becomes zero corresponding to a stiffness of infinity. Figure 4.3 compares the force-extension relationships produced by uniform and hyperbolic distributions of stop

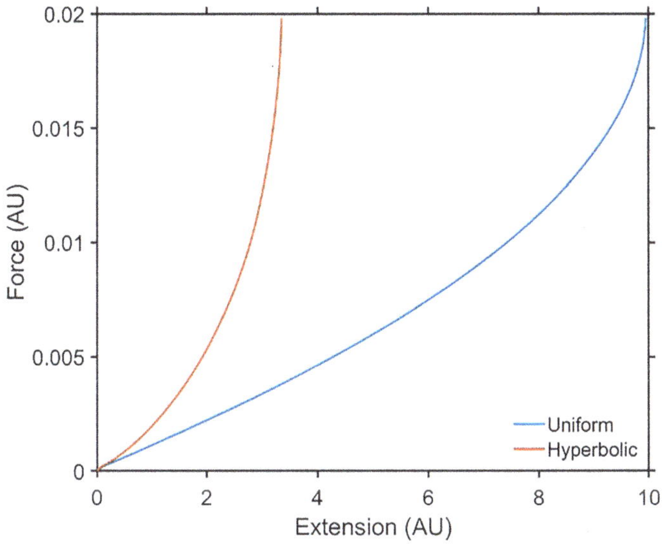

Fig. 4.3 Force-extension curves predicted by the discrete spring-string model (Eq. 4.10) for uniform and hyperbolic distributions of stop extensions

extensions. Force increases with total extension significantly faster for the hyperbolic distribution, which has more units with low stop extensions compared to the uniform distribution.

The spring-string model can also be formulated in terms of continuous rather than discrete distributions of stop extensions. The defining equation is [16]

$$n(l_s) = -k^2 \frac{d^2x}{dF^2} \tag{4.12}$$

from which the stop extension distribution $n(l_s)$ can be calculated directly from a measured force-length relationship. Equation 4.12 can be conceptually adapted to describe progressive collagen fiber recruitment in alveolar walls to give rise to a distribution $n(v_s)$ of stop volumes v_s in a homogeneous isotropic lung. Replacing x with V, F with P, and k with lung elastance E_L gives

$$n(v_s) = -E_L^2 \frac{d^2V}{dP^2} \tag{4.13}$$

which can be applied to the strain-stiffening portion of any P-V curve such as the upper half of the comprehensive relationship described by Eq. 4.1. An algebraically more straightforward example [16] is provided by the celebrated Salazar-Knowles equation [4]

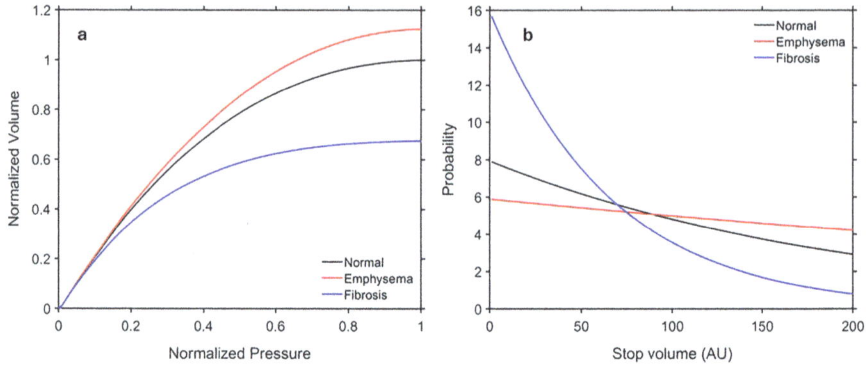

Fig. 4.4 (**a**) Pressure-volume curves calculated using Eq. 4.15 applied to a typical normal (blue), emphysematous (red) and fibrotic (brown) lung. (**b**) Corresponding stop volume distributions. The parameter χ in the exponential distribution was multiplied by 3 to create the emphysema model and divided by 3 to create the fibrosis model

$$V = C - De^{-KP} \tag{4.14}$$

which yields an exponential stop volume distribution given by

$$n(v_s) = \frac{KE_L D}{\chi} e^{-\frac{v_s}{\chi}} \tag{4.15}$$

where C, D and K are empirical constants and $\chi = \frac{1}{KE_L}$.

In addition to a distribution of string lengths, nonlinear P-V behavior in the spring-string model (Fig. 4.2) can also be achieved by having identical string lengths in each unit along with an appropriate distribution of spring constants [16], or some combination of both.

Figure 4.4a shows hypothetical P-V curves for normal, emphysematous and fibrotic lungs [17] based on Eq. 4.15. These curves exhibit the relative softening typical of emphysema and the stiffening seen in fibrosis. The corresponding stop volume distributions are shown in Fig. 4.4b, which demonstrate a greater likelihood of finding high stop volumes in an emphysematous lung. This is consistent with the fact that force-induced rupture of enzymatically weakened fibrils and fibers [18] in the emphysematous lung [19] can lead to the characteristic increases in airspace size that are seen both histologically and radiologically [20, 21]. The stop volume distribution shifts lower in fibrosis (Fig. 4.4b), suggesting that the collagen fibers have remodeled to become less wavy with reduced stop extensions so that they straighten at lower lung volumes. Thus, despite their simplicity, these models provide ample inspiration for creative analytic and computational modeling.

The spring-string model is, of course, a greatly simplified representation of a complex reality that neglects numerous important factors that link P and V, such as surface tension. The predictions of this model should thus not be taken to imply that anatomic correlates of $n(l_s)$ or $n(v_s)$ can be identified in actual lung tissue. Rather, the power of the model lies in the mechanistic insight that it provides because it demonstrates how macroscopic behavior such as the P-V relationship can be determined as much by how the constituents of a complex tissue are arranged as by what those constituents are made of.

4.4 Continuum Mechanical Models

Although the alveolar parenchyma with its interconnected septal walls naturally lends itself to being described as a 3D network, when modeling mechanical behavior at a level of scale well above that of a single alveolus, it can be convenient to resort to a description of the parenchyma as an elastic continuum [22–24]. Modeling the lung as a solid continuum requires the use of tensor analysis, which allows stresses and strains to be related to each other in 3D as follows. External forces acting on a body produce a net force vector $\Delta\boldsymbol{F}$ acting on a small surface element located within the body. This surface has area ΔA and an outward pointing unit normal vector $\boldsymbol{n}$. The stress vector at the location of $\Delta\boldsymbol{F}$ is defined as

$$\boldsymbol{T} = \lim_{\Delta A \to 0} \frac{\Delta \boldsymbol{F}}{\Delta A} = \sigma \boldsymbol{n}. \tag{4.16}$$

The Cauchy theorem states that $\boldsymbol{T}$ on the left-hand side is a linear function of $\boldsymbol{n}$ on the right-hand side, with σ being the Cauchy stress tensor. A tensor is a mathematical object with an existence in 3D space that is independent of coordinate system, meaning that its coordinates with respect to any particular coordinate system transform to another coordinate system according to the same type of mathematical rule that transforms the coordinates of a vector (a vector is a rank-1 tensor). The stress tensor is a rank-2 tensor comprised of 9 components that are, in general, functions of the three space coordinates and time. The inner product of $\boldsymbol{n}$ with σ in Eq. 4.16 is expressed as

$$T_x = \sigma_{xx} n_x + \sigma_{yx} n_y + \sigma_{zx} n_z = \sigma_{ix} n_i \tag{4.17}$$

where the Einstein notation has been employed on the right-hand side; the expression is summed over each value of any index that appears twice (i in this case), while the statement is true for each value of an index appearing only once (x in this case). The unit normal vector $\boldsymbol{n} = (1, 0, 0)$ lies parallel to the x axis, so $\sigma_{xx} = T_x$ is the normal stress (the ratio of ΔF_x acting on a surface ΔA perpendicular to the x axis). The shear stress σ_{yx} is the ratio of ΔF_x acting on a surface perpendicular to the y axis. σ_{xx} is a tensile stress when it is positive, and is a compressive stress otherwise. σ_{yx} is positive when $\boldsymbol{n}$ and ΔF_x point in the positive y and positive x directions, respectively. At equilibrium, σ is symmetric ($\sigma_{xy} = \sigma_{yx}$) and has only 6 independent components. This may not be true under non-equilibrium conditions, however.

The conditions for mechanical equilibrium of an elementary volume ΔV within a solid body are provided by the Navier equations which, again in Einstein notation, are

$$\frac{\partial \sigma_{ij}}{\partial x_i} + \rho b_j = \rho \frac{\partial^2 u_j}{\partial t^2} \tag{4.18}$$

where ρ is the material density within ΔV and u_j is the jth component of the displacement of ΔV from its original position before a force is applied. The first term on the left-hand side of Eq. 4.18 expresses the net force acting across ΔV, where differentiation is carried out with respect to coordinates in the deformed state. The second term represents the total body force acting on ΔV, such as that due to gravity. The term on the right-hand side is the acceleration of ΔV, so Eq. 4.18 is simply a statement of Newton's second law of motion. For a given a set of boundary conditions, the solution to Eq. 4.18 provides the stress distribution throughout the body. If the constitutive equations connecting stress and strain are known, the displacements of each point within the body may also be determined.

In addition to Eq. 4.18, the most general case of linear elasticity involves 15 independent quantities - 3 displacements, 6 strains and 6 stresses—along with 6 equations expressing the so-called compatibility conditions. Also involved are a set of constitutive equations, the number of which depends on the symmetry of the material properties. Appropriate displacement and/or traction boundary conditions must be specified to completely solve for the stresses and configuration of the solid. Because of the complexities involved in solving this problem for the general case, it is often only the static equilibrium configuration that is sought, in which case the right-hand side of Eq. 4.18 is set to zero. This approach can potentially be used to determine the configuration of the lung inside the thoracic cavity at a fixed volume. However, due to the irregular shape of the chest cavity, analytic solutions are not possible, so it is necessary to resort to numerical techniques to find the solution. A unique solution to this problem exists provided the bulk and shear moduli are positive. Consequently, a network model of the parenchyma can only be replaced by an equivalent continuum model if the unit cells within the network exert some degree of resistance to shape change (i.e., the network has a finite shear modulus). Even so, finding an equivalent continuum model can be computationally time consuming, and converging to a unique solution may be challenging.

In 1971, West argued that since mechanical stresses in the lung are nonuniform because of its weight, these stresses must contribute to disease development. He noted that "If an engineer is asked where a structure such as a bridge is likely to break first, he will probably want to know where it is stressed most. It is curious that, although the lung is a structure which often fails mechanically, little thought has been given to the distribution of stresses within it" [25]. West used the Finite Element Method to solve Eqs. 4.18 for the mechanical stresses due to the weight of the human lung in the upright position (Fig. 4.5a), an impressive achievement at a time when the computational tools we take for granted today were not available. His

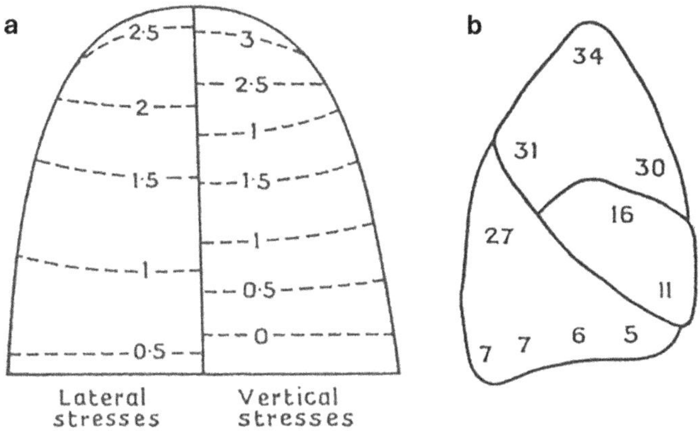

Fig. 4.5 (**a**) Finite Element calculation of the stress distribution in the lung due to its weight. (**b**) Distribution of centrilobular emphysema in a random series of necropsies in 138 patients in which disease was found. (Adopted with permission from Ref. [25])

calculations showed that the vertical stresses (due to the force of gravity) and the lateral stresses (due to compression against the chest wall) in the lung are at least 5 times greater at the apex than the base. These predictions are in accord with the fact that pleural pressure at the apex of the upright lung is significantly more negative than at the base [26]. Since local pleural pressure is transmitted laterally via septal walls to all alveoli at a given height [27], the alveoli at the apex experience larger stresses than those at the base and are thus correspondingly more expanded [28]. Even more importantly, West found a strong spatial correlation between the vertical stress gradient and the apical predominance of centrilobular emphysema (Fig. 4.5b). This correlation is modulated by the fact that the lung is divided into discrete lobes, so the vertical dependence of disease on height is not entirely monotonic. Also, as the lung changes volume, the lobes slide relative to each other along the fissures, which helps to reduce stress in the tissue, particularly in the lower lobes as shown in Fig. 4.6 [29, 30]. Nevertheless, the hand of gravity in emphysema pathogenesis and progression seems unmistakable.

4.5 Constitutive Relations and Large Deformations

There is a large body of literature on the constitutive relations of the lung parenchyma [31]. These are needed for an analysis of both small and large tissue deformations, for which the general definition of strain is also required. Strain is loosely defined as relative change in length. A more precise definition requires the consideration of displacements. If the position of a point in a continuous body at time t is denoted by the vector $\boldsymbol{X}$ (with respect to some arbitrary origin) before deformation, and by the vector $\boldsymbol{x}$ after deformation, then the displacement of the point is

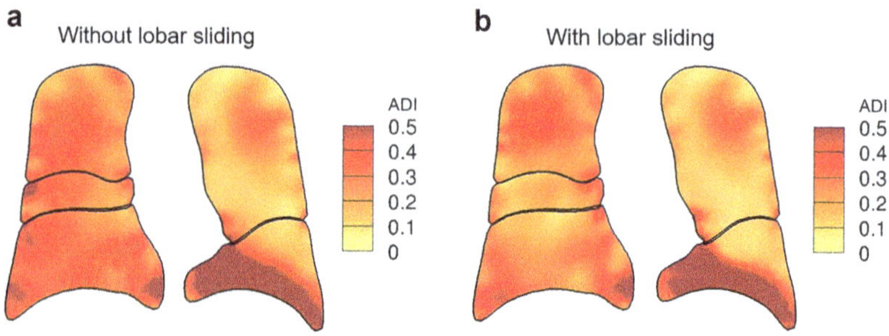

Fig. 4.6 Distribution of parenchymal distortion in a representative subject. Contour plots of anisotropic deformation index (ADI) in the non-sliding (**a**) and sliding (**b**) models. In both models, the largest distortion occurs near the diaphragm, which might explain why the lower lobe receives the most benefit from lobar sliding, especially in the right lung which is composed of 3 lobes. (Figure kindly provided by A. E. Galloy, J. M. Reinhardt and M. L. Raghavan)

$$\boldsymbol{u}(\boldsymbol{X},t) = \boldsymbol{x}(\boldsymbol{X},t) - \boldsymbol{X}. \tag{4.19}$$

The rank-2 linear strain tensor, ϵ, has diagonal elements that define the normal strains $\epsilon_{ii} = \frac{\partial u_i}{\partial x_i}$, and symmetric off-diagonal elements that define the shear strains $\epsilon_{ij} = \frac{1}{2}\left(\frac{\partial u_j}{\partial x_i} + \frac{\partial u_i}{\partial x_j}\right)(i \neq j)$. The normal strains represent uniaxial extension when they are positive and compression when they are negative, whereas the shear strains account for changes in angle between the original and deformed edges of a small element. This interpretation of the linear strain tensor is only valid for small deformations because rigid body rotations without deformation produce artificial normal components of the linear strain tensor. To avoid such artifacts, it is necessary to introduce nonlinear strain measures (see below).

Constitutive equations are mathematical relationships linking the stress and strain tensors. For small strains, the general form of Hooke's law links the stress and the strain tensors according to

$$\sigma_{ij} = C_{ijkl}\epsilon_{kl} \tag{4.20}$$

where C_{ijkl} is a rank-4 elasticity tensor. In general, C_{ijkl} is comprised of 81 material constants, but it becomes greatly simplified when the material exhibits symmetry. For example, when a material is isotropic, the deforming effect of an applied force is the same regardless of the direction in which the force is applied. In this case, two independent elastic constants suffice to link the stress and strain tensors thus:

$$\sigma_{ij} = \frac{Y}{1+\nu}\left[\epsilon_{ij} + \frac{\nu}{2(1-2\nu)}\epsilon_{kk}\delta_{ij}\right] \tag{4.21}$$

where ν is the Poisson's ratio that accounts for the lateral contraction of the material under uniaxial tensile stretching, and δ_{ij} is the Kronecker delta that equals 1 if $i = j$ and is zero otherwise. The Young's modulus (Y), the shear modulus (G) and ν are related to each other via $Y = 2G(1 + \nu)$. Equation 4.21 is the simplest example of a linear isotropic and ideal (no dissipation) constitutive relationship and is often invoked to describe lung elasticity [22, 24, 32]. Nevertheless, Y, G and ν are functions of lung volume so when given fixed values, they apply only to small variations in volume around a given inflation state of the lung. More complicated constitutive equations are required to account for nonlinear and/or time-dependent tissue mechanics (see Chap. 5).

A common measure of nonlinear deformation is the Green strain, ϵ_G, defined as

$$\epsilon_G = \frac{1}{2}\frac{l^2 - l_0^2}{l_0^2} = \frac{1}{2}\frac{(l - l_0)(l + l_0)}{l_0^2} = \frac{1}{2}\left(\lambda^2 - 1\right) \tag{4.22}$$

where l_0 and l are the stress-free and deformed lengths, respectively, and $\lambda = l/l_0$ is the stretch ratio. ϵ_G reduces to the linear strain when the displacement $\Delta l = l - l_0$ is small compared to l_0. Indeed, when $\lambda \to 1$, Eq. 4.22 simplifies to $\epsilon_G = \frac{1}{2}(\lambda + 1)(\lambda - 1) \approx \frac{1}{2}(2)(\lambda - 1) = \epsilon$. To extend Eq. 4.22 to a general 3D deformation, consider a material point $\boldsymbol{X}$ prior to deformation becoming $\boldsymbol{x}$ after deformation, as in Eq. 4.19. Provided the body does not rupture during deformation, $\boldsymbol{x}$ will be a continuous function of $\boldsymbol{X}$: $\boldsymbol{x} = \boldsymbol{x}(\boldsymbol{X}, t)$. The deformation gradient tensor F_d is defined as the rate of change of $\boldsymbol{x}$ with respect to $\boldsymbol{X}$ thus:

$$F_{d,ij} = \frac{\partial x_i}{\partial X_j}. \tag{4.23}$$

Let two neighboring points in the undeformed body be separated by an infinitesimal vector $d\boldsymbol{X}$. The distance $d\boldsymbol{x}$ between the corresponding points after deformation can be expressed using Eq. 4.23 as $d\boldsymbol{x} = F_d d\boldsymbol{X}$, which is invertible if the determinant of F_d, called the Jacobian, is positive ($J = \det F_d > 0$). If 3 vectors span a small volume dV_0 before deformation, the volume after deformation becomes $dV = JdV_0$. The change in the squared distance between the two points following deformation is then given by

$$|d\boldsymbol{x}|^2 - |d\boldsymbol{X}|^2 = d\boldsymbol{x}^T d\boldsymbol{x} - d\boldsymbol{X}^T d\boldsymbol{X} = d\boldsymbol{X}^T\left(F_d^T F_d - I\right)d\boldsymbol{X} \tag{4.24}$$

where the superscript T denotes the matrix transpose and I is the identity matrix. The general form of the Green strain tensor is thus

$$\epsilon_G = \frac{1}{2}\left(F_d^T F_d - I\right) \tag{4.25}$$

where the factor ½ ensures that, for small displacements, Eq. 4.25 reduces to the infinitesimal strain tensor. The Green strain is a true measure of deformation as it is free of the rotation artifact that affects the linear strains mentioned above [33].

Finally, the Green strain and the stress tensor are connected via the strain energy. The general treatment of nonlinear solid mechanics is beyond the scope of this book, so the present discussion is limited to the theory of linear elasticity and a class of materials called hyperelastic whose behavior does not exhibit plasticity or viscoelasticity, so stress does not depend on strain history. A complete account on nonlinear elasticity can be found in Chap. 3 of reference [34], or the nonlinear continuum mechanics textbook by Holzapfel [35].

For hyperelastic materials, the work done on the body by an externally applied force is stored as elastic potential energy that is returned when the force is removed. In such cases, the stress can be obtained from the so-called strain energy density function defined as the strain energy per unit volume. For isotropic materials, the strain energy is independent of the direction in which the material is deformed and thus is not a direct function of the strain tensor components. Since the invariants of a tensor do not change after a coordinate transformation, the best way to set up a strain energy function is to express it using the invariants of the deformation tensor. The Cauchy-Green deformation tensor is defined as

$$C_G = F_d^T F_d. \tag{4.26}$$

Since F_d can be decomposed into the matrix product of a pure rotation (R) and a pure stretch (U), it can be written $F_d = RU$. Substituting this into Eq. 4.26 and using $R^{-1} = R^T$ gives $C_G = U^2$, which demonstrates that C_G, and hence ϵ_G, are invariant to rigid body rotation, unlike ϵ.

Following a given deformation, the principal directions and the corresponding eigenvectors of F_d are computed. Rotating the coordinate system into the principal directions diagonalizes F_d:

$$F_d = \begin{bmatrix} \lambda_1 & 0 & 0 \\ 0 & \lambda_2 & 0 \\ 0 & 0 & \lambda_3 \end{bmatrix}. \tag{4.27}$$

Substituting this diagonalized matrix into Eq. 4.26 diagonalizes C_G so that $C_{G,ij} = \lambda_i \lambda_j \delta_{ij}$, giving its invariants the following simple forms:

$$\begin{aligned} I_1 &= \lambda_1^2 + \lambda_2^2 + \lambda_3^2, \\ I_2 &= \lambda_1^2\lambda_2^2 + \lambda_2^2\lambda_3^2 + \lambda_3^2\lambda_1^2, \\ I_3 &= \lambda_1^2\lambda_2^2\lambda_3^2. \end{aligned} \tag{4.28}$$

A general formulation of the strain energy density, e, is then given by

$$e(I_1, I_2, I_3) = \sum_{i+m+n=1}^{N} D_{kmn}(I_1 - 3)^i (I_2 - 3)^m (I_3 - 1)^n \tag{4.29}$$

where D_{kmn} are material constants. If the material is incompressible then $J = 1$, since a small volume transforms into itself (i.e., $dV = dV_0 = JdV_0$), which in turn implies that $I_3 = J^2 = 1$. Thus, e cannot depend on I_3 and Eq. 4.29 is rewritten as

$$e(I_1, I_2) = \sum_{m+n=1}^{N} D_{mn}(I_1 - 3)^m (I_2 - 3)^n. \tag{4.30}$$

For a Hookean spring, the elastic energy is $E = \frac{1}{2}k\Delta l^2$ which, when differentiated with respect to Δl gives the force $F = k\Delta l$. A similar formalism can be derived that links the strain energy density to stress. However, the stress is not the Cauchy stress in Eq. 4.18 that refers to a reference area in the deformed body because Eq. 4.30 contains strain invariants rather than strain values. Instead, the stress involved in this case is force acting over a differential area in the undeformed body. This is called the nominal stress, engineering stress or the first Piola-Kirchhoff stress denoted by S_I, and is related to σ via the transformation

$$S_I = JF_d^{-1}\sigma. \tag{4.31}$$

For the general case of a 3D deformation (see Chapter 3 of Ref. [34]), S_I can be computed as the derivative of the strain energy density with respect to the components of the deformation gradient tensor, that is

$$S_I = \frac{\partial e}{\partial F_d}. \tag{4.32}$$

Unfortunately, S_I is not symmetric in general, so yet another stress is needed, the second Piola-Kirchhoff stress which is a transformation of S_I defined as

$$S_{II} = S_I F_d^T = JF_d^{-1}\sigma F_d^T. \tag{4.33}$$

Similar to S_I in Eq. 4.32, S_{II} is related to the Green strain tensor via

$$S_{II} = \frac{\partial e}{\partial \epsilon_G}. \tag{4.34}$$

The conventional approach to modeling solid mechanics is to formulate the strain energy in terms of the invariants of F_d or ϵ_G, calculate S_I or S_{II}, and then transform these stresses back to the Cauchy stress via Eqs. 4.31 or 4.33, respectively. For an incompressible material, a pressure term is also added to reflect the constraint $J = 1$. This results in principal stresses given by

$$\sigma_i = \lambda_i \frac{\partial e}{\partial \lambda_i} - P. \tag{4.35}$$

As a special case of Eq. 4.30, consider an incompressible neo-Hookean material that has a strain energy density given by

$$e(I_1) = D_{10}(I_1 - 3). \tag{4.36}$$

To understand the difference between the behavior of a neo-Hookean material and a Hookean material (Eq. 4.21), consider a bar with a square cross-section of area A_0 and length l_0 that is stretched in the x direction by a force F. If the axis of the bar coincides with the x axis, then $\lambda_1 = \lambda = \frac{l}{l_0}$. Due to the symmetry of the situation, the λ values in the y and z directions are equal (i.e., $\lambda_2 = \lambda_3 = \lambda'$). Incompressibility implies that $\det(F_d) = \lambda\lambda'^2 = 1$ so that $\lambda' = \frac{1}{\sqrt{\lambda}}$. Thus, Eq. 4.36 can be written

$$e(I_1) = D_{10}(\lambda^2 + \frac{2}{\lambda} - 3). \tag{4.37}$$

Equation 4.32 gives $S_{I,xx} = S_I = 2D_{10}\left(\lambda - \frac{1}{\lambda^2}\right)$ and so the corresponding Cauchy stress from Eq. 4.31 is given by

$$\sigma = 2D_{10}(\lambda^2 - \frac{1}{\lambda}) = 2D_{10}(1 + 2\epsilon + \epsilon^2 - \frac{1}{1+\epsilon}). \tag{4.38}$$

Unlike Eq. 4.21, this is a nonlinear relationship. Expanding the last term up to second order gives.

$$\sigma = 6D_{10}\epsilon. \tag{4.39}$$

Setting $D_{10} = \frac{Y}{6}$ makes the neo-Hookean formulation consistent with linear elasticity. The neo-Hookean model has also been implemented with the assumption that the lung tissue is compressible, which makes nonlinear mechanical models of lung tissue suitable for finite element analysis [36, 37].

An alternative incompressible strain energy density function is that due to Ogden [38], given by

$$e(\lambda_1, \lambda_2) = \sum_{i=1}^{N} \frac{\mu_i}{\alpha_i}(\lambda_1^{\alpha_i} + \lambda_2^{\alpha_i} - \lambda_1^{-\alpha_i}\lambda_2^{-\alpha_i} - 3). \tag{4.40}$$

This function describes nonlinear material behavior that exhibits stiffening for $\alpha_i > 2$, which is consistent with the behavior of biological tissue. When $N = 1$ and $\alpha_1 = 2$, the Ogden formulation describes a neo-Hookean material. Both the neo-Hookean and the Ogden models as well as other strain energy formulations, such as the exponential strain energy function [39], have been used to model the nonlinear elastic behavior of the lung parenchyma [29, 36, 37, 39–46] and the airways [47, 48]. It is possible, in principle, to include the contribution of collagen

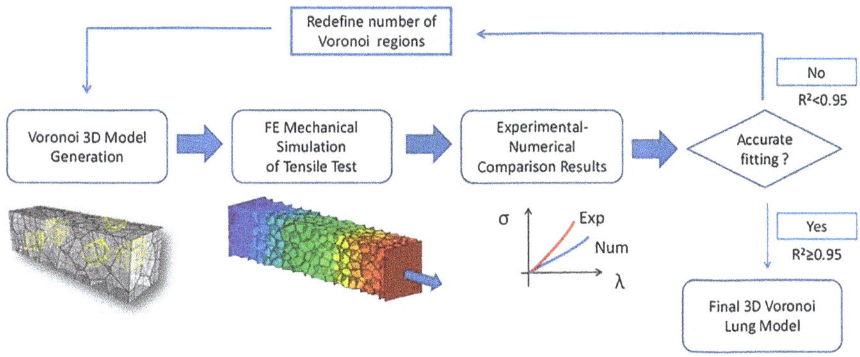

Fig. 4.7 Algorithm used for creating a 3D model of a lung tissue strip based on Voronoi tessellation. An iterative process was employed with three steps: (1) Generation of the 3D model for a fixed number of alveoli, (2) Simulation of the tensile test, and (3) Comparison of the stress-strain curve against experimental data. If the fitting error is too high, a new 3D Voronoi model is generated, and the process is repeated until a satisfactory fit is achieved. (Reproduced with permission from Ref. [44])

fibers with a given directionality in the strain energy function [49], but this is not usually implemented with respect to the lung because of a lack of specific fiber orientation in the parenchyma. A constituent-specific approach has nevertheless been proposed in which elastin, collagen or both are digested and then a nonlinear inverse model is fit to the simulated data [50].

The above theory has been applied to the modeling of lung tissue strip mechanics [44]. The model consists of a collection of packed alveoli represented by 3D Voronoi tessellation of a stretched cube (Fig. 4.7). Each septal wall has a strain energy function implemented according to the so-called Yeoh model originally formulated to describe the elasticity of rubber [51]. An iterative process was used to find the Voronoi structure that represents the required number of alveoli while best mimicking experimental stress-strain curves [44], as illustrated in Fig. 4.7. Applying a finer finite element mesh to a small section of this model, septal wall stiffness was calculated because this is a key biophysical attribute of parenchymal tissue that is sensed by resident alveolar cells. A small radial displacement of 0.1 μm was imposed on the mesh nodes on a region of the alveolar surface having an area similar to the footprint of a cell. This was designed to mimic the local distortion of the ECM caused by contraction of the cytoskeleton as cells probe their mechanical environment [52]. The distributions of radial displacement and maximum principal stress were calculated for both a Type I alveolar epithelial cell residing on a flat surface and a Type II alveolar epithelial cell sitting in the corner of the alveolus (Fig. 4.8). This model indicates that stress is much more regionally heterogenous and that stiffness is approximately twice as high in the alveolar corners where septa join compared to on the surface of a single septum, which has potentially significant implications for regional cell behavior [53].

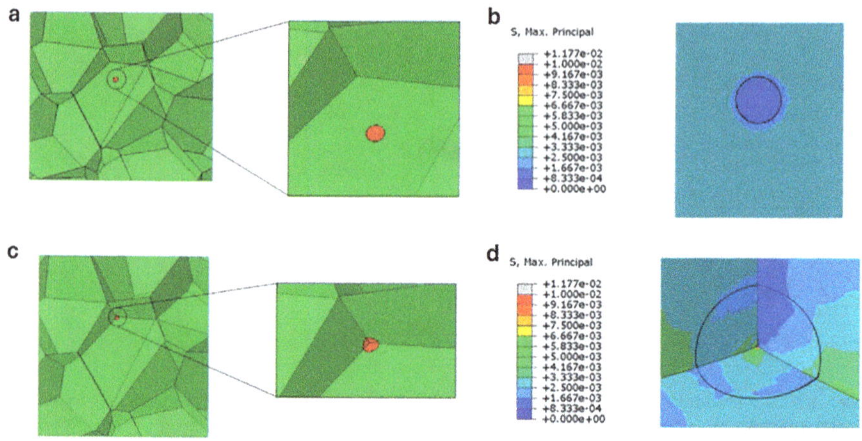

Fig. 4.8 (**a**) Circumference projected on a flat region of the alveolus for modeling the Type I alveolar epithelial cell. (**b**) Maximum principal stress [MPa] distribution obtained for a biaxial strain of 14.3% for the Type I cell case. (**c**) Circumference projected on a corner region of the alveolus for modeling the Type II alveolar epithelial cell. (**d**) Maximum principal stress distribution [MPa] obtained for biaxial strain of 14.3% for the type II cell case. (Reproduced with permission from Ref. [44])

The continuum approach can also be coupled with imaging. For example, CT imaging has been used to determine the exact boundaries of the lung, from which a personalized continuum model can be created [29] as exemplified in Fig. 4.6. Images of tissue in the baseline state can also be warped to match an image taken after the tissue has been deformed. When this is combined with a solution to the inverse elasticity problem, a stiffness map can be estimated [54]. This has been done with images collected using the crystal ribcage method in mouse lungs to determine stiffness maps of subpleural regions [55] that agree with direct measurements made using atomic force microscopy [41]. These maps have revealed that while a healthy lung exhibits substantial heterogeneity in alveolar scale stiffness, heterogeneity is increased in a lung with primary cancer (Fig. 4.9). Also, normal alveoli show strain-stiffening behavior with local tissue stiffness increasing by a factor of ~1.75 as P_{tp} is elevated, while stiffness in the cancerous lung can increase by sixfold over the same range of P_{tp}. Simulations in Fig. 4.3 indicate that such stiffening must be related to altered waviness of collagen, and the data in Fig. 4.9 suggest this phenomenon must occur at the level of single septal walls, which motivates the discussion of the role of wavy collagen in single alveolar mechanics.

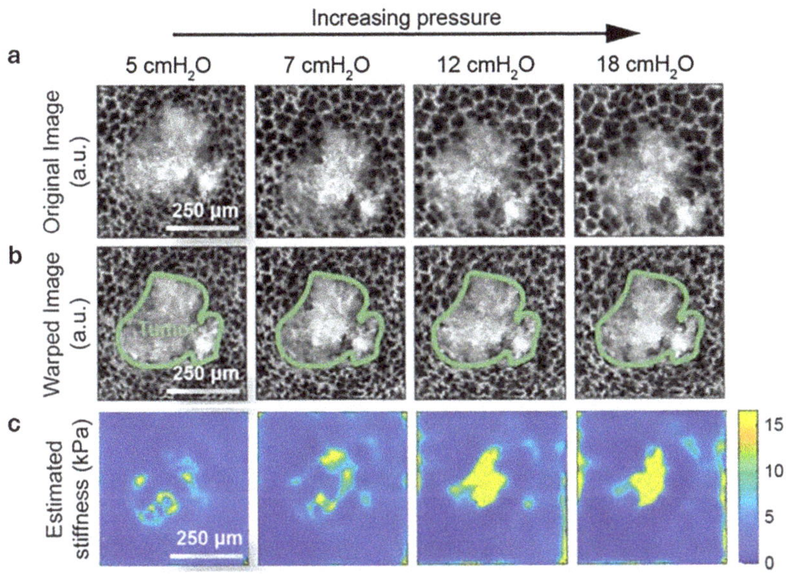

Fig. 4.9 (**a**) Alveolar region with a primary tumor from a transgenic mouse expressing the tdTomato fluorescent label at four different distending pressures. These images were used as inputs to a multiscale computational model to determine the distribution of stiffnesses throughout the tumor and its surrounding tissue. (**b**) Images that are computationally deformed back to their geometries at 2 cmH_2O. (**c**) Corresponding stiffness maps obtained by solving the inverse elasticity problem. (Reproduced with permission from Ref. [54])

4.6 An Analytic P-V Model of a Single Alveolus

It is possible to marry the spring-string model in Fig. 4.2 with a continuum mechanical description of an expanding alveolus in 3D [56]. An example of such an alveolar model is shown in Fig. 4.10a, which consists of a spherical thick-walled sphere having inner radius R_i and outer radius R_e in the baseline state when the transmural pressure across the wall is zero. During isotropic expansion, R_i and R_e increase to r_i and r_e, respectively (Fig. 4.10b), such that the total cross-sectional area of the wall is conserved. A spherical shell within the wall at radius $R_i \leq R \leq R_e$ at baseline thus achieves radius $r_i \leq r \leq r_e$ in the inflated state.

The wall is composed of a tissue matrix within which circumferentially oriented wavy collagen fibers and non-wavy elastin fibers are embedded (Fig. 4.10c). Expansion of the alveolus, driven by transmural alveolar pressure (P_{alv}), causes the elastic fibers to elongate and the collagen fibers to progressively straighten in what is essentially a circular version of the linear spring-string model shown in Fig. 4.2. As strain increases within the alveolar wall, stress builds up commensurately, but because the wall has finite thickness, stress and strain cannot be accounted for in terms of the single parameters that are appropriate for lumped parameter models.

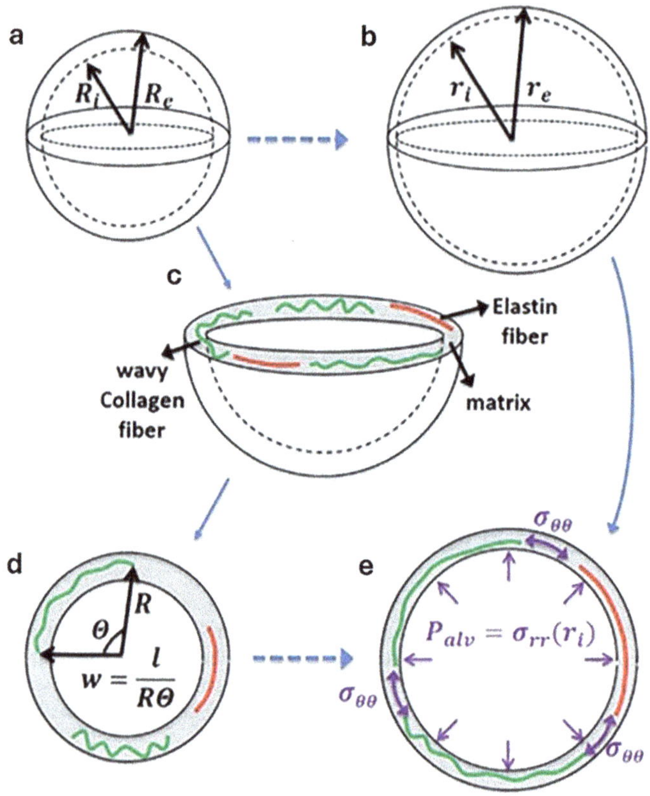

Fig. 4.10 Model concept of alveolar inflation. (**a**) A thick-walled spherical alveolus of initial inner and outer radii r_i and r_e, respectively. (**b**) As the alveolus expands isotropically, the wall thickness decreases as the stretched inner and outer radii R_i and R_e, respectively, increase. (**c**) Wavy collagen fibers (green) and the non-wavy elastin fibers (red) embedded in a matrix (gray) lie circumferentially within a plane slicing through the great circle of the sphere. (**d**) The waviness of a collagen fiber is defined as its contour length (l) over its end-to-end distance ($R\theta$) in the undeformed sphere. (**e**) As the alveolus expands due to alveolar pressure (P_{alv}), the wavy fibers become less wavy. Once fully straightened, they contribute together with the already elongated elastin fibers to the circumferential stress ($\sigma_{\theta\theta}$). Note that P_{alv} is equal to the radial stress (σ_{rr}) at r_i. (Reproduced with permission from Ref. [56])

In the alveolar model illustrated in Fig. 4.10d, the wavy collagen fibers become progressively less wavy as volume starts to increase above baseline. Only when a collagen fiber is fully straightened, however, does it contribute to the circumferential wall stress $\sigma_{\theta\theta}$ (Fig. 4.10e). The radial stress σ_{rr} also increases with inflation and at the inner radius r_i, is always equal to P_{alv}. The elastic modulus of straightened collagen, Y_c, is not infinite in this model (as it was in the 1D spring-string model above), but is nevertheless much larger than the modulus, Y_e, of elastin.

The total fiber area fractions, given by the product of fiber number and average fiber cross-sectional area divided by the alveolar wall cross-sectional area, are $\delta_{A,\ c}$ and $\delta_{A,\ e}$ for collagen and elastin, respectively. The effective fiber elastic moduli are thus $Y_{ce} = \delta_{A,\ c}Y_c$ and $Y_{ee} = \delta_{A,\ e}Y_e$, respectively. Collagen fiber waviness (w) is defined as the fiber contour length divided by its end-to-end distance in the undeformed state (Fig. 4.10d). The parameter w is a measure of how much the tissue can be strained before the fiber becomes straight. Based on waviness measurements [56], it can be assumed that w within a population of fibers conforms to the beta probability distribution function (p)

$$p(w) = \frac{(w - w_1)^{\alpha-1}(w_2 - w)^{\beta-1}}{B(\alpha,\beta)(w_2 - w_1)^{\alpha+\beta-1}} \tag{4.41}$$

where α and β are shape parameters, $B(\alpha,\beta)$ is the beta function, and w_1 and w_2 are the minimum and maximum measured waviness values, respectively, in the unstretched tissue.

Next, define the circumferential stretch ratio $\lambda = \frac{r}{R}$, the inner stretch ratio $\lambda_i = \frac{r_i}{R_i}$ and a thickness mapping parameter $S = \frac{R}{R_i}$ that spans the distance through the septal wall thickness such that $1 \le S \le \frac{R_e}{R_i} = S_e$. By accounting for conservation of wall tissue mass, and using the deformed area in calculating stress, the Cauchy circumferential stress in the alveolar wall due to elastin at any radial position S in the wall is

$$\sigma_{e,\theta\theta}(\lambda, S) = Y_{ee}\frac{\lambda}{S}(\lambda - 1). \tag{4.42}$$

The Cauchy stress due to collagen fibers is

$$\sigma_{c,\theta\theta}(\lambda, S) = \frac{Y_{ce}}{B(\alpha,\beta)\alpha(\alpha+1)}\frac{\lambda}{S}\frac{(\lambda - w_1)}{w_1}\left(\frac{\lambda - w_1}{w_2 - w_1}\right)^{\alpha}((\alpha+1)A_1 - \alpha B_1),\ w_1 \le \lambda < w_2 \tag{4.43}$$

$$\sigma_{c,\theta\theta}(\lambda, S) = \frac{Y_{ce}}{B(\alpha,\beta)\alpha(\alpha+1)}\frac{\lambda}{S}\frac{(\lambda - w_1)}{w_1}((\alpha+1)A_2 - \alpha B_2), \lambda \ge w_2 \tag{4.44}$$

where A_1, B_1, A_2 and B_2 are calculated from the Appell hypergeometric function of α and β (see [56] for derivation details). The total circumferential stress is the sum of the stresses due to collagen and elastin, $\sigma_{\theta\theta}(\lambda, S) = \sigma_{e,\ \theta\theta} + \sigma_{c,\ \theta\theta}$, which result in a nonlinear constitutive equation. Because of the spherical symmetry of the model, there are no shear stresses within the alveolar wall. Applying the equations of equilibrium for a continuum (setting the right-hand side of Eq. 4.18 to zero) then gives the contribution to alveolar pressure from collagen and elastin fibers as

$$P_{alv,f} = \frac{2}{\lambda_i^2}\int_1^{S_e} \frac{S^2}{(S^3+\lambda_i^3-1)^{\frac{1}{3}}}\sigma_{\theta\theta}(\lambda,S)dS. \tag{4.45}$$

Finally, if the alveolus is thin walled with initial thickness h_0, the stress in the wall is essentially independent of r and equal to the mean stress ($\overline{\sigma}_{\theta\theta}$) evaluated at the middle layer, in which case Eq. 4.45 becomes

$$\begin{aligned} P_{alv,f}(\lambda_i) &\cong \bar{\sigma}_{\theta\theta}\left[\frac{(\lambda_i^3+S_e^3-1)^{\frac{2}{3}}}{\lambda_i^2}-1\right] \\ &\cong \bar{\sigma}_{\theta\theta}\left[\frac{\left(\lambda_i^3+3\left(\frac{h_0}{R_i}\right)+3\left(\frac{h_0}{R_i}\right)^2+\left(\frac{h_0}{R_i}\right)^3\right)^{\frac{2}{3}}}{\lambda_i^2}-1\right]. \end{aligned} \tag{4.46}$$

The corresponding alveolar air volume (V_{alv}) is

$$V_{alv} = V_o\lambda_i^3 \tag{4.47}$$

where V_0 is the baseline volume.

Alveolar pressure also has a component due to surface tension γ at the air-liquid interface [57] given by

$$P_{alv,s} = 2\frac{\gamma(r_i)}{r_i} = \frac{2\gamma\left(\frac{\Delta V}{V_o}\right)}{R_i\left(\frac{\Delta V}{V_o}+1\right)^{1/3}} \tag{4.48}$$

where $\Delta V = V_{alv} - V_o$. The value of γ can be estimated as a function of ΔV by fitting the sigmoidal curve

$$\gamma\left(\frac{\Delta V}{V_o}\right) = \frac{33}{1+\exp\left(-\frac{\frac{\Delta V}{V_o}-3}{0.5}\right)} \tag{4.49}$$

to experimental data [58].

Total P_{alv} is the sum of $P_{alv,f}$ (Eq. 4.46) and $P_{alv,s}$ (Eq. 4.48), and is thus defined by 8 parameters, those related to the fibers (Y_{ce}, Y_{ee}, α, β, w_1, w_2) and those related to alveolar geometry (S_e, V_0). This total number reduces to 6 if alveolar geometry is specified.

Using second harmonic generation images, it was determined that the distribution of w for bovine lung tissue is skewed to the right with values ranging from $w_1 = 1.12$ to $w_2 = 2.53$. Equation 4.18 describes these data well with $\alpha = 1.89$ and $\beta = 3.60$ [56]. The model of P_{alv} (Eq. 4.46 plus Eq. 4.48) was fit to published deflation P-V curves from human lungs [59–61] using a fixed value for the difference $w_2 - w_1$.

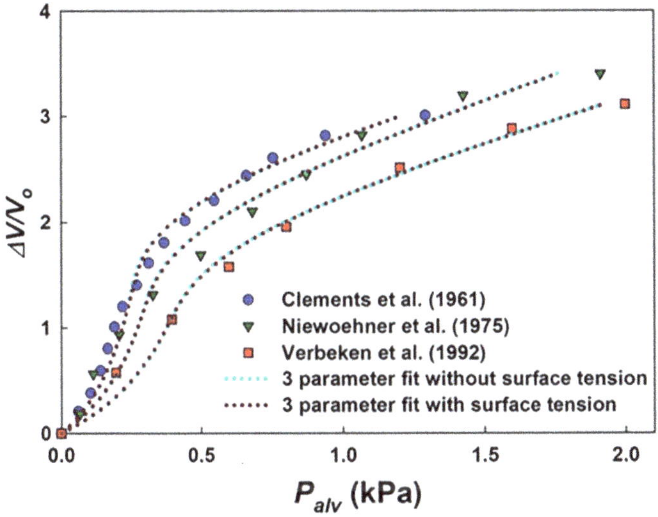

Fig. 4.11 Published deflation P-V curves from human lungs and the fits provided by the model of P_{alv} (Eq. 4.46 plus Eq. 4.48). The fits were achieved by adjusting the values of Y_{ce}, Y_{ee}, and w_1 while the values of α, β and w_2 were fixed based on experimental measurements of collagen waviness. The fits with and without surface tension are almost indistinguishable. The y-axis is the relative volume change. (Reproduced with permission from Ref. [56])

Figure 4.6 demonstrates that the model describes the data well, with minimal improvement made by the inclusion of surface tension (Eq. 4.48). The effective values estimated for collagen and elastin stiffness are 1419 ± 164 kPa and 18.9 ± 4.3 kPa, respectively. Using the measured volume fractions of 0.08 and 0.05 for collagen and elastic fibers, respectively [62], gives Young's moduli for the fibers themselves of approximately 18 MPa and 380 kPa, respectively. These numbers are about 3 and 2 times smaller, respectively, than measured values reported for collagen [18] and elastin [63], but this may reflect the limitations of the model and/or the challenges inherent in making the experimental measurements. Nevertheless, alveolar wall stiffness predicted by the model is in good agreement with direct measurements made using Atomic Force Microscopy [41].

An unexpected feature of the P-V curve of the alveolar model in Fig. 4.10 becomes apparent when the model is inflated to non-physiologically high volumes (i.e., beyond those shown in Fig. 4.11). It can be seen that P_{alv} reaches a maximum value beyond which it rapidly decreases even though $\Delta V/V_0$ keeps increasing (Fig. 4.12a). For a range of $\Delta V/V_0$ starting at 7.3 corresponding to $P_{alv} = 5.5$ kPa, the specific elastance (bulk modulus) of the model, $E_s = \frac{V_0 dP_{alv}}{dV}$, is decreasing but still positive. This regime corresponds to geometric instability. Eventually, however, at $\Delta V/V_0 = 43.8$ and $P_{alv} = 10.4$ kPa, E_s becomes negative. This is essentially a negative feedback mechanism that causes runaway inflation (as happens in a badly run economy!). The result is a progressively increasing alveolar size while P_{alv} approaches zero (Fig. 4.12b). The appearance of this inflation instability can occur

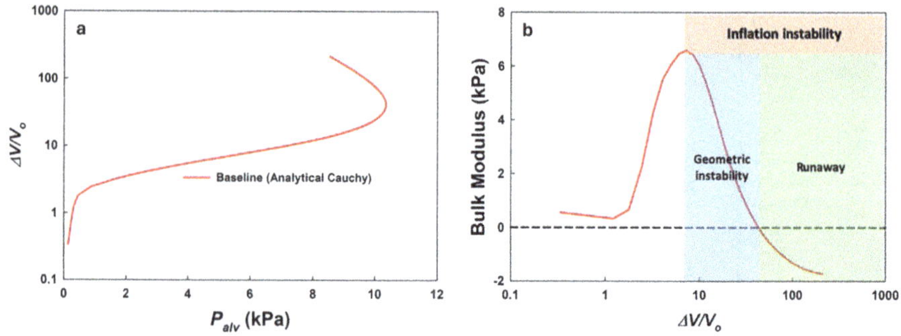

Fig. 4.12 (**a**) Inflation instability in the P-V curve of the lung occurring above P_{alv}= 10 kPa. (**b**) Alveolar bulk modulus exhibits a peak value, an adjacent region of geometric instability (blue), and finally a runaway region (green). (Reproduced with permission from Ref. [56])

at much lower values of P_{alv} if Y_{ce}, S_e and the rate of collagen recruitment with alveolar volume are reduced sufficiently. This has possible implications for diseases in which collagen fiber stiffness and/or degree of waviness are altered. If local runaway inflation of the parenchymal tissue occurs at physiologic volumes, it could potentially lead to tissue rupture resulting in pneumothorax, as is sometimes seen in emphysema.

Alveolar instability is actually inherent in the Laplace Law for the alveolus, which is

$$P_{alv} = \frac{2(T_f + \gamma)}{r_i} \tag{4.50}$$

where T_f is fiber tension in the ECM. During deflation, the increasing surfactant concentration at the interface causes γ to fall at a rate faster than r_i, causing P_{alv} to also decrease. If γ remains fixed, a decreasing r_i causes P_{alv} to increase, which leads to smaller alveoli emptying into adjacent larger ones. This process is known as deflation instability. One of the key functions of pulmonary surfactant is to prevent this from happening so that alveoli of different sizes can coexist stably. The situation is different during inflation, however. Equation 4.50 shows that as r_i increases, P_{alv} decreases, suggesting the possibility of runaway inflation. In general, alveolar volume asymptotes toward a finite value as P_{alv} continues to increase [64]. It has been pointed out [65] that such behavior cannot be due to surfactant because although γ increases with volume as the surfactant molecules become more sparsely distributed in the expanding interface area [66, 67], γ cannot increase above the surface tension of pure water [59].

Alveolar stability must ultimately rest, therefore, on the volume dependence of T_f. Specifically, since $\Delta V/V_0 \sim r_i^3$, Eq. 4.50 shows that T_f/r_i must be proportional to r_i^{3+q}, where $q > 0$. Fitting a measured collagen waviness distribution with the beta distribution gives $q = 1.89$ [56], indicating that inflation stability is guaranteed, at

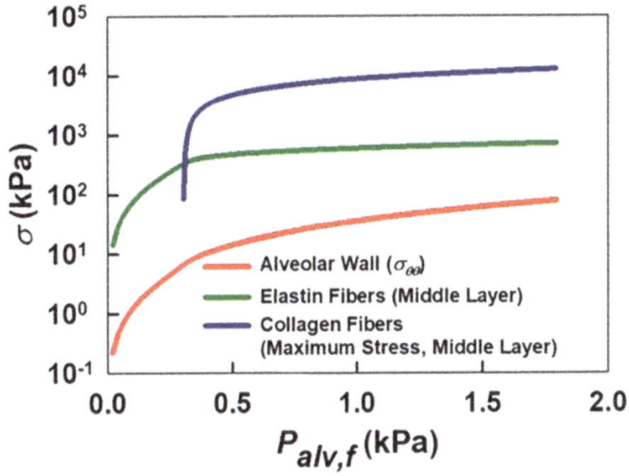

Fig. 4.13 Multiscale stresses in the alveolus. The alveolar wall, elastic fiber and collagen fiber stresses can be related to the alveolar pressure ($P_{alv,f}$) due to both collagen and elastic fibers. The total circumferential stress in the wall due to fibers is shown by the red line. The collective elastin and collagen circumferential stresses in the middle layer are shown in green and blue, respectively. Note that the stress for collagen represents the maximum stress in the middle layer. (Reproduced with permission from Ref. [56])

least until all collagen fibers have become straight. Figure 4.13 shows the estimated multiscale inflation stresses on the key structural components of the alveolar wall. The model predicts that the stresses on the septal walls are much smaller than those on the elastin fibers, which themselves are much smaller than those on the collagen fibers. The progressive recruitment of collagen into the stress-bearing role is thus what protects alveoli from inflation instability. This raises the question, however, as to whether inflation instability could ever be reached by inflating the alveolus enough, as suggested by (Fig. 4.12). One possibility is that the alveolar wall always undergoes catastrophic failure (e.g., pneumothorax) before this regime is reached. On the other hand, careful slow inflation of the mouse lung has shown it to be capable of sustaining inflation pressures up to 80 cmH_2O, far higher than is usually accepted as the physiologic maximum of 30–40 cmH_2O [68]. Furthermore, the mouse P-V relationship is concave downwards (decreasing stiffness with increasing pressure) at the upper end of the pressure range, which might be explained by Eq. 4.50. As with all models, however, the above 3D model of alveolar inflation has limitations [56]. Key among these is that it is a model of only a single alveolus and thus neglects the mechanical effects exerted by neighboring alveoli with which it shares septal walls. The second limitation is related to the fact that the human lung is large, spanning a linear dimension of roughly 30 cm. Lung tissue has weight, and even though the septal wall thickness is extremely thin (~6 μm), there are hundreds of millions of alveoli in the lung. In aggregate, these tiny alveoli add up to create a noticeable gravitational influence on lung deformation and ventilation.

4.7 Network Models of the Parenchyma

As pointed out by Mead et al. in a highly influential study [27], "Since, statically, gas pressure in all interconnected airways and air spaces is the same, there are no pressure differences across walls, and all forces distending airways and air spaces must arise from tissue attachments...This mechanical interdependence has a bearing on airspace expansion and stability, the relative dryness of air space, and on the forces distending airways and blood vessels". This *mechanical interdependence* between neighboring alveoli arises because of their shared septal walls, which means that pleural pressure (P_{pl}) is transmitted from the pleural surface all the way through the interconnected parenchymal network. If the parenchymal network is homogeneous and alveolar gas pressure (P_{alv}) is uniform, the stress at any point in the parenchyma equals transpulmonary pressure ($P_{tp}= P_{alv} - P_{pl}$). In a heterogeneous lung, non-uniform local expansion of alveoli can be caused by tissue stresses that differ from the average P_{tp}. This can arise, for example, when a septal wall becomes remodeled by a fibrotic process, causing it to contract and thus distort neighboring septal walls. But because the wall tissue is nonlinear, sufficient stretching of the neighboring walls will increase their stiffness and thus affect their degree of stretch. This nonlinear neighbor-neighbor interaction, a phenomenon sometimes called a "network effect", may have long-range spatial influence on alveolar distension and distortion.

The criterion for parenchymal stability is that the P-V curve must have a positive slope, which is equivalent to saying that the lung has a positive bulk modulus [69]. Since the recoil pressure of the parenchyma comes from all tissue and surface forces transmitted across a given plane per unit area, the rate of increase of these forces during lung expansion must exceed the rate of decrease in the area density of linear force-bearing elements [70–72]. Parenchymal stability is thus not something that can be understood by considering a single alveolus in isolation, but rather is related to the interconnectivity of the peripheral septal network [65, 73]. This applies not only to individual alveoli but also to regions of lung parenchyma that are prone to collapse [39, 72]. Of course, this can only occur at positive P_{tp} in heterogeneous lungs [57], and is mitigated by mechanical interdependence. The conclusion that arises from these analytical studies is that collapse in a region, which impairs local gas exchange, is naturally prevented by network effects in the lung parenchyma. Further insight can be obtained when network phenomena are studied via computational models.

Parenchymal stability is also conferred to some degree by the way in which adjoining alveolar septal walls resist a change in the angle between them. Septal angles change when parenchymal strips are immersed in fluids of different tonicity (a situation devoid of surface tension) [74]. This is at odds with the common perception that resistance to parenchymal shear is conferred entirely by the prestress due to P_{tp}. This perception has arisen because parenchymal shear modulus appears to be linearly related to P_{tp} [32], implying that there should be no resistance to shear deformation when P_{tp} is zero [75]. Furthermore, structures composed in 2D of packed hexagons or in 3D of packed truncated octahedra (see Chap. 2) are unstable

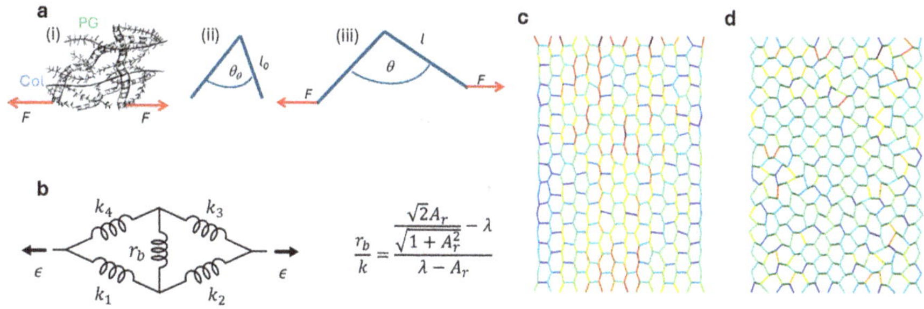

Fig. 4.14 Role of proteoglycans in stabilizing the alveolar septal network. (**a**) A model of resilience to bond-bending: (i) Forces (F) pull on wavy collagen fibers (Col) embedded in a proteoglycan matrix. (ii) A model of the collagen-proteoglycan matrix consisting of two elastic line elements with initial lengths l_0 and intervening angle θ_0. (iii) F causes a change in θ and an increase in line element length l. (**b**) Network of linear springs k_i representing septal walls with a bond-bending spring r_b that opposes compression. When $k_i = k$ ($i = 1, \ldots, 4$), the ratio r_b/k can be expressed as a function of the change in the overall model aspect ratio A_r and the stretch ratio λ. (**c**) A 2D network of nonlinear springs with $r_b = 0.01$. (**d**) The same network with $r_b = 100$. Both networks are stretched to 30% uniaxial strain in the vertical direction. The line colors denote relative wall elastic energy (red and blue correspond to high and low energy, respectively). The pathways of high energy (red) that percolate across the network in (**c**) are replaced by isolated islands of high energy when r is increased in (**d**). (Panels **c** and **d** are reproduced with permission from Ref. [74])

without prestress because they collapse by simply folding flat. The resistance to folding in the parenchyma is thought to be due to the stabilizing influence of proteoglycans in the septal walls that can sequester water molecules and thus change their bulk as a function of environmental tonicity. Thus, proteoglycans interfere with the ability of adjacent septal walls to fold into perfect alignment in the direction of macroscopic strain [74].

The proteoglycans can be viewed as springs that elastically oppose changes in septal angle (Fig. 4.14a). A simple 2D model of a pair of septa with a bond-bending spring is illustrated in Fig. 4.14b. In a network of alveoli composed of such septa, the total elastic energy, E, is given by the sum of two terms:

$$E = \frac{1}{2}\sum_i k_i \Delta l_i^2 + \frac{1}{2}\sum_j r_{b,j} \Delta\theta_j^2 \tag{4.51}$$

where $\Delta\theta$ is the change in angle between adjacent septa, $r_{b,\,j}$ are the corresponding bond-bending spring constants, k_i are the tensile spring constants of the septal walls (assumed linear), and Δl_i is extension of spring i. The summations with respect to i and j in Eq. 4.51 pertain to all links (septal walls) and all the nodes (wall junctions) of the network, respectively. The equilibrium configuration of the spring network subject to a uniaxial strain can be determined using an optimization method such as simulated annealing [74]. Configurations of this 2D alveolar network model are shown in Figs. 4.14c, d for two different values of the bond-bending parameter r_b (the $r_{b,\,j}$ and k_i are constant throughout the network) [74].

The networks in Figs. 4.14c, d reveal that the lateral contraction of the model during uniaxial stretch is determined by the strength of r_b relative to k. In other words, the Poisson's ratio is influenced significantly by mechanical interactions between the collagen and elastin fiber networks and the proteoglycans. Also, the model is non-affine because its macroscopic deformation is not reflected in the deformation patterns of the individual alveoli, in accord with experimental observation [76]. Interestingly, when the model is adapted to the emphysematous lung, the proteoglycans appear to be protective by reducing the rate of septal wall rupture [77].

Force does not typically transit uniformly across an elastic network but rather follows discrete pathways that percolate from one side to the other [78]. Figures 4.14c, d show that these percolating patterns of force transmission are influenced by impediments to bond-bending, which has potential implications for mechanotransduction within the cells that inhabit the parenchyma, and which are affected by its mechanical environment. Too much geometric regularity in the network can lead to preferential directions of force transmission, which is probably not good for the uniform maintenance of the parenchyma (see Fig. 2.8). Random variation in the proteoglycan load at each vertex of the network makes the tissue architecture more isotropic beyond the scale of a single alveolus, which may be beneficial for mechanotransduction in general [79].

The parenchymal model above considers only the alveolar contribution to parenchymal mechanics and does not include mesoscopic structures such as alveolar ducts and small airways. These structures can be included by modeling the parenchyma as a set of truncated octahedral alveoli surrounding an open tubular structure representing the alveolar duct (see Fig. 2.6) [80–83]. Elastic recoil is generated in this model by ascribing distinct constitutive stress-strain relationships to bundles of elastin and collagen fibers. Each alveolar septal face is imbued with a surface tension having a nonlinear area-tension relationship. P-V curves simulated with this model compare well with experimental curves obtained from both saline-filled and air-filled lungs, behavior that is attributed largely to the mechanical behavior of the alveolar duct. Investigations with this model indicate that for it to exhibit mechanical isotropy, parenchymal regions with smaller than average alveoli must have a higher density of fibers along the duct but a lower density in septal walls compared with regions containing larger alveoli [81]. The model is isotropic for small strains, but becomes anisotropic at large strains for which it requires five strain-dependent elastic moduli to fully characterize its mechanical behavior [83]. In particular, the Young's modulus in the direction of uniaxial strain increases significantly with strain while the shear modulus remains essentially constant, in fairly good agreement with data from triaxially prestressed lung tissue [84]. Lung deformation during breathing exhibits regional and directional variations that are more pronounced during uniaxial stretching compared to uniform triaxial stretching [85]. Nonuniform parenchymal distortion also occurs in the immediate vicinity of an airway during bronchoconstriction. Accounting for these phenomena in detail will likely require a model incorporating fibers that are arranged in bundles at specific spatial orientations [81] superimposed on an amorphous network [79].

Before discussing the influence of gravity, it is instructive to examine the relationship between network and continuum representations of the parenchyma. Continuum analysis predicts that parenchymal distortion caused by a contracting airway remains locally confined [22–24]. When an airway is embedded in a 2D network of hexagonal alveolar cells, however, the distortion propagates much further in an anisotropic manner [86, 87]. This is in contrast to the case when the parenchyma is represented as a triangular network [86], or as a 3D network in which the alveoli also resist shape change [88]. Since a hexagonal network representation of the alveoli is anatomically more realistic than either a triangular network or a continuum, these findings suggest that alveoli resist shape change, likely due to the presence of proteoglycans as discussed above, with a reaction force approximately 10% of that arising from changes in fiber length in the walls [88]. Different model predictions arise because of differences in shear moduli. The triangular network can be viewed as a hexagonal network reinforced with the addition of 6 additional springs inside each hexagon, making it resist shape change even in the absence of prestress. This is similar to the effect of the bond-bending springs that oppose angle change in Fig. 4.14b, which also attenuates long-range force transmission (Fig. 4.14c versus Fig. 4.14d).

4.8 What About Gravity?

The mechanical properties of the human lung as manifest in the P-V curve have been studied extensively under both normal and diseased conditions [4–7, 10, 16, 56, 89–94]. The P-V curve is, however, a measure of global physiologic behavior that does not reflect what may be substantial regional heterogeneity in specific ventilation [95]. These heterogeneities are the result of many factors that operate over a wide range of spatial scales. Factors intrinsic to the lung such as spatial variation in alveolar compliance [96] determine how the alveolus expands due to a change in P_{tp} [56]. The P-V curve for a single alveolus, as modeled in Fig. 4.10, can vary substantially even between neighbors [54]. Extrinsic factors are also critical, however, key among which is the force of gravity. Breathing is driven by cyclic variations in pleural pressure which, in the human lung, exhibits a marked rostral-caudal gradient (in the upright position) due to the weight of the lung [95, 97–99]. This results in a strong effect on regional lung expansion [100]. Together with geometrically complex boundary conditions due to the shape of the lung and its confinement within the thoracic cavity, the net effect of all relevant intrinsic and extrinsic factors on regional ventilation in the lung is a complex matter that can only be understood properly with the help of computational modeling.

Continuum mechanical modeling has been implemented in a gravitational field to predict regional ventilation and perfusion in the lung [101], showing that regional variations in tissue mechanical properties in the healthy lung are more dominant than variations in airway resistance in causing preferential ventilation of the base of the lung. The weight of the blood in the lung also plays an important role in the

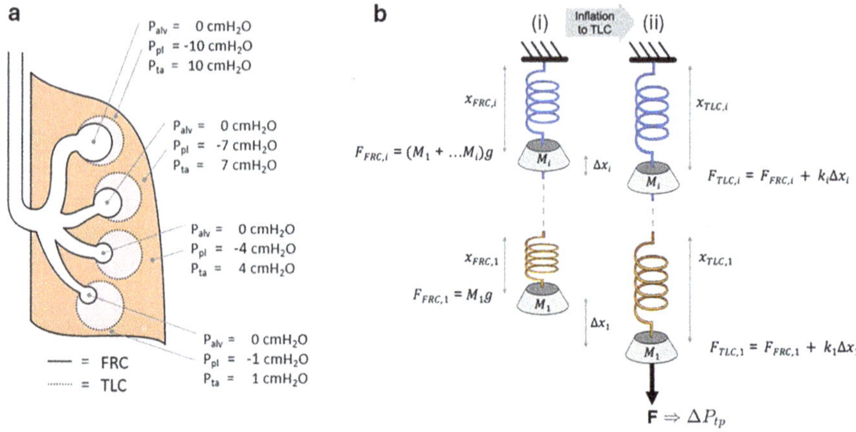

Fig. 4.15 (**a**) schematic representation of the upright human lung with pleural (P_{pl}), alveolar (P_{alv}) and transalveolar (P_{ta}) pressure at FRC. Dotted lines show how individual alveoli expand when the lung is inflated to TLC. (**b**) A spring-mass in-series representation of the lung in panel **a**. Subpanels (i) and (ii) show the system at FRC and TLC, respectively. Springs are extended by different amounts due to the mass elements M_i representing regional tissue weight. At the apex (blue), all subtended mass elements act to extend the top spring whereas at the base (orange), only one mass element extends the spring. When the system is stretched to TLC, a common force F, representing the change in transpulmonary pressure (ΔP_{tp}), is applied to the bottom spring. (Reproduced with permission from Ref. [103])

distribution of perfusion, with important implications for gas exchange [102]. The results of this modeling suggest that, in the supine human lung during normal breathing, ventilation-perfusion matching is driven predominantly by tissue deformation influenced by the weight of the lung tissue and the blood it contains, with a lesser role played by the close anatomic alignment of the airway and vascular trees.

Nevertheless, continuum modeling does not account for the role played by the collagen and elastic fiber networks that are so important for determining the strain-stiffening nature of the P-V curve of a single alveolus as illustrated by the model in Fig. 4.10 [56]. This model can be extended to the lung as a whole by adding the effect of a gravitationally dependent pleural pressure [103]. This provides alveolar stiffness as a function of vertical position within the lung, and how this stiffness translates into regional distributions of mechanical stress in the septal walls and the collagen and elastin fibers. Figure 4.15a illustrates how the alveoli expand at the top, middle and base of the lung during a quasi-static deep inhalation from FRC to TLC. At FRC, pleural pressure is -10 cmH_2O at the apex and only -1 cmH_2O at the base, resulting in transalveolar pressures of 10 and 1 cmH_2O, respectively.

A mechanical analogue of this situation consisting of springs and mass elements is depicted in Fig. 4.15b. In this model, an upright-oriented lung is divided from base to apex into $n = 20$ regions of equal height. The average alveolus within the ith region exhibits a unique alveolar pressure-volume (P_{alv}-V) relationship due to its

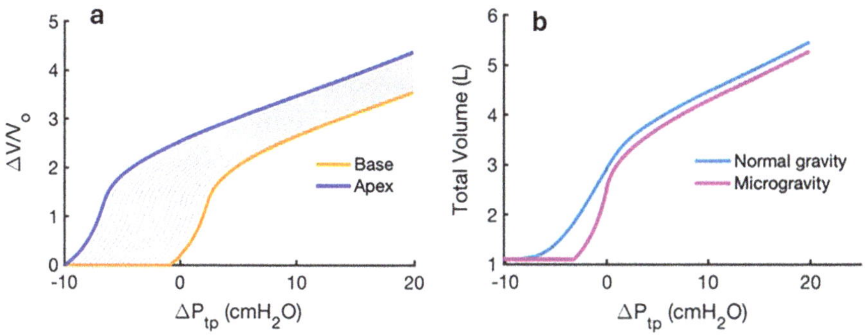

Fig. 4.16 (**a**) Regional alveolar P-V curves that include the effects of gravity. P-V curves at the base and apex of the lung are shown in orange and blue, respectively, while P-V curves for the intermediate regions are shown in gray. The y-axis is relative volume change, as in Fig. 4.11, while the x-axis is the change in transpulmonary pressure, ΔP_{tp}, relative to FRC. At FRC, pleural pressure at the lung base is -1 cmH_2O, so $P_{tp} = 1$ cmH_2O and $\Delta P_{tp} = -1$ cmH_2O. This creates $P_{ta} = 0$ for which $\Delta V = 0$. (**b**) Comparison of the deflation limb of the model PV curve under normal gravity and microgravity conditions (the latter assuming a homogenous distribution of pleural pressure of 3.25 cmH_2O)

particular value of the pleural pressure ($P_{pl,\ i}$). Since transalveolar pressure, $P_{ta,\ i}$, in the ith region is equal to the negative of $P_{pl,\ i}$, regional P_{alv}-V curves initiated from FRC for each of the n regions can be generated from a universal alveolar P_{ta}- V curve similar to Eqs. 4.45 through 4.48 and the curves in Fig. 4.11 by defining the values $P_{pl,\ i}$ $(i = 1, \ldots, n)$ appropriately. Based on the reported effects of gravity on P_{pl} [97], the alveoli at the base of the upright lung in region $i = 1$ at FRC are subject to $P_{pl,\ 1} = -1$ cmH_2O. This decreases linearly up to the alveoli at the apex in region $i = 20$ which experience $P_{pl,\ 20} = -10$ cmH_2O [104], although this gradient varies depending on body position [97, 105]. The volume of a given lung region also depends on the number of alveoli in that region [106].

Figure 4.16a demonstrates that the regional P-V curves simulated by this model differ substantially from lung apex to base. Not surprisingly, the P-V curves are also affected if the lung is placed in microgravity conditions as is achieved in low earth orbit (Fig. 4.16b) where all alveoli are exposed to an essentially uniform pleural pressure, assumed here to be $P_{pl,\ i} = -3.25$ cmH_2O at FRC. This corresponds to a lung volume of 2.55L, which is consistent with the 15% reduction in FRC in microgravity that has been reported [107]. TLC is also reduced to 4.28 L in microgravity at $\Delta P_{tp} = 20$ cmH_2O, while residual volume (RV) remains unchanged. Based on this discussion, it is easy to see that the circumferential septal wall and fiber stresses for both collagen and elastin would have a similar relationship as in Fig. 4.13, but these stresses are larger at the apex than the base of the lung. Inevitably, this model has limitations [103], but it serves to link the macroscopic forces of gravity to microscale stresses in the various components of lung tissue and their spatial distributions in a way that may provide insight into mechanisms of disease. Emphysema, for example, occurs preferentially in the lung apex [25] whereas fibrosis is more common in subpleural and basilar regions [108]. Tissue stresses and strains likely play an important role in these pathologies (see Chaps. 8 and 9). Computational modeling has the potential to elucidate exactly how.

4.9 Conclusions

The lung is a mechanical organ with elastic properties that are crucial determinants of its function. The link between lung structure and its elastic function is highly complex, involving mechanisms operating at many levels of scale from that of a single molecule up to the entire organ. Understanding how the lung functions mechanically thus requires detailed knowledge of the intricacies associated with its multiscale features that include the structural fibers in the tissue and whether they are under tension or wavy, the dynamic nature of surface tension, regional variations in tissue stiffness, and the effects of gravity. The overall elastic behavior of the lung is also an emergent property arising not only from the properties of its individual components but most especially from the way these components are arranged with respect to each other. Changes in this arrangement alone can lead to major changes in function, which raises the intriguing possibility that evolution might operate at the level of component arrangement as well as at the individual component level. Lung elasticity in all its manifestations is thus impossible to address in a quantitative fashion without the use of rather sophisticated mathematical and computational models. Different modeling methods are best suited to addressing different aspects of lung elasticity. In every case and at each scale, however, these methods must be closely guided by relevant experimental data to ensure that the resulting models link lung structure and mechanical function in a way that serves the purpose at hand. Predictions provided by such computational models can provide important insights into the nature of fundamental physiologic and biologic processes as well as how these processes become deranged in disease.

4.10 Problems

1. Write a code to simulate the force-extension model of Eq. 4.10 and reproduce the result shown in Fig. 4.3. Notice that the numerical results somewhat depend on the choice of model parameters. Try setting the total number of springs to 2000, the number of bins of the stop extension distribution to 200, $\Delta l_s = 0.0001$ and $k=1$.

   (a) Implement a lognormal distribution for the stop extensions. Use a mean of 1 and vary the standard deviation between 0.5 and 10. How does the force-

extension curve compare to that of the hyperbolic model as you vary the standard deviation of the lognormal distribution?
(b) Plot the stop extension distributions separately for each case. Given the macroscopic force-extension curve, can you infer the underlying stop extension distribution?

2. This problem deals with the hyperelastic material properties of the lung tissue.

   (a) Derive the Cauchy stress-strain relation for the incompressible Ogden material given by Eq. 4.40 for the case of uniaxial stretching of a uniform bar. Simplify the general case by assuming that $N = 1$, which should be written in a form similar to that of Eq. 4.38. Starting with the last equation, derive the incremental Young's modulus (Y), the derivative of the Cauchy stress with respect to linear strain (ϵ), as a function of $\alpha_1 = \alpha$. Show that consistency requires that $\mu = 2D_{10}$ when $\alpha = 2$, which also means that $\mu = Y/3$ in agreement with linear elasticity theory. Calculate the modulus at zero strain and determine how it depends on α. Under what condition does $Y = 0$ at $\epsilon = 0$? Is this condition consistent with linear elasticity?
   (b) Write a code to simulate the stress-strain curve of a block of tissue using your Ogden model from part (a). Set $Y = 1$ kPa and $N = 1$ and calculate the stress strain curves for ϵ between 0 and 1 and for α= 2, 3, 4, and 5. Plot these curves on the same set of axes. Also plot the stress-strain curve corresponding to linear elasticity. Does the latter agree with the case when $\alpha = 2$? If not, how does the deviation depend on strain? Explain why.
   (c) To calculate the incremental modulus, take the derivatives of each of the stress-strain curves and plot them as a function of strain. Also plot the theoretical values of Y at $\epsilon = 0$. Do these theoretical values agree with the numerically calculated moduli?

3. The pressure-volume curve of a single alveolus can be modeled using Eqs. 4.40 through 4.49. Assume that the elastic and collagen fibers have absolute moduli of $Y_e = 400$ kPa and $Y_c = 20$ MPa, and their volume fractions are 0.05 and 0.07, respectively. The normalized thickness at the outer layer of the thick-walled sphere is $S_e = 1.05$.

   (a) Using Eq. 4.42, calculate the stress due to elastin in the middle layer $S_m = (1 + S_e)/2$ under the thin wall assumption for which $\lambda = \lambda_i$, where λ_i is the stretch ratio at the inner surface of the alveolus. Also calculate the stress due to collagen alone assuming the fiber is linear with the above modulus. Next, utilize the incremental modulus as a function of λ from Problem 2c and calculate the stress due to nonlinear collagen for several values of α. Use a range of λ between 1 and 1.7. Also calculate the total stress due to elastin and nonlinear collagen. Compare these curves on a single plot.
   (b) Based on Eq. 4.46, calculate the alveolar pressure-volume curve for pure elastin, pure linear collagen, pure nonlinear collagen and the combination of elastin and nonlinear collagen. Assume that your stresses are the effective stresses in the middle layer ($\overline{\sigma}$). Plot these stresses as a function of the relative

change in volume using λ. How do these curves compare with those in Fig. 4.11. If the magnitudes of the pressures are different from those in Fig. 4.11, try to find the reason why. Try different α values. If this doesn't help, move on to part c.

(c) Come up with a solution for the discrepancy in Problem 3b. Hint: the stress due to collagen may rise too much and too early with λ. The recruitment of wavy collagen effectively delays the contribution of collagen to the stress. Equations 4.43 and 4.44 can account for this phenomenon. Implement a similar but simpler approach without explicitly considering wavy collagen.

(d) Add surface tension (Eqs. 4.48 and 4.49) to your model. Compare the pressure-volume curves with and without surface tension.

(e) What argument can you make that neither pure elastin nor pure linear collagen model are able to support life? Why is collagen waviness crucial for life?

4. As an advanced exercise, use simulated annealing to determine the mechanical equilibrium configuration of the hexagonal spring network implemented in Problem 2.3. Look up the details of the algorithm in Reference [74]. The expression for the energy of the network will contain only the first term on the right-hand side of Eq. 4.51. Assign a stiffness value of $k = 1$ to each spring and an initial length that is half of the side length of the hexagons. Randomize the positions of the internal nodes and fix the positions of the boundary nodes. If the simulated annealing works correctly, then the energy minimization should restore the uniform hexagonal network. Now the model is ready for free exploration. For example, you can carve out an airway and stiffen up the airway walls and solve the system or you can calculate the stiffness of the network.

References

1. Froese AR, Shimbori C, Bellaye PS, Inman M, Obex S, Fatima S, Jenkins G, Gauldie J, Ask K, Kolb M. Stretch-induced activation of transforming growth factor-beta1 in pulmonary fibrosis. Am J Respir Crit Care Med. 2016;194:84–96.
2. Neergaard K. Neue Auffasungen über einn Grundbergriff der Atemtechnik. Die Retraktionskraft der unge, abhangig von den Oberflachenspannung in den Alveolen (new interpretations of basic concepts of respiratory mechanics correlation of pulmonary recoil force with surf). Zeitschrift Fur Gesamte Experi Medizin. 1929;66:373–94. https://doi.org/10.1007/bf02621963.
3. Harris RS. Pressure-volume curves of the respiratory system. Respir Care. 2005;50:78–98. discussion 98–79
4. Salazar E, Knowles JH. An analysis of pressure-volume characteristics of the lungs. J Appl Physiol. 1964;19:97–104.
5. Murphy BG, Engel LA. Models of the pressure-volume relationship of the human lung. Respir Physiol. 1978;32:183–94.
6. Paiva M, Yernault JC, Eerdeweghe PV, Englert M. A sigmoid model of the static volume-pressure curve of human lung. Respir Physiol. 1975;23:317–23.
7. Venegas JG, Harris RS, Simon BA. A comprehensive equation for the pulmonary pressure-volume curve. J Appl Physiol (1985). 1998;84:389–95.

8. Albaiceta GM, Luyando LH, Parra D, Menendez R, Calvo J, Pedreira PR, Taboada F. Inspiratory vs. expiratory pressure-volume curves to set end-expiratory pressure in acute lung injury. Intensive Care Med. 2005;31:1370–8.
9. Albaiceta GM, Garcia E, Taboada F. Comparative study of four sigmoid models of pressure-volume curve in acute lung injury. Biomed Eng Online. 2007;6:7.
10. Alencar AM, Arold SP, Buldyrev SV, Majumdar A, Stamenovic D, Stanley HE, Suki B. Physiology: dynamic instabilities in the inflating lung. Nature. 2002;417:809–11.
11. Suki B, Barabasi AL, Hantos Z, Petak F, Stanley HE. Avalanches and power-law behaviour in lung inflation. Nature. 1994;368:615–8.
12. Gaver DP 3rd, Samsel RW, Solway J. Effects of surface tension and viscosity on airway reopening. J Appl Physiol. 1990;69:74–85.
13. Toshima M, Ohtani Y, Ohtani O. Three-dimensional architecture of elastin and collagen fiber networks in the human and rat lung. Arch Histol Cytol. 2004;67:31–40.
14. Gosline J, Lillie M, Carrington E, Guerette P, Ortlepp C, Savage K. Elastic proteins: biological roles and mechanical properties. Philos Trans R Soc Lond Ser B Biol Sci. 2002;357:121–32.
15. Suki B, Stamenovic D, Hubmayr RD. Lung parenchymal mechanics. In: Fredberg JJ, Sieck GC, Gerthoffer WT, editors. Comprehensive physiology, the respiratory system, respiration mechanics: organ, cell, molecule, vol. 1. Wiley-Blackwell; 2011. p. 1317–51.
16. Maksym GN, Bates JH. A distributed nonlinear model of lung tissue elasticity. J Appl Physiol. 1997;82:32–41.
17. Pride NB, Macklem PT. Lung mechanics in disease. In: Comprehensive physiology. Wiley; 2011. p. 659–92.
18. Deng Y, Herrmann J, Wang Y, Nguyen M, Hall JK, Kim JH, Smith ML, Lutchen KR, Bartolak-Suki E, Suki B. Physiological mechanical forces accelerate the degradation of bovine lung collagen fibers by bacterial collagenase. Sci Rep. 2024;14:29008.
19. Kononov S, Brewer K, Sakai H, Cavalcante FS, Sabayanagam CR, Ingenito EP, Suki B. Roles of mechanical forces and collagen failure in the development of elastase-induced emphysema. Am J Respir Crit Care Med. 2001;164:1920–6.
20. Greaves IA, Colebatch HJ. Elastic behavior and structure of normal and emphysematous lungs post mortem. Am Rev Respir Dis. 1980;121:127–36.
21. Wright JL, Churg A. Smoke-induced emphysema in Guinea pigs is associated with morphometric evidence of collagen breakdown and repair. Am J Phys. 1995;268:L17–20.
22. Lai-Fook SJ. A continuum mechanics analysis of pulmonary vascular interdependence in isolated dog lobes. J Appl Physiol Respir Environ Exerc Physiol. 1979;46:419–29.
23. Lai-Fook SJ, Hyatt RE, Rodarte JR. Effect of parenchymal shear modulus and lung volume on bronchial pressure-diameter behavior. J Appl Physiol Respir Environ Exerc Physiol. 1978;44:859–68.
24. Wilson TA. A continuum analysis of a two-dimensional mechanical model of the lung parenchyma. J Appl Physiol. 1972;33:472–8.
25. West JB. Distribution of mechanical stress in the lung, a possible factor in localisation of pulmonary disease. Lancet. 1971;1:839–41.
26. Krueger JJ, Bain T, Patterson JL Jr. Elevation gradient of intrathoracic pressure. J Appl Physiol. 1961;16:465–8.
27. Mead J, Takishima T, Leith D. Stress distribution in lungs: a model of pulmonary elasticity. J Appl Physiol. 1970;28:596–608.
28. Milic-Emili J, Henderson JA, Dolovich MB, Trop D, Kaneko K. Regional distribution of inspired gas in the lung. J Appl Physiol. 1966;21:749–59.
29. Galloy AE, Reinhardt JM, Raghavan ML. Role of lung lobar sliding on parenchymal distortion during breathing. J Appl Physiol (1985). 2023;135:534–41.
30. Galloy AE, Reinhardt JM, Raghavan ML. Lobar sliding reduces parenchymal distortion more in the right lung than the left lung. J Biomech Eng. 2025;147:1–15.
31. Bhana RH, Magan AB. Lung mechanics: a review of solid mechanical elasticity in lung parenchyma. J Elast. 2023;153:53–117.

32. Lai-Fook SJ, Wilson TA, Hyatt RE, Rodarte JR. Elastic constants of inflated lobes of dog lungs. J Appl Physiol. 1976;40:508–13.
33. Humphrey JD, O'Rourke SL. An introduction to biomechanics : solids and fluids, analysis and design. 2. edition. pp. 1 online resource (xxvi, 692 pages). New York: Springer; 2015:1 online resource (xxvi, 692 pages).
34. Holzapfel GA, Ogden RW. Biomechanics : trends in modeling and simulation. In: Studies in mechanobiology, tissue engineering and biomaterials volume 20. pp. 1 online resource (vii, 316 pages). Cham: Springer; 2017:1 online resource (vii, 316 pages).
35. Holzapfel GA. Nonlinear solid mechanics : a continuum approach for engineering. Chichester/New York: Wiley; 2000.
36. Birzle AM, Martin C, Uhlig S, Wall WA. A coupled approach for identification of nonlinear and compressible material models for soft tissue based on different experimental setups - exemplified and detailed for lung parenchyma. J Mech Behav Biomed Mater. 2019;94:126–43.
37. Birzle AM, Martin C, Yoshihara L, Uhlig S, Wall WA. Experimental characterization and model identification of the nonlinear compressible material behavior of lung parenchyma. J Mech Behav Biomed Mater. 2018;77:754–63.
38. Ogden RW. Large deformation isotropic elasticity—on the correlation of theory and experiment for incompressible rubberlike solids. Proc R Soc Lond A Math Phys Sci. 1972;325:565–84.
39. Fung YC. Stress, deformation, and atelectasis of the lung. Circ Res. 1975;37:481–96.
40. Sadeghi Naini A, Patel RV, Samani A. Measurement of lung hyperelastic properties using inverse finite element approach. IEEE Trans Biomed Eng. 2011;58:2852–9.
41. Jorba I, Beltran G, Falcones B, Suki B, Farre R, Garcia-Aznar JM, Navajas D. Nonlinear elasticity of the lung extracellular microenvironment is regulated by macroscale tissue strain. Acta Biomater. 2019;92:265–76.
42. Singh D, Slutsky AS, Cronin DS. Alveolar wall hyperelastic material properties determined using alveolar cluster model with experimental stress-stretch and pressure-volume data. J Mech Behav Biomed Mater. 2024;159:106685.
43. Lee GC, Frankus A. Elasticity properties of lung parenchyma derived from experimental distortion data. Biophys J. 1975;15:481–93.
44. Beltran G, Navajas D, Garcia-Aznar JM. Mechanical modeling of lung alveoli: from macroscopic behaviour to cell mechano-sensing at microscopic level. J Mech Behav Biomed Mater. 2022;126:105043.
45. Rausch SM, Martin C, Bornemann PB, Uhlig S, Wall WA. Material model of lung parenchyma based on living precision-cut lung slice testing. J Mech Behav Biomed Mater. 2011;4:583–92.
46. Concha F, Sarabia-Vallejos M, Hurtado DE. Micromechanical model of lung parenchyma hyperelasticity. J Mech Phys Solids. 2018;112:126–44.
47. Eskandari M, Nordgren TM, O'Connell GD. Mechanics of pulmonary airways: linking structure to function through constitutive modeling, biochemistry, and histology. Acta Biomater. 2019;97:513–23.
48. LaPrad AS, Lutchen KR, Suki B. A mechanical design principle for tissue structure and function in the airway tree. PLoS Comput Biol. 2013;9:e1003083.
49. Holzapfel GA, Gasser TC, Ogden RW. A new constitutive framework for arterial wall mechanics and a comparative study of material models. J Elast Phys Sci Solids. 2000;61:1–48.
50. Birzle AM, Hobrack SMK, Martin C, Uhlig S, Wall WA. Constituent-specific material behavior of soft biological tissue: experimental quantification and numerical identification for lung parenchyma. Biomech Model Mechanobiol. 2019;18:1383–400.
51. Yeoh OH. Some forms of the strain energy function for rubber. Rubber Chem Technol. 1993;66:754–71.
52. Engler AJ, Sen S, Sweeney HL, Discher DE. Matrix elasticity directs stem cell lineage specification. Cell. 2006;126:677–89.

53. Tschumperlin DJ. Mechanotransduction. Compr Physiol. 2011;1:1057–73.
54. LeBourdais R, Grifno GN, Banerji R, Regan K, Suki B, Nia HT. Mapping the strain-stiffening behavior of the lung and lung cancer at microscale resolution using the crystal ribcage. Front Netw Physiol. 2024;4:1396593.
55. Banerji R, Grifno GN, Shi L, Smolen D, LeBourdais R, Muhvich J, Eberman C, Hiller BE, Lee J, Regan K, et al. Crystal ribcage: a platform for probing real-time lung function at cellular resolution. Nat Methods. 2023;20:1790–801.
56. Bou Jawde S, Karrobi K, Roblyer D, Vicario F, Herrmann J, Casey D, Lutchen KR, Stamenovic D, Bates JHT, Suki B. Inflation instability in the lung: an analytical model of a thick-walled alveolus with wavy fibres under large deformations. J R Soc Interface. 2021;18: 20210594.
57. Stamenovic D, Wilson TA. Parenchymal stability. J Appl Physiol. 1992;73:596–602.
58. Smith JC, Stamenovic D. Surface forces in lungs. I. Alveolar surface tension-lung volume relationships. J Appl Physiol. 1986;60:1341–50.
59. Clements JA, Hustead RF, Johnson RP, Gribetz I. Pulmonary surface tension and alveolar stability. J Appl Physiol. 1961;16:444–50.
60. Niewoehner DE, Kleinerman J, Liotta L. Elastic behavior of postmortem human lungs: effects of aging and mild emphysema. J Appl Physiol. 1975;39:943–9.
61. Verbeken EK, Cauberghs M, Mertens I, Clement J, Lauweryns JM, Van de Woestijne KP. The senile lung. Comparison with normal and emphysematous lungs. 2. Functional aspects. Chest. 1992;101:800–9.
62. Mercer RR, Crapo JD. Spatial distribution of collagen and elastin fibers in the lungs. J Appl Physiol (1985). 1990;69:756–65.
63. Berquand A, Wahart A, Henry A, Gorisse L, Maurice P, Blaise S, Romier-Crouzet B, Pietrement C, Bennasroune A, Sartelet H, et al. Revealing the elasticity of an individual aortic fiber during ageing at nanoscale by in situ atomic force microscopy. Nanoscale. 2021;13: 1124–33.
64. Mercer RR, Laco JM, Crapo JD. Three-dimensional reconstruction of alveoli in the rat lung for pressure-volume relationships. J Appl Physiol (1985). 1987;62:1480–7.
65. Wilson TA. Parenchymal mechanics at the alveolar level. Fed Proc. 1979;38:7–10.
66. Schurch S, Goerke J, Clements JA. Direct determination of surface tension in the lung. Proc Natl Acad Sci USA. 1976;73:4698–702.
67. Wilson TA, Bachofen H. A model for mechanical structure of the alveolar duct. J Appl Physiol. 1982;52:1064–70.
68. Soutiere SE, Mitzner W. On defining total lung capacity in the mouse. J Appl Physiol (1985). 2004;96:1658–64.
69. Mead J. Mechanical properties of lungs. Physiol Rev. 1961;41:281–330.
70. Wilson TA. Surface tension-surface area curves calculated from pressure-volume loops. J Appl Physiol. 1982;53:1512–20.
71. Wilson TA. The mechanics of lung parenchyma. In: Chang HK, Paiva M, editors. Respiratory physiology – an analytical approach, vol. 40. New York: Dekker; 1989. p. 317–42. [Lung Biol Health Dis Ser].
72. Fung YC. Does the surface tension make the lung inherently unstable? Circ Res. 1975;37:497–502.
73. Reifenrath R. The significance of alveolar geometry and surface tension in the respiratory mechanics of the lung. Respir Physiol. 1975;24:115–37.
74. Cavalcante FS, Ito S, Brewer KK, Sakai H, Alencar AM, Almeida MP, Andrade JS Jr, Majumdar A, Ingenito EP, Suki B. Mechanical interactions between collagen and proteoglycans: implications for the stability of lung tissue. J Appl Physiol. 2005;98:672–9.
75. Stamenovic D. Micromechanical foundations of pulmonary elasticity. Physiol Rev. 1990;70: 1117–34.

76. Brewer KK, Sakai H, Alencar AM, Majumdar A, Arold SP, Lutchen KR, Ingenito EP, Suki B. Lung and alveolar wall elastic and hysteretic behavior in rats: effects of in vivo elastase treatment. J Appl Physiol. 2003;95:1926–36.
77. Takahashi A, Majumdar A, Parameswaran H, Bartolak-Suki E, Suki B. Proteoglycans maintain lung stability in an elastase-treated mouse model of emphysema. Am J Respir Cell Mol Biol. 2014;51:26–33.
78. Maksym GN, Fredberg JJ, Bates JH. Force heterogeneity in a two-dimensional network model of lung tissue elasticity. J Appl Physiol. 1998;85:1223–9.
79. Hall JK, Bates JHT, Casey DT, Bartolak-Suki E, Lutchen KR, Suki B. Predicting alveolar ventilation heterogeneity in pulmonary fibrosis using a non-uniform polyhedral spring network model. Front Netw Physiol. 2023;3:1124223.
80. Denny E, Schroter RC. The mechanical behavior of a mammalian lung alveolar duct model. J Biomech Eng. 1995;117:254–61.
81. Denny E, Schroter RC. Relationships between alveolar size and fibre distribution in a mammalian lung alveolar duct model. J Biomech Eng. 1997;119:289–97.
82. Denny E, Schroter RC. Viscoelastic behavior of a lung alveolar duct model. J Biomech Eng. 2000;122:143–51.
83. Denny E, Schroter RC. A model of non-uniform lung parenchyma distortion. J Biomech. 2006;39:652–63.
84. Hoppin FG Jr, Lee GC, Dawson SV. Properties of lung parenchyma in distortion. J Appl Physiol. 1975;39:742–51.
85. Amelon R, Cao K, Ding K, Christensen GE, Reinhardt JM, Raghavan ML. Three-dimensional characterization of regional lung deformation. J Biomech. 2011;44:2489–95.
86. Ma B, Bates JH. Continuum vs. spring network models of airway-parenchymal interdependence. J Appl Physiol (1985). 2012;113:124–9.
87. Ma B, Breen B, Bates JH. Influence of parenchymal heterogeneity on airway-parenchymal interdependence. Respir Physiol Neurobiol. 2013;188:94–101.
88. Ma B, Smith BJ, Bates JH. Resistance to alveolar shape change limits range of force propagation in lung parenchyma. Respir Physiol Neurobiol. 2015;211:22–8.
89. Demory D, Arnal JM, Wysocki M, Donati S, Granier I, Corno G, Durand-Gasselin J. Recruitability of the lung estimated by the pressure volume curve hysteresis in ARDS patients. Intensive Care Med. 2008;34:2019–25.
90. Ma B, Lutchen KR. An anatomically based hybrid computational model of the human lung and its application to low frequency oscillatory mechanics. Ann Biomed Eng. 2006;34:1691–704.
91. Oliveira CL, Araujo AD, Bates JH, Andrade JS Jr, Suki B. Entropy production and the pressure-volume curve of the lung. Front Physiol. 2016;7:73.
92. Salmon RB, Primiano FP Jr, Saidel GM, Niewoehner DE. Human lung pressure-volume relationships: alveolar collapse and airway closure. J Appl Physiol Respir Environ Exerc Physiol. 1981;51:353–62.
93. Sansores RH, Ramirez-Venegas A, Perez-Padilla R, Montano M, Ramos C, Becerril C, Gaxiola M, Pare P, Selman M. Correlation between pulmonary fibrosis and the lung pressure-volume curve. Lung. 1996;174:315–23.
94. Wilson TA. Mechanics of the pressure-volume curve of the lung. Ann Biomed Eng. 1981;9:439–49.
95. Frerichs I, Dargaville PA, Rimensberger PC. Regional respiratory inflation and deflation pressure-volume curves determined by electrical impedance tomography. Physiol Meas. 2013;34:567–77.
96. Parameswaran H, Bartolak-Suki E, Hamakawa H, Majumdar A, Allen PG, Suki B. Three-dimensional measurement of alveolar airspace volumes in normal and emphysematous lungs using micro-CT. J Appl Physiol. 2009;107:583–92.
97. D'Angelo E, Bonanni MV, Michelini S, Agostoni E. Topography of the pleural pressure in rabbits and dogs. Respir Physiol. 1970;8:204–29.

98. Lai-Fook SJ, Rodarte JR. Pleural pressure distribution and its relationship to lung volume and interstitial pressure. J Appl Physiol (1985). 1991;70:967–78.
99. Bigler AH, Renzetti AD Jr, Watanabe S, Begin R, Clark J. Regional lung expansion and vertical pleural pressure gradients in normal human subjects. Bull Eur Physiopathol Respir. 1979;15:773–88.
100. Washko GR, O'Donnell CR, Loring SH. Volume-related and volume-independent effects of posture on esophageal and transpulmonary pressures in healthy subjects. J Appl Physiol (1985). 2006;100:753–8.
101. Swan AJ, Clark AR, Tawhai MH. A computational model of the topographic distribution of ventilation in healthy human lungs. J Theor Biol. 2012;300:222–31.
102. Kang W, Clark AR, Tawhai MH. Gravity outweighs the contribution of structure to passive ventilation-perfusion matching in the supine adult human lung. J Appl Physiol (1985). 2018;124:23–33.
103. Shi L, Herrmann J, Bou Jawde S, Bates JHT, Nia HT, Suki B. Modeling the influence of gravity and the mechanical properties of elastin and collagen fibers on alveolar and lung pressure-volume curves. Sci Rep. 2022;12:12280.
104. Casha AR, Caruana-Gauci R, Manche A, Gauci M, Chetcuti S, Bertolaccini L, Scarci M. Pleural pressure theory revisited: a role for capillary equilibrium. J Thorac Dis. 2017;9: 979–89.
105. Agostoni E, D'Angelo E. Comparative features of the transpulmonary pressure. Respir Physiol. 1970;11:76–83.
106. Diallo MH, Guenard H, Laurent F, Carles P, Giron J. Distribution of lung density and mass in patients with emphysema as assessed by quantitative analysis of CT. Chest. 2000;118:1566–75.
107. Elliott AR, Prisk GK, Guy HJ, West JB. Lung volumes during sustained microgravity on spacelab SLS-1. J Appl Physiol (1985). 1994;77:2005–14.
108. Wang JM, Robertson SH, Wang Z, He M, Virgincar RS, Schrank GM, Smigla RM, O'Riordan TG, Sundy J, Ebner L, et al. Using hyperpolarized (129)Xe MRI to quantify regional gas transfer in idiopathic pulmonary fibrosis. Thorax. 2018;73:21–8.

Chapter 5
Models of Tissue Viscoelasticity

Abstract Lung tissue is viscoelastic because it dissipates energy, in addition to storing energy, when stretched. This gives rise to the phenomenon of stress adaptation and creep and makes lung resistance and elastance depend on the frequency of volume cycling. Viscoelasticity is predicted in general terms by simple lumped-parameter models such as the Maxell and Kelvin bodies governed by ordinary differential equations. Lung tissue viscoelasticity is more accurately predicted by the constant-phase model, which is governed by a differential equation of fractional order. The constant-phase model predicts that stress relaxation following a step increase in strain will conform to a power-law function of time, as is observed experimentally. The combination of tissue viscoelasticity and nonlinear quasi-static stress-strain behavior observed in lung tissue can be accurately represented by quasi-linear viscoelasticity in which stress results from strain first being transformed by a static nonlinear function followed by a linear dynamic system. Although the origin of the power law stress adaptation is not fully understood, quasi-linear viscoelasticity can potentially be explained as arising from a sequence of micro-yield events that take place over time as local tissue regions of high stress break apart and pass the stress they were bearing on to other regions.

Keywords Energy storage and dissipation · Maxwell and Kelvin bodies · Frequency dependence of resistance and elastance · Power-law stress relaxation · Quasi-linear viscoelasticity

5.1 Introduction

It is natural to think of biological soft tissue as being elastic in nature. Pull down on an earlobe and the tissue will lengthen, but then let go and it will recoil upward. Similarly, the elastic recoil of the lung parenchymal tissue is what returns its volume to functional residual capacity at the end of every expiration (Chap. 4). However, not all the energy used to extend an earlobe or inflate a lung is returned to its supplier when the distending force is removed. Some of this energy is dissipated within the tissues as heat. In other words, tissues have resistive as well as elastic properties. The genesis of tissue resistance is complex and somewhat less well understood than

B. Suki, J. H. T. Bates, *Mathematical Modeling of the Healthy and Diseased Lung*,
https://doi.org/10.1007/978-3-031-95322-4_5

tissue elastance (Chap. 4), but its source can be appreciated in general terms by considering how airway resistance arises. Friction is generated within the pulmonary airways when the ordered motion of the molecules in a flowing stream of viscous gas is converted into random motion as a result of molecular collisions (Sect. 3.2.2). This randomly oriented molecular kinetic energy manifests as heat and cannot be utilized to do useful work in any particular direction, so we say the energy has been *dissipated*. A corresponding dissipative process takes place in the lung tissues during breathing except that it involves the molecular constituents of tissue rather than gas. Thus, when lung tissue is deformed as the lung inflates and deflates, its constituents move relative to each other, just as occurs in a flowing stream of gas, causing friction which generates heat. Lung tissue thus dissipates energy during breathing.

Unlike a flowing gas, however, lung tissue also stores energy when it is deformed, and it releases this energy in returning to its original undeformed configuration. The stored energy can be harnessed to perform useful work in the same way that the potential energy stored in a stretched spring can, for example, be converted to kinetic energy if the spring is attached to a projectile. Materials that store potential energy when deformed are called elastic, while materials that dissipate energy are called viscous. Since lung tissue (in fact, all biologic tissues) does both, it is called *viscoelastic*.

Viscoelastic materials exhibit two canonical properties that can be taken to define them as being viscoelastic. One of these is *stress adaptation* (Fig. 5.1), a phenomenon readily observed in the lung by applying a sudden increase in volume and then

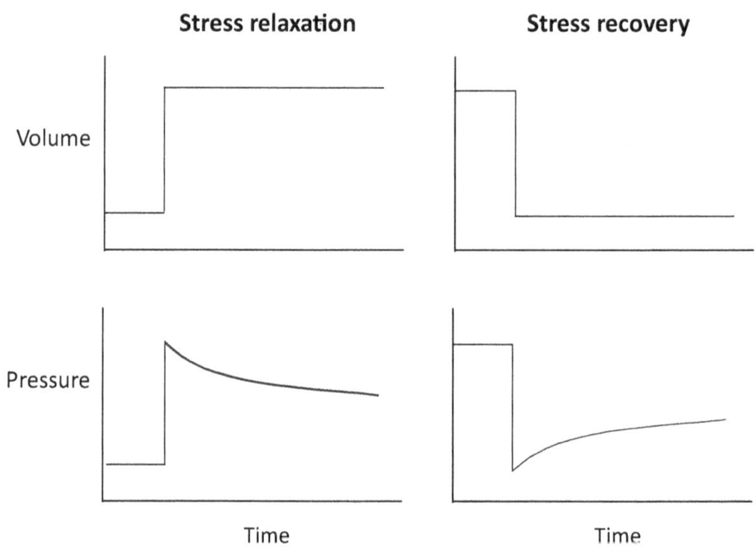

Fig. 5.1 Manifestations of stress adaptation in the lung. Stress relaxation (left-hand side) is the transient decrease in airway pressure following a sudden step increase in lung volume. The converse is stress recovery (right-hand side) where pressure increases following a sudden step decrease in volume

observing what happens to inflation pressure as the new volume is maintained. Immediately following the sudden increase in pressure generated by the step change in volume, pressure will decrease gradually, appearing to asymptote toward some intermediate level between its pre- and immediately post-step levels (Fig. 5.1a, b). This relaxation in pressure reflects the adjustments taking place among the tissue constituents as they seek to find a more comfortable juxtaposition compatible with the new level of volume. The converse of stress relaxation is stress recovery, which occurs when volume suddenly changes from a higher to a lower level (Fig. 5.1c, d). The other canonical signature of viscoelasticity is *creep*. If a sudden change in pressure is applied to the lungs, the result will be a transient change in volume (Fig. 5.2). Stress adaptation and creep are thus counterparts of each other.

This chapter covers models of lung tissue viscoelasticity that are rationalized on the basis of the above discussion. These models are empirical in the sense that their viscoelastic properties are produced by idealized constructs that do not correspond to identifiable structures within the tissue. The ultimate source of lung tissue viscoelasticity itself remains elusive.

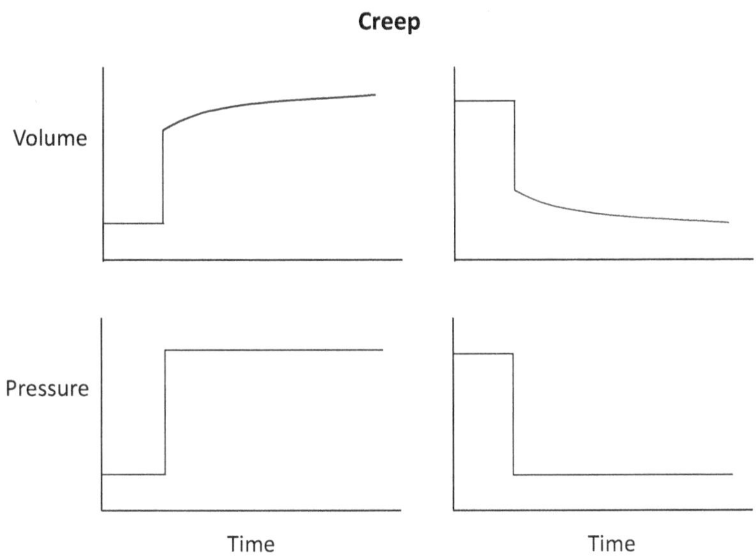

Fig. 5.2 Manifestations of creep in the lung. Lung volume increases transiently following a sudden step increase in airway pressure (left-hand side). Conversely, volume decreases transiently following a sudden step decrease in pressure

5.2 Lumped-Parameter Models of Viscoelasticity

5.2.1 *The Maxwell Body*

Mathematical models of viscoelastic materials date back to the nineteenth Century. Such models were first applied to the lung by Mount in 1955 [1] and then revisited in detail several decades later by Hildebrandt who made detailed studies of the nature of lung tissue viscoelasticity in cats [2, 3]. The advent of the alveolar capsule technique in the 1980's [4, 5], which allowed alveolar pressure to be measured directly in anesthetized animals, confirmed that lung tissue viscoelasticity is an important contributor to the dynamic mechanical behavior of the lungs when lung volume is oscillated at frequencies pertinent to spontaneous breathing [6, 7].

The overall features of viscoelastic behavior, namely stress adaptation and creep, can be recapitulated by simple lumped-parameter models. The simplest of these is named for the nineteenth Century Scottish physicist James Clerk Maxwell. The Maxwell model consists of a Hookean (linear) spring connected in series to a resistive element known as a dashpot that is fixed at its left-hand end in Fig. 5.3. The dashpot is represented by a plunger fitting snugly inside a barrel, much like a syringe, so its resistance arises from the friction generated as the plunger slides along the inside of the barrel (which is assumed long enough that the two components never come apart). Let the length of the spring be expressed by the variable x, and define $x = 0$ when the spring is at its relaxed length (Fig. 5.3a). If the spring is suddenly stretched by an amount x_0 at $t = 0$ (Fig. 5.3b) the elastic force in the spring will rise immediately to $F = Ex_0$, where E is the elastance of the spring. (Note that, in the context of the lung, x and F can be replaced by lung volume and airway pressure, respectively, in which case E corresponds to lung elastance.) However, the elastic force across the spring must equal the resistive force across the dashpot at all times because the two elements are connected in series. The resistive force is given by $F = Rdx/dt$, where R is the dashpot resistance, and is generated as the dashpot barrel

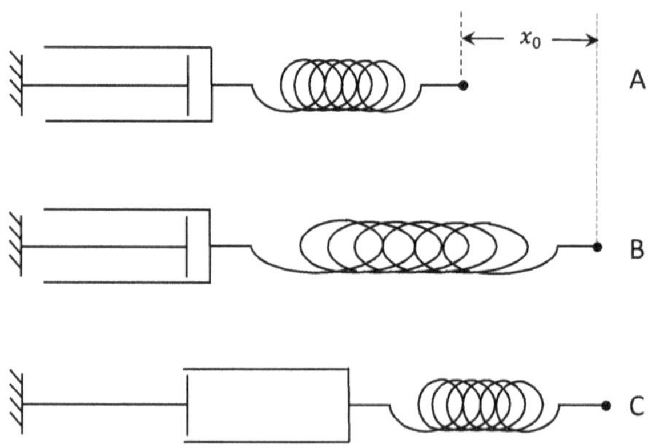

Fig. 5.3 (**a**) A Maxwell body in its relaxed configuration. (**b**) The configuration immediately after the spring has been stretched by amount x_0 to the right. (**c**) The configuration after the dashpot has moved and relaxed the spring again

moves to the right in Fig. 5.3c. This reduces the stretch, and thus the force, in the spring. In other words, the Maxwell model exhibits stress relaxation following a step change in length.

To analyze this situation mathematically, note that the rate at which the dashpot changes length for $t > 0$ is equal and opposite to the rate at which the spring changes length because the sum of their individual lengths is fixed. This situation is described by the first-order linear ordinary differential equation

$$F(t) = -R\dot{x}(t) = Ex(t) \tag{5.1}$$

(the dot notation denoting a time derivative). If E and R are constants independent of x and/or dx/dt, the solution to Eq. 5.1 is

$$x(t) = x_0 e^{-\frac{Et}{R}}. \tag{5.2}$$

The force (or stress) in the model is then

$$F(t) = Ex(t) = Ex_0 e^{-\frac{Et}{R}}. \tag{5.3}$$

The Maxwell model thus predicts that stress relaxes in a viscoelastic material in an exponential manner following a step change in length. The stress asymptotes toward zero as time increases, however, which is characteristic of a viscous fluid rather than a solid. One usually expects stress in a viscoelastic solid such as lung tissue to approach a finite level of elastic recoil as depicted in Fig. 5.1.

If a constant force F_0 is applied to a Maxell body at $t = 0$, the length of the body immediately increases by an amount $x_0 = F_0/E$ due to the elongation of the spring. If the force is maintained unchanged, the dashpot continues to elongate at a constant rate $\dot{x}(t) = F_0/R$. The length of the Maxwell element beyond its resting length as a function of time is given by

$$x(t) = \frac{F_0}{E} + \frac{F_0}{R}t. \tag{5.4}$$

Thus, the Maxell body also predicts creep, although again in a way that continues indefinitely at a constant rate rather than approaching a finite asymptote as depicted in Fig. 5.2.

5.2.2 The Kelvin Body

The addition of a second spring in parallel with the Maxwell body leads to the Kelvin body shown in Fig. 5.4. The elastance of the parallel spring is denoted E_{stat} because it represents the static elastance of the model, which pertains when the Maxwell body

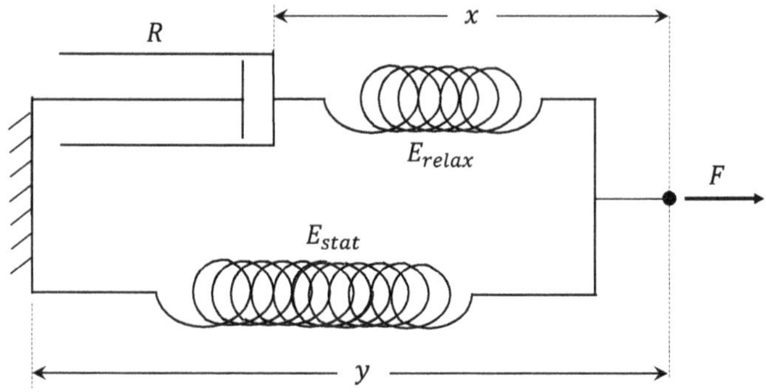

Fig. 5.4 The Kelvin body consisting of a Maxwell body (R and E_{relax}) in parallel with a second spring (E_{stat})

is fully relaxed. The elastance of the spring in the Maxwell body is denoted E_{relax} because it represents the loss in elastance that manifests following a step change length due to stress relaxation. The Kelvin body has two mechanical degrees of freedom that can be specified as the extensions of its two springs (as opposed to the single degree of freedom of the Maxwell body). Let the extension of the parallel spring (i.e., the extension of the entire model) be y, and the extension of the Maxwell body spring be x, so the extension of the dashpot is $y - x$ (Fig. 5.4). If the total force across the Kelvin body is F, the behavior of the Kelvin body is accounted for by the following two coupled ordinary differential equations (provided R, E_1 and E_2 are constants):

$$F(t) = E_{stat}y(t) + E_{relax}x(t) \tag{5.5}$$

and

$$E_{relax}x(t) = R[\dot{y}(t) - \dot{x}(t)]. \tag{5.6}$$

Eliminating $x(t)$ and its derivative between the above two equations yields the second-order differential equation of motion of the Kelvin body given by

$$R\dot{F}(t) + E_{relax}F(t) = R(E_{stat} + E_{relax})\dot{y}(t) + E_{stat}E_{relax}y(t). \tag{5.7}$$

Following a step change in total length of y_0, while $\dot{y}(t) = 0$, Eq. 5.7 reduces to

$$\dot{F}(t) = -\frac{E_{relax}}{R}F(t) + \frac{E_{stat}E_{relax}}{R}y_0 \tag{5.8}$$

which has solution

$$F(t) = E_{stat}y_0 + E_{relax}y_0e^{-\frac{E_{relax}t}{R}}. \tag{5.9}$$

The Kelvin body, like the Maxwell body, thus predicts an exponential progression for stress relaxation following a step increase in length. In contrast to the Maxwell model, however, the stress relaxes asymptotically toward the finite level of $E_{stat}y_0$ instead of zero. This is more in keeping with the subjective impression gained from experimental data, that lung tissue has a non-zero static elastance.

Although stress adaptation (Fig. 5.1) and creep (Fig. 5.2) are phenomena that manifest over time, they have frequency-domain counterparts revealed in plots of force versus length (or pressure versus volume, in the case of the lung) when length (or volume) is cycled sinusoidally. Such plots comprise closed loops with areas determined by the relative importances of elastic versus resistive behavior. This can be appreciated by letting $y(t) = y_0 \sin \omega t$ in the equation for the Kelvin body (Eq. 5.7), so $\dot{y} = 2\pi f y_0 \cos \omega t$, where $\omega = 2\pi f$. The right-hand side of Eq. 5.7 is thus the sum of sine and cosines functions both having the same frequency f. The result is a sinusoidal function also with frequency f but with a non-zero phase, meaning that that $F(t)$ must be a sinusoidal function with frequency f. Plotting two sinusoidal functions against each other produces an ellipse with an area determined by the extent to which the two sinusoids differ in phase. The area of this ellipse has units of energy and, in fact, is equal to the energy that is dissipated within the tissues during a single complete oscillatory cycle. The ratio of tissue resistance to tissue elastance is called *hysteresivity*. An important observation is that the hysteresivity of many biologic soft tissues, including lung parenchyma, is virtually independent of the frequency and amplitude at which lung tissue length (or lung volume) is cycled [8].

Frequency independence of hysteresivity is not a characteristic of the Kelvin body, however. This can be seen by taking the Fourier transform of Eq. 5.7, which is

$$i\omega R F(f) + E_{relax} F(t) = R(E_{stat} + E_{relax})\dot{y}(f) - \frac{iE_{stat}E_{relax}}{\omega}\dot{y}(f) \tag{5.10}$$

where $F(f)$ and $\dot{y}(f)$ are the Fourier transforms of $F(t)$ and $\dot{y}(t)$, respectively, and $i = \sqrt{-1}$. Note that Eq. 5.10 uses the fact that since the Fourier transform of a time-domain function, say $y(t)$, is defined as

$$y(f) = \int_{-\infty}^{\infty} e^{-i\omega t} y(t)dt \tag{5.11}$$

then integration by parts shows the Fourier transform of its time-derivative to be

$$\int_{-\infty}^{\infty} e^{-i\omega t}\dot{y}(t)dt = \left[-i2\pi f e^{-i\omega t}\dot{y}(t)\right]_{-\infty}^{\infty} + \int_{-\infty}^{\infty} i2\pi f e^{-i\omega t} y(t)dt = i\omega y(f) \tag{5.12}$$

under the assumption that $y(t)$ is square-integrable, meaning that both $y(t)$ and $\dot{y}(t)$ tend to zero as t tends to either $-\infty$ or ∞. Conversely, the Fourier transform of $y(t)$ equals the Fourier transform of $\dot{y}(t)$ multiplied by $1/i\omega$.

Rearranging Eq. 5.10 gives the impedance (see Sect. 3.5) of the Kelvin body as

$$Z(f) = \frac{F(f)}{\dot{y}(f)} = \frac{RE_{relax}^{\ 2}}{E_{relax}^{\ 2} + \omega^2 R^2} - i\left[\frac{\omega^2 R^2 (E_{stat} + E_{relax}) + E_{stat}E_{relax}^{\ 2}}{\omega E_{relax}^{\ 2} + \omega^3 R^2}\right]. \tag{5.13}$$

Inspection of Eq. 5.13 shows that, in the limit as ω tends to ∞, the real part of $Z(f)$ tends to zero. The real part of impedance corresponds to the resistance of the Kelvin body, which makes intuitive sense because as the oscillation frequency increases, the dashpot (Fig. 5.4) has progressively less time to slide back and forth, until eventually it does not move at all. At the same time, the imaginary part of impedance, which describes the elastic behavior of the Kelvin body, tends to the sum of E_{stat} and E_{relax}, again as one would expect since both springs are stretched equally in the limit of high frequencies. Conversely, at ω approaches zero, the real part of $Z(f)$ tends to R while the imaginary part tends to E_{stat}/ω, reflecting the fact that the dashpot now has time to follow the changes in y(t) faithfully because the Maxwell element spring never gets stretched at all. The stiffness at zero frequency is thus due solely to E_{stat}.

Further inspection of Eq. 5.13 also shows that the phase of $Z(f)$, equal to the ratio of its imaginary part to its real part, is a function of frequency. This means that the force-length hysteresis, observed when the Kelvin body is cyclically stretched in a sinusoidal fashion, has an area that varies with frequency.

5.2.3 A Viscoelastic Lung Model

The Kelvin body can be used to represent the viscoelastic properties of the lung parenchymal tissues in a model of the entire lung. The simplest version of this model is an extension of the model shown in Fig. 3.2. In the new model (Fig. 5.5) the representation of tissue resistance by a single dashpot has been replaced by a Maxwell element, and the dashpot in the Maxwell element is now denoted R_{tis} to denote the fact that it accounts for the resistance of the tissues alone and not the resistance of the entire lung. At oscillation frequencies below about 2 Hz, the inertance of the airway gas can be neglected, in which case the differential equation of motion of the model relating pressure (P) applied at the airway opening to the flow ($\dot{V}$) entering the airway is

$$R_{tis}\dot{P}(t) + E_{relax}P(t) = R_{aw}R_{tis}\ddot{V}(t) + [R_{tis}(E_{stat} + E_{relax}) + R_{aw}E_{relax}]\dot{V}(t) \\ + E_{stat}E_{relax}V(t). \tag{5.14}$$

The impedance of the model is

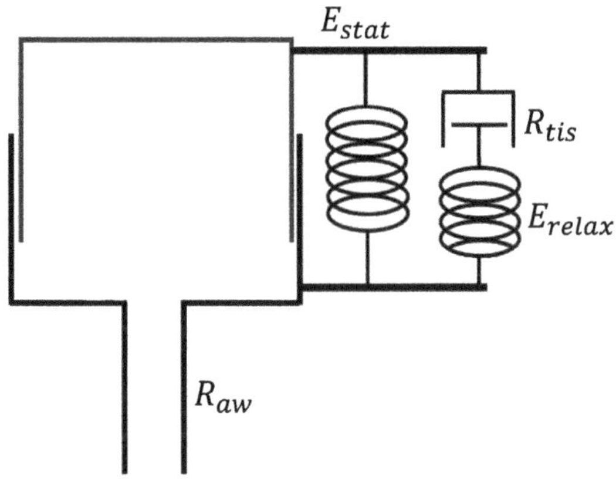

Fig. 5.5 Single-compartment model of the lung consisting of a single airway (resistance R_{aw}) serving a single alveolar compartment composed of viscoelastic tissue represented by a Kelvin body

$$Z(f) = \frac{(R_{aw} + R_{tis})E_{relax}^{\ 2} + \omega^2 R_{aw} R_{tis}^{\ 2}}{E_{relax}^{\ 2} + \omega^2 R_{tis}^{\ 2}} - i\left[\frac{\omega^2 R_{tis}^{\ 2}(E_{stat} + E_{relax}) + E_{stat} E_{relax}^{\ 2}}{\omega E_{relax}^{\ 2} + \omega^3 R_{tis}^{\ 2}}\right]. \tag{5.15}$$

This viscoelastic model of the lung is thus governed by a second-order linear ordinary differential equation. The same applies to the series and parallel two-compartment models of the lung encountered in Chap. 3 (Eqs. 3.15 and 3.16, respectively). Because the equations of these three models are identical in mathematical form, differing only in how the coefficients of the equations relate to the parameters of the respective models, it is impossible to distinguish them from measurements of P and $\dot{V}$ made at the airway opening. In other words, given P and $\dot{V}$ alone, there is no maneuver one can perform that will discriminate between a two-compartment lung that is heterogeneously ventilated in either a serial or parallel fashion or a homogeneous lung surrounded by viscoelastic tissue. Making this discrimination requires the measurement of a signal somewhere else in the system, and was achieved through direct measurement of alveolar pressure [6]. These measurements showed that, of the three contenders, the viscoelastic model captures reality the best. Of course, it is possible to make the lung heterogeneous through appropriate interventions, in which case none of the models individually does a particularly good job [7].

5.3 The Constant-Phase Model of Tissue Impedance

The Kelvin body provides a very useful and often acceptably accurate representation of the dynamic relationship between stress and strain in a viscoelastic tissue. In particular, it predicts that the mechanical impedance of a viscoelastic tissue should conform to Eq. 5.13. Careful experimental investigation reveals, however, that the mechanical impedance of lung tissue is significantly better described by the function

$$Z(f) = \frac{G - iH}{\omega^{\alpha}} \tag{5.16}$$

where G, H and α are constants, i is the imaginary unit, and α is determined by G and H through the expression

$$\alpha = \frac{2}{\pi} \tan^{-1} \frac{H}{G}. \tag{5.17}$$

Inspection of Eq. 5.16 shows that the ratio of the imaginary part to the real part of $Z(f)$ is simply the ratio $\frac{H}{G}$. This ratio is the tangent of the phase of $Z(f)$, and it is independent of ω, unlike the impedance of the Kelvin body (Eq. 5.13). Accordingly, Eqs. 5.16 and 5.17 together define what is known as the *constant-phase model* of tissue impedance [9, 10]. The constant-phase model also predicts that force-length hysteresis will be independent of ω, in keeping with experimental observation [8].

The parameters G and H are known as *tissue damping* and *elastance*, respectively, and have been widely used to characterize parenchymal mechanics under a wide range of experimental conditions [11–13]. Curiously, since G and H are linked through the parameter α via Eq. 5.17, their units depend on the value of α and hence on their values relative to each other. This strange state of affairs can be avoided if frequency is defined relative to some standard frequency ω_0 [14]. This normalized version of the constant-phase model, given by

$$Z(f) = \frac{G - iH}{\left(\frac{\omega}{\omega_0}\right)^{\alpha}}, \tag{5.18}$$

results in G and H both having the same units as Z, regardless of the value of α. By conveniently choosing $\omega_0 = 1$, the numerical values of G and H do not change when fitting Eq. 5.18 to measured data.

The Kelvin body also predicts that stress relaxation will follow an exponential time course after a step change in length (Eq. 5.9). Again, experimental studies reveal this not to be precisely the case. Early studies by Hildebrandt [3] indicated that stress relaxation is linear with the logarithm of time, that is

$$F(t) = A + B \ln t \tag{5.19}$$

where A is a constant. It was subsequently shown that a more accurate representation is given by the power law expression [15]

$$F(t) = At^{-\beta} \tag{5.20}$$

where β is a small constant (typically in the order of 0.1). In fact, Eq. 5.19 is a good approximation to Eq. 5.20 when β is small because the exponential series expansion

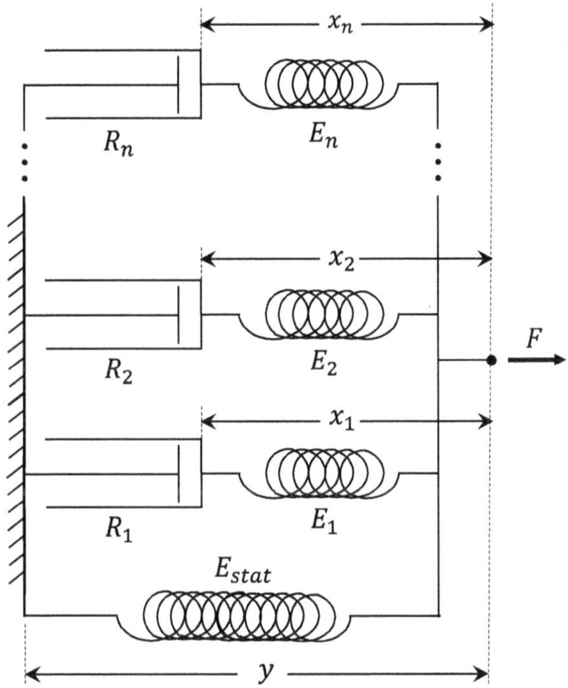

Fig. 5.6 Multiple parallel Maxwell bodies with the appropriate distribution of time constants can be made to exhibit any monotonically decreasing stress relaxation profile

$$t^{-\beta} = e^{-\beta \ln t} = 1 - \beta \ln t + \frac{(\beta \ln t)^2}{2!} - \cdots \tag{5.21}$$

is then accurately approximated by only its first two terms, which is equivalent to Eq. 5.19.

Equation 5.20 suffers from the same issue with units as occurs with the constant-phase model of impedance, since both sides of the equation must have units of force but the factor $t^{-\beta}$ has fractional units of time. This means that the units of A depend on the value of β. Again, a convenient way to sidestep the awkwardness this causes [15] is to normalize time so that $(\omega_0 t)^{-k}$ is now unitless, allowing A itself to assume units of force while its numerical value remains unchanged.

The question now arises as to how lumped-parameter models like those shown in Figs. 5.3 and 5.4 can be used to represent a viscoelastic material that behaves according to Eq. 5.21. One possibility is to extend the Kelvin body with additional parallel Maxwell bodies as illustrated in Fig. 5.6. If the time constants of these Maxwell bodies (i.e., the ratios R_i/E_i) are chosen appropriately, it is possible to recapitulate any monotonically decreasing stress relaxation profile, including one that decreases as a power-law function of time [16].

This raises another question, however; why should Nature have chosen a very specific distribution of Maxwell element time constants to direct the course of stress relaxation? There is no obvious answer to this question, suggesting that the use of multiple Maxwell elements is simply an empirical strategy to account for a complex phenomenon in a way that has nothing to do with its underlying mechanistic basis.

5.4 Quasilinear Viscoelasticity

A detailed examination of viscoelasticity in lung tissue, or any biologic soft tissue for that matter, reveals it to be significantly nonlinear. The models of viscoelasticity considered so far in this chapter are all linear, meaning that their responses in pressure (volume) scale precisely with the amplitude of an imposed volume (pressure) signal. The elastic behavior of lung tissue, however, is highly nonlinear. This is most obviously apparent in the quasi-static pressure-volume behavior of tissue seen when length changes are imposed very slowly on a strip of tissue while the force across the tissue is recorded. (Truly static behavior requires an infinitely slow rate of length change, so all that can be done experimentally is to change length slowly compared to the rate of viscoelastic stress adaption and hope that this allows most of the stress adaptation to take place while length is changing, a problematic strategy for tissues conforming to Eq. 5.21). A plot of force versus length starting from the relaxed length of the tissue is far from linear. Instead, it exhibits a marked upward concavity indicative of progressive strain stiffening, a behavior that is seen all the way down to the level of a single alveolar septal wall [17]. Similarly, the quasi-static pressure-volume relationship of the lung is only approximately linear over the tidal volume range, becoming markedly nonlinear as volume approaches total lung capacity (e.g., Eq. 3.11 or 3.12 in Sect. 3.3.2).

This suggests that if a Kelvin body is to be used as a model of lung tissue viscoelasticity then, at the very least, the spring E_{stat} in Fig. 5.4 should become stiffer with increasing strain. Let F_{stat} be the force across the spring E_{stat}, and suppose that its force-length behavior is given by a suitably nonlinear function such as the exponential expression

$$F_{stat} = E_{stat} y e^{qy} \tag{5.22}$$

where q is a constant defining the strength of the nonlinearity. In general, F_{stat} and y are functions of time so, in analogy with Eq. 5.5, the total force across this nonlinear Kelvin body is

$$F(t) = E_{stat} y(t) e^{qy(t)} + E_{relax} x(t). \tag{5.23}$$

Because the Maxwell body in this model is linear, Eq. 5.6 still applies. Eliminating $x(t)$ and its derivative between Eqs. 5.6 and 5.23 yields

$$\dot{F(t)} + \frac{E_{relax}}{R}F(t) = \left\{E_{relax} + E_{stat}[1 + qy(t)]e^{qy(t)}\right\}\dot{y}(t) + \frac{E_{stat}E_{relax}}{R}e^{qy(t)}y(t). \tag{5.24}$$

(Note that Eq. 5.24 reduces to Eq. 5.7 when $q = 0$.)

Now impose a step increase in length on the nonlinear Kelvin body such that $y(t) = 0$ for $t < 0$ and $y(t) = y_0$ for $t \geq 0$. For $t > 0$, Eq. 5.24 then reduces to

$$\dot{F}(t) + \frac{1}{\tau}F(t) = \frac{E_{stat}}{\tau}e^{qy_0}y_0 \tag{5.25}$$

where $\tau = \frac{R}{E_{relax}}$ is a time constant. This is a linear nonhomogeneous ordinary differential equation in $F(t)$ with the solution

$$F(t) = E_{relax}e^{-t/\tau} + E_{stat}y_0e^{qy_0}. \tag{5.26}$$

Thus, stress relaxation is still exponential following a step change in length when only the parallel spring in the Kelvin body is nonlinear, even when the quasi-static stress-strain behavior is highly nonlinear.

Nevertheless, experiments indicate that the stiffness and resistance giving rise to viscoelasticity (i.e., those of the series spring and dashpot in Fig. 5.5) are also nonlinear [18]. In particular, the amount of stress adaptation taking place following a dynamic change in volume increases with the quasi-static stiffness of the tissue, implying that the stiffness of the spring E_{relax} should also depend on its extension. But if E_{relax} is not constant, the Maxwell body no longer has a well-defined time constant (given by the ratio R/E_{relax}) unless R is constrained to have the same strain dependence as E_{relax}. A potential rationale for such a constraint can be made on the basis that the Kelvin body, being a lumped parameter model, attributes the resistive and elastic properties of tissue to separate entities. In fact, no such identifiable entities exist. If they did, it would be possible to subdivide tissue into portions small enough to isolate either the purely resistive or elastic behavior. There is no evidence that this can be done, which means that the resistive and elastic properties responsible for viscoelastic behavior are co-located within the same physical structures and are thus "coupled at the level of the stress bearing element" [8].

Experimental studies have shown that nonlinear viscoelastic behavior in numerous disparate soft tissues, including lung parenchyma, adheres closely to a very particular and perhaps peculiar functional form in which all the nonlinearities are confined to the static force-length (pressure-volume) properties of the tissue while the dynamic properties are linear, an example of which is Eq. 5.26. This behavior has been termed *quasi-linear viscoelasticity* by Fung [16]. Quasi-linear viscoelasticity can be expressed in system theoretic terms with respect to the lung as illustrated in Fig. 5.7, which shows lung volume, $V(t)$, being transformed into pressure, $P(t)$, in two distinct steps. First, the nonlinear function g is applied to $V(t)$. The result is then convolved with the impulse response function, $h(t)$, of a linear dynamic system. This type of block structure is known as a Hammerstein model.

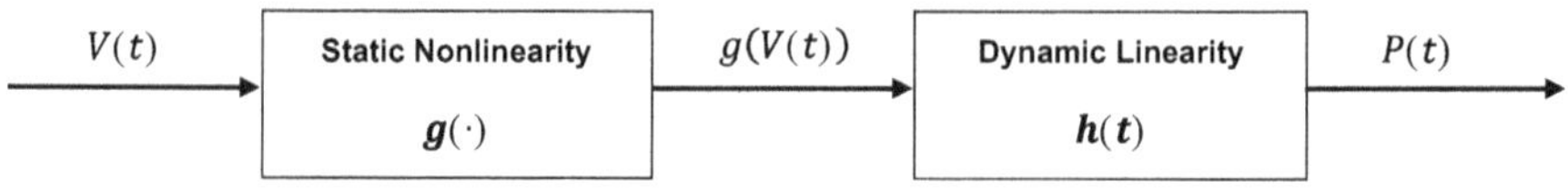

Fig. 5.7 Quasi-linear viscoelasticity. Volume is converted to pressure first via a static nonlinear function and then via a linear dynamic system

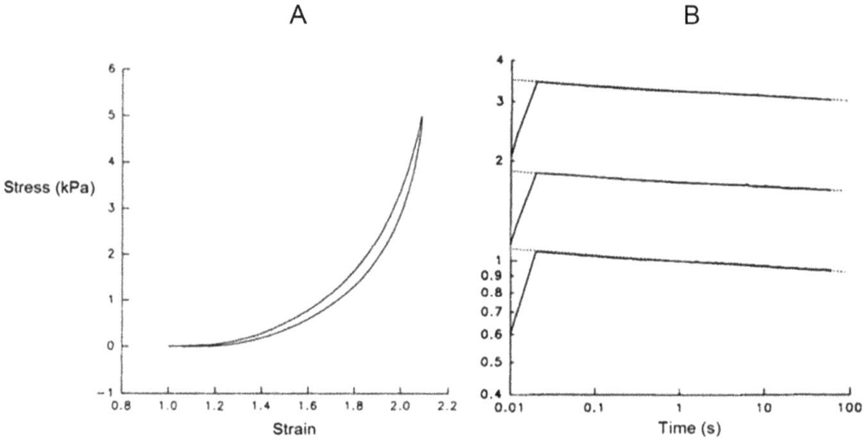

Fig. 5.8 (**a**) Quasi-static force-length behavior of a strip of lung parenchyma. (**b**) Stress relaxation when the strip is stretched by 10% from three different starting lengths. (Adapted with permission from [15])

The two steps in Fig. 5.7 are combined in the expression

$$P(t) = \int_0^t g(V(u))h(t-u)du \tag{5.27}$$

where u is a dummy variable of integration. A dramatic illustration of quasi-linear viscoelasticity is shown by the data in Fig. 5.8, which were obtained from a strip of lung parenchyma in an organ bath. One end of the strip was attached to a force sensor and at the other end to a linear actuator. Beginning from a length at zero stress, the strips were stretched by 100% and then returned to resting length again over a period of 1 min, which is very slow compared to the cycling rate of normal breathing (hence this rate of cycling can be considered quasi-static). The force-length relationship is highly nonlinear (Fig. 5.8a), which reflects the static function g in Eq. 5.27. This relationship is well approximated by the exponential relationship $g(V) = Ae^V$, $V > 0$. There is also a slight degree of hysteresis reflecting some residual stress adaptation that had not been given a chance to manifest even at this slow cycling rate.

The dynamic behavior of the parenchymal strips is evident in the behavior they exhibited when they were suddenly (within about 10 ms) stretched by 10% and force recorded for the subsequent 100 s. The resulting force-time behavior was linear in a log-log plot (Fig. 5.8b) indicative of stress relaxation behavior conforming to Eq. 5.20. In addition, the slope of the relationship, equal to $-\beta$ in Eq. 5.20, was independent of the point on the nonlinear stress-strain relationship that the 10% strain was applied from.

Putting these two steps together gives the following model for the viscoelastic behavior of a strip of lung parenchyma subjected to a sudden step in length:

$$P(t) = \int_0^t Ae^{V(u)}(t-u)^{-\beta}du. \tag{5.28}$$

There are two intriguing aspects of the data shown in Fig. 5.8 that beg for explanation: 1) the fact that stress relaxation follows a power law of time as opposed to an exponential or multi-exponential time course, and 2) the fact that the power law exponent, β, is independent of where along a highly nonlinear stress-strain curve the stretch is applied. These two observations do not at the outset appear to be related to each other, so it is temping to think they reflect different underlying physics. It turns out, however, that there may be a unifying explanation that accounts for both simultaneously.

First, consider the power law form of the stress relaxation, the existence of which is not entirely surprising given how ubiquitously power law processes appear throughout the natural world as codified in Zipf's law [19]. Examples include the distributions of earthquake magnitudes [20], words in language [21] and frequencies in music [22], the degree of notoriety of internet sites [23], and even the distribution of airway opening cascades in the lung [24]. Although a universal explanation for this ubiquity remains unclear, several mechanisms have been proposed for specific cases. A common theme among these explanations is that they have a serial character to them, meaning that the underlying processes involved take place as a series of sequentially dependent events rather than as multiple simultaneous events occurring in parallel [25].

It is difficult to see how this could apply to the multiple Maxwell body model in Fig. 5.6 in which all dashpots move together, albeit at different rates. If the Maxwell bodies are configured as shown in Fig. 5.9, however, the serial nature of their respective contributions to stress relaxation becomes apparent. Specifically, stress is initially born by a single spring-dashpot pair (Fig. 5.6a) and energy is dissipated as the dashpot piston slides along its barrel. In this case, however, barrel is very short barrel so the piston and barrel soon separate. When this happens, a second initially unstressed dashpot is recruited to share the stress with the first spring, as shown in Fig. 5.9b. Since the two springs are each stretched by half as much as the first spring was initially, the stress across the model is reduced. The second dashpot now

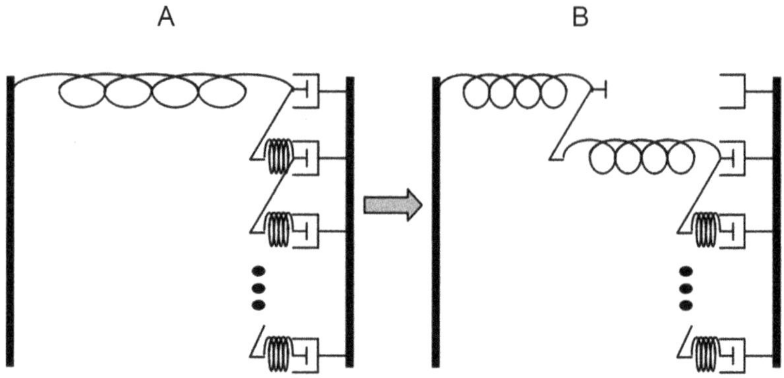

Fig. 5.9 A sequential spring recruitment model of quasi-linear viscoelasticity showing (**a**) a single Maxwell element bearing the entire stress initially, and (**b**) a second spring having been recruited to share the stress and the strain with the first spring

dissipates energy but, like the first, soon comes apart to recruit a third spring-dashpot pair, and so on. The stress across the model thus decreases in ever diminishing steps.

The mathematical description of this model begins with a specification of the constitutive properties of the spring and dashpot, which are the same for all the Maxwell bodies in the model. To begin with, assume that the springs are Hookean (each has a spring constant, E, that is independent of strain), and the dashpot resistances are also constants (independent of strain rate). When there are n springs recruited to share the total length, L, of the model, the extension of each of the recruited springs is L/n, so the force (F) in each stretched spring is

$$F = E\frac{L}{n}. \tag{5.29}$$

The decrease, ΔF, in force when the $n + 1^{th}$ spring is recruited into the lineup is, via a binomial expansion,

$$\Delta F = E\left(\frac{L}{n+1}\right) - E\left(\frac{L}{n}\right) = E\left(\frac{L}{n}\right)\left(1+\frac{1}{n}\right)^{-1} - E\left(\frac{L}{n}\right) = E\left(\frac{L}{n}\right)\left[\frac{-1}{n} + \frac{2}{2!n^2}\right] - \cdots \tag{5.30}$$

For large n, this approximates to

$$\Delta F \approx -\frac{EL}{n^2} = -\frac{1}{LE}\left[\frac{EL}{n}\right]^2 = -\frac{1}{LE}F^2. \tag{5.31}$$

The same force F is also experienced by the dashpot in series with the most recently recruited spring, for which

$$F = R\dot{x} \tag{5.32}$$

where R is the dashpot resistance. If the length of the dashpot barrel is D, then the time taken for the dashpot piston to reach the end of its travel and come out of the barrel is

$$\Delta t \approx \frac{D}{\dot{x}} = DRF \tag{5.33}$$

where $\dot{x}$ is the velocity of the piston in the barrel. Combining Eqs. 5.31 and 5.33 gives, in the limit of large n,

$$\frac{\Delta F}{\Delta t} \to \frac{dF}{dt} = \frac{1}{LEDR} F(t)^3 \tag{5.34}$$

which has solution

$$F(t) = \left[\frac{LEDR}{2}\right]^{\frac{1}{2}} \left[\frac{Dn_0{}^2}{2LER} + t\right]^{-\frac{1}{2}} \tag{5.35}$$

where n_0 is the initial number of stress-bearing springs in series prior to the application of force.

As t increases, this expression asymptotes to

$$F(t) = At^{-\frac{1}{2}} \tag{5.36}$$

where $A = \left[\frac{LEDR}{2}\right]^{\frac{1}{2}}$ is a constant of the model. In other words, rather than resulting in exponential-type stress relaxation like conventional spring-and-dashpot models, the sequential recruitment of identical springs automatically leads to power law stress relaxation, with the exponent of the power law having a magnitude of 1/2. Interestingly, this model behaves similarly to models of polymer physics, which have themselves been invoked as useful representations of lung tissue viscoelasticity [26]. Such models are described by differential equations involving fractional derivatives [26, 27].

The question remains, however, as to whether this sequential recruitment mechanism still gives rise to power law behavior when the constitutive properties of the springs and dashpots are nonlinear. To demonstrate that this is so, let force (F) in a spring be given by the relationship

$$F = Ex^a. \tag{5.37}$$

where a is a positive constant that determines an arbitrary degree of nonlinear stress-strain behavior. With this modification, Eq. 5.31 becomes, in the limit of large n,

$$\begin{aligned}\Delta F &= E\left(\tfrac{L}{n+1}\right)^{\alpha} - E\left(\tfrac{L}{n}\right)^{\alpha} = E\left(\tfrac{L}{n}\right)^{\alpha}\left[\left(1+\tfrac{1}{n}\right)^{-\alpha} - 1\right] \\ &= E\left(\tfrac{L}{n}\right)^{\alpha}\left[\frac{-\alpha}{n} + \frac{\alpha(\alpha+1)}{2!n^2} - \cdots\right] \approx -\frac{aE}{n}\left(\frac{L}{n}\right)^{a} \\ &= -\frac{a}{LE^{\frac{1}{a}}}\left[E\left(\frac{L}{n}\right)^{a}\right]^{\frac{a+1}{a}} = -\frac{a}{LE^{\frac{1}{a}}}F^{1+\frac{1}{a}}.\end{aligned} \tag{5.38}$$

Similarly, let the force across the dashpot be

$$F = R\dot{x}^{b} \tag{5.39}$$

where b is another positive constant permitting an arbitrary degree of nonlinearity. In this case, Eq. 5.33 becomes

$$\Delta t \approx \frac{D}{\dot{x}} = DR^{1/b}F^{1/b}. \tag{5.40}$$

Equation 5.34 then becomes

$$\frac{dF}{dt} = \frac{a}{LE^{\frac{1}{a}}DR^{1/b}}F^{1+\frac{1}{a}+\frac{1}{b}}. \tag{5.41}$$

This has solution

$$F = \left[\frac{bLE^{\frac{1}{a}}DR^{1/b}}{a+b}\right]^{\frac{ab}{a+b}}\left[\frac{bL^{-\frac{a}{b}}E^{-\frac{1}{b}}DR^{\frac{1}{b}}n_0{}^{\left(1+\frac{a}{b}\right)}}{a+b} + t\right]^{-\frac{ab}{a+b}} \tag{5.42}$$

which asymptotes for large t to

$$F = At^{-\frac{ab}{a+b}}. \tag{5.43}$$

Thus, even with potentially extreme nonlinear elastic and resistive behavior of the springs and dashpots, the stress relaxation response of the model shown in Fig. 5.9 to a step increase in length is still a power law function of time, as found experimentally for lung tissue (Eq. 5.20). What is more, the exponent of the power law, $\beta = \frac{ab}{a+b}$, does not contain any reference to the size of the step change in length or the initial strain from which the step was initiated. Qualitatively, this is qualitatively like the quasi-linear viscoelastic behavior inherent in the data shown in Fig. 5.8.

The construct in Fig. 5.9 is only a toy model, of course, so it can hardly be taken as a detailed description of exactly what goes on inside a quasi-linear viscoelastic tissue during stress relaxation. It also has some significant limitations in terms of the value of the power law exponent it predicts via Eq. 5.43, which is substantially greater in magnitude than that observed experimentally [25]. To have this model predict more realistic values of the exponent, it seems necessary to invoke a multi-

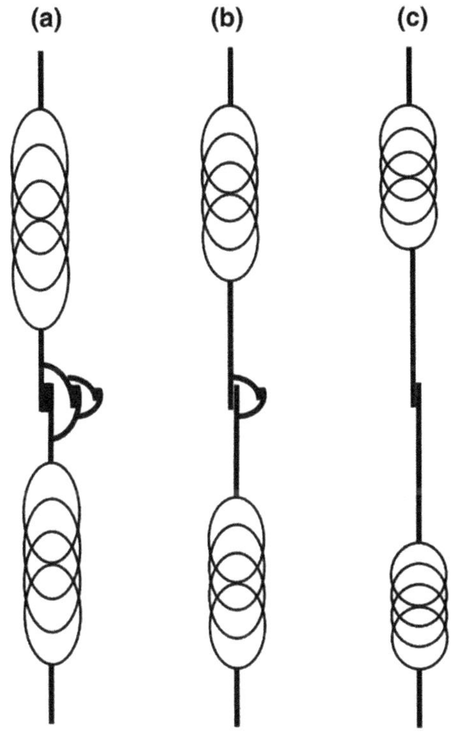

Fig. 5.10 Schematic representation of fiber slippage occurring over multiple length scales. When the first slip occurs, progressing from (**a**) to (**b**), the retraction of the fibers is halted temporarily by the tension in the previously slack but smaller secondary fibers. These secondary fibers then slip, only to be saved by the straightening of an even smaller set of tertiary fibers, and so on. The relaxed length of the entire assembly thus increases over time by ever decreasing increments. (Reproduced with permission from [25])

scale cascade of yield events. For example, a two-dimensional numerical version of this model composed of a network of fibers exhibits power law stress relaxation if overlapping fibers are allowed to suddenly slip with a certain probability once the force pulling them apart exceeds a set threshold [25]. In order to have the exponent of the power law assume a value close to that observed experimentally, however, it is necessary to invoke a multiscale process in which multiple slips between overlapping fibers are possible, where each successive slip is smaller in magnitude than the previous one (Fig. 5.10). There thus appears to be a fundamental link between tissue structure and viscoelastic function. On the other hand, this hierarchical process does not extend down to the smallest length scales since single elastin molecules are essentially purely elastic, so the viscoelastic properties of lung tissue are not a function of its individual components but rather of how these components are arranged [28].

Another limitation of the model is that it does not explain how stress recovery occurs, since the yield events in Figs. 5.9 and 5.10 would appear to be irreversible. On the other hand, if the basis of stress adaptation is slippage of overlapping fibers, then there is no reason to think this could not happen when the fibers move relative to each other in either direction. Indeed, both stress relaxation and recovery can be predicted by a model in which the fibers interact transiently with each other via cross-linkers that are broken when subjected to sufficient stress, and then reestablished with a certain probability when the fibers have moved relative to each other to a position of lesser stress [29].

Regardless of the precise details, however, the model depicted in Fig. 5.9 does suggest a mechanistic view of viscoelasticity at the micro level that is fundamentally different to the notion represented in Fig. 5.6. Specifically, rather than sliding friction being generated continuously throughout a strained tissue as stress relaxes, the model in Fig. 5.9 suggests that stress relaxes as a result of micro yield events occurring sequentially at points of high stress. That is, as soon as the stress born by one of these high stress points is released, other regions in the tissue must pick up the stress until they too yield, and so on. The precise nature of these yield events may not be important; they may be transient rupture of cross-linkers between adjacent fibers as alluded to above, or perhaps the step-by-step unravelling of intertwined fibers or molecules as occurs in polymer reptation [30]. Regardless of the details, the separability of the static and dynamic aspects of lung tissue mechanics (Fig. 5.7) may be an inevitable consequence of such a sequence of events.

Finally, while the quasi-linear viscoelastic behavior of lung tissue can be interpreted in terms of the block-structured Hammerstein model shown in Fig. 5.7, the order of the two blocks can be reversed to create what is known as a Weiner model. In this case, an input $V(t)$ is first passed through a linear dynamic system and then processed by a static nonlinearity to produce the output $P(t)$. This causes the dynamics to be inherently nonlinear, making the Weiner model more complex than the Hammerstein model. The subtle differences in behavior between the two model structures only becomes apparent, however, with inputs that are richer than a simple step change, such as white noise or sums of sine waves. Dynamic testing of tissue using large-amplitude broad-band inputs has shown that both model structures describe the measured dynamics quite accurately [31–33]. Whether the Hammerstein or Weiner model is better is unclear, however, so our understanding of the nonlinear dynamic nature of lung tissue mechanics remains incomplete.

5.5 Conclusions

The viscoelastic behavior of lung parenchymal tissue is an important aspect of overall lung function, contributing more to the dynamic mechanical behavior of the normal lung over the breathing frequency range than the effects of ventilation heterogeneity. Assemblies of springs and dashpots provide convenient and often accurate empirical models of viscoelastic behavior. The underlying mechanistic basis of viscoelasticity is nevertheless complex and still not completely understood. The phenomenon known as quasi-linear viscoelasticity, where the nonlinear static stress-strain behavior of lung tissue is separable from its linear dynamic behavior, can be explained in general terms as resulting from micro-yield events taking place throughout the tissue in a sequentially dependent manner. The exact nature of these yield events remains to be elucidated, and indeed there may be numerous varieties of such events taking place over a wide range of length scales. It does seem clear, however, that viscoelasticity is a consequence of how the components of a tissue are arranged relative to each other and how they interact, rather than their individual identities.

5.6 Problems

1. Express Eqs. 5.16 and 5.17 as an ordinary differential equation in the time domain. Explain why this equation involves fractional integration. (Hint: Illustrate $Z(f)$ in Eq. 5.16 as an Argand diagram in the complex plane.)
2. Equation 5.20 gives the response in force to a step change in length. Why is this expression problematic as the step response for a real system?
3. Equation 5.27 expresses the relationship between input and output for a Hammerstein model (Fig. 5.7). Write down the equivalent equation for a Wiener model in which the order of the blocks is reversed (i.e., dynamic linearity followed by static nonlinearity). Suppose the nonlinear function in both cases is a second-order polynomial function $g(\cdot) = a_1(\cdot) + a_2(\cdot)^2$. Use this function in the equations for both the Hammerstein and Wiener models and compare. Is it possible to model the data shown in Fig. 5.8b using the Wiener model formula?
4. The micro-yield model of quasi-linear viscoelasticity (Fig. 5.9) introduces nonlinear behavior into its springs and dashpots through Eqs. 5.37 and 5.39, respectively. The values of α in Eq. 5.37 obviously must be greater than 1 if the model is to exhibit strain stiffening as seen in the data in Fig. 5.8a. Assuming $\alpha = 2$, what would the value of β in Eq. 5.39 have to be to predict rates of stress-relaxation similar to those seen in Fig. 5.8b? Is there any way that this value of β could make physical sense?

References

1. Mount LE. The ventilation flow-resistance and compliance of rat lungs. J Physiol. 1955;127: 157–67.
2. Hildebrandt J. Comparison of mathematical models for cat lung and viscoelastic balloon derived by Laplace transform methods from pressure-volume data. Bull Math Biophys. 1969;31:651–67.
3. Hildebrandt J. Pressure-volume data of cat lung interpreted by a plastoelastic, linear viscoelastic model. J Appl Physiol. 1970;28:365–72.
4. Fredberg JJ, Keefe DH, Glass GM, Castile RG, Frantz ID 3rd. Alveolar pressure nonhomogeneity during small-amplitude high-frequency oscillation. J Appl Physiol Respir Environ Exerc Physiol. 1984;57:788–800.
5. Fredberg JJ, Ingram RH Jr, Castile RG, Glass GM, Drazen JM. Nonhomogeneity of lung response to inhaled histamine assessed with alveolar capsules. J Appl Physiol. 1985;1985(58): 1914–22.
6. Bates JH, Ludwig MS, Sly PD, Brown K, Martin JG, Fredberg JJ. Interrupter resistance elucidated by alveolar pressure measurement in open-chest normal dogs. J Appl Physiol. 1985;1988(65):408–14.
7. Ludwig MS, Romero PV, Sly PD, Fredberg JJ, Bates JH. Interpretation of interrupter resistance after histamine-induced constriction in the dog. J Appl Physiol. 1985;1990(68):1651–6.
8. Fredberg JJ, Stamenovic D. On the imperfect elasticity of lung tissue. J Appl Physiol. 1985;1989(67):2408–19.
9. Hantos Z, Daroczy B, Suki B, Nagy S, Fredberg JJ. Input impedance and peripheral inhomogeneity of dog lungs. J Appl Physiol. 1985;1992(72):168–78.

10. Hantos Z, Suki B, Csendes T, Daroczy B. Constant phase modelling of pulmonary tissue impedance. Bull Eur Pathophysiol Respir. 1987;12:326s.
11. Wagers S, Lundblad LK, Ekman M, Irvin CG, Bates JH. The allergic mouse model of asthma: normal smooth muscle in an abnormal lung? J Appl Physiol. 1985;2004(96):2019–27.
12. Petak F, Hantos Z, Adamicza A, Asztalos T, Sly PD. Methacholine-induced bronchoconstriction in rats: effects of intravenous vs. aerosol delivery. J Appl Physiol. 1985;1997(82):1479–87.
13. Tomioka S, Bates JH, Irvin CG. Airway and tissue mechanics in a murine model of asthma: alveolar capsule vs. forced oscillations. J Appl Physiol. 1985;2002(93):263–70.
14. Ito S, Ingenito EP, Arold SP, Parameswaran H, Tgavalekos NT, Lutchen KR, Suki B. Tissue heterogeneity in the mouse lung: effects of elastase treatment. J Appl Physiol. 1985;2004(97): 204–12.
15. Bates JH, Maksym GN, Navajas D, Suki B. Lung tissue rheology and 1/f noise. Ann Biomed Eng. 1994;22:674–81.
16. Fung YC. Biomechanics. In: Mechanical properties of living tissues. New York: Springer; 1981.
17. Jorba I, Beltran G, Falcones B, Suki B, Farre R, Garcia-Aznar JM, Navajas D. Nonlinear elasticity of the lung extracellular microenvironment is regulated by macroscale tissue strain. Acta Biomater. 2019;92:265–76.
18. Bates JH, Donoso FA, Peslin R. Airway and tissue impedances of canine lungs after step volume changes. J Appl Physiol. 1985;1993(75):1460–6.
19. Zipf G. Human behavior and the principle of last effort. Cambridge, MA: Addison-Wesley; 1949.
20. Gutenberg B, Richter CF. Seismicity of the earth and associated phenomena. Princeton: Princeton University Press; 1949.
21. Piantadosi ST. Zipf's word frequency law in natural language: a critical review and future directions. Psychon Bull Rev. 2014;21:1112–30.
22. Voss RF, Clarke J. 1/f noise in music: Music from 1/f noise. J Acoust Soc Am. 1978;63:258–63.
23. Barabasi AL, Albert R. Emergence of scaling in random networks. Science. 1999;286:509–12.
24. Suki B, Barabasi AL, Hantos Z, Petak F, Stanley HE. Avalanches and power-law behaviour in lung inflation. Nature. 1994;368:615–8.
25. Bates JH, Ma B. A progressive rupture model of soft tissue stress relaxation. Ann Biomed Eng. 2013;41:1129–38.
26. Suki B, Barabasi AL, Lutchen KR. Lung tissue viscoelasticity: a mathematical framework and its molecular basis. J Appl Physiol. 1985;1994(76):2749–59.
27. Ionescu C, Lopes A, Copot D, Machado JAT, Bates JHT. The role of fractional calculus in modeling biological phenomena: a review. Commun Nonlinear Sci Numer Simul. 2017;51: 141–59.
28. Suki B. Structure and function of the extracellular matrix. A multiscale quantitative approach. London: Academic; 2022. p. 105.
29. Donovan GM, Bullimore SR, Elvin AJ, Tawhai MH, Bates JHT, Lauzon A-MS. J.: A continuous-binding cross-linker model for passive airway smooth muscle. Biophys J. 2010;99:3164–71.
30. Gennes D. Reptation of a polymer chain in the presence of fixed obstacles. J Chem Phys. 1971;55:572–9.
31. Suki B, Zhang Q, Lutchen KR. Relationship between frequency and amplitude dependence in the lung: a nonlinear block-structured modeling approach. J Appl Physiol. 1995;79:660–71.
32. Maksym GN, Bates JH. Nonparametric block-structured modeling of rat lung mechanics. Ann Biomed Eng. 1997;25:1000–8.
33. Maksym GN, Kearney RE, Bates JH. Nonparametric block-structured modeling of lung tissue strip mechanics. Ann Biomed Eng. 1998;26:242–52.

Part II
Modeling the Diseased Lung

Chapter 6
Homeostasis and Repair

Abstract Living systems are able to maintain internal order throughout life, a phenomenon known as homeostasis, by importing energy from the environment while exporting entropy. Homeostasis is achieved by the actions of cells that continually maintain their tissue environments and repair them when injured. Agent-based modeling demonstrates that homeostasis requires that the cells responsible for tissue maintenance receive feedback from the tissue concerning the results of their labors. It also demonstrates that a system capable of ongoing tissue maintenance automatically has the capacity to repair tissue injury, although temporary mechanisms that accelerate repair may be required by the demands of life. Agent-based models also provide insight into the conditions that predispose to the development of pulmonary fibrosis. Living systems can also be viewed as complex dynamic networks, raising the general question as to how such networks can fail to function normally. Three possibilities are identified—structural damage, a persistent external insult, and the existence of multiple stable network states (attractors). A Hopfield net model illustrates the conditions under which multiple stable states may exist when only one of these states corresponds to normality. Mixture theory allows tissue remodeling to be simulated over continuous time and space.

Keywords Homeostasis · Tissue maintenance · Injury repair · Agent-based model · Pulmonary fibrosis · Hopfield net · Multiple stable states · Tissue remodeling · Mixture theory

6.1 Introduction

Despite the inexorable tendency for disorder in any system to increase over time, the form and function of the human body remain largely intact over the entire lifespan despite ongoing abuse from the environment. This is not a violation of the *second law of thermodynamics*—that law is sacrosanct. Staying healthy from birth to old age is a consequence of the fact that living systems are *thermodynamically open*, meaning they import *energy* from their surroundings while exporting *entropy* in the form of heat and waste products [1–4]. The energy imported by a living system drives biological mechanisms that maintain internal order in the face of forces that

B. Suki, J. H. T. Bates, *Mathematical Modeling of the Healthy and Diseased Lung*,
https://doi.org/10.1007/978-3-031-95322-4_6

work to disrupt it. The "normal" internal order represents a set-point to which a healthy system always tries to return when its state is perturbed, a situation known as *homeostasis* [5]. Homeostatic mechanisms operate continually to deal with the rigors of daily living as well as occasionally repairing tissues that suffer injury.

Self-repair of injured tissues requires a mechanism for detecting injury and initiating appropriate reparative processes. Equally important, however, is the need to terminate those processes when repair is complete; failure to do so leads to chronic diseases associated with aberrant tissue remodeling such as pulmonary fibrosis (Chap. 8). The initiation and termination of self-repair arise from the ensemble behavior of the various cell types that take part in self-repair along with the signals they receive from the extracellular matrix (ECM). Failure of one or more homeostatic mechanisms underlies most chronic diseases, including chronic lung diseases such as asthma, emphysema and pulmonary fibrosis. Understanding the pathophysiology of chronic lung disease thus begins with the question of how homeostasis is maintained in the normal lung, followed by a consideration of the various ways in which it can fail. These are questions that require the analysis of complex nonlinear dynamic systems and are thus dependent on the use of mathematical and computational modeling.

6.2 Homeostatic Tissue Maintenance

Living tissues are able to maintain internal order and to restore that order after injury without direction from any sentient controller. This appears to depend on an automatic program of continual tissue replacement. For example, epithelial layers such as skin are continually being sloughed off from the surface of the body by daily wear and tear while being replaced from below [6], and bone is continually being broken down by osteoclasts and rebuilt by osteoblasts [7]. A similar program of continual tissue maintenance is also an important component of lung tissue health as evidenced by the fact that the mechanical stiffness of the lung parenchyma is maintained within normal limits throughout most of life [8, 9]. The homeostatic maintenance of lung tissue in its normal healthy state is thus a matter of having an appropriate balance between the removal of existing tissue and its replacement with new material. When these processes become out of balance, the homeostatic setpoint shifts from its normal position toward one corresponding to pathology. For example, there is strong evidence that emphysema can develop as a result of an imbalance between the activities of proteases that effect tissue breakdown and the anti-proteases that regulate it [10, 11]. A corresponding imbalance in the opposite direction likely plays a key role in pulmonary fibrosis [12].

The primary effectors of tissue maintenance are the cells that manufacture the extracellular matrix and the cells that manufacture chemicals that degrade the matrix [13]. These different cell types can be viewed as autonomous agents comprising a cooperating community engaged in a common task that, when working properly, benefits the host. The behaviors of the cells are governed by sets of evolutionarily codified rules specifying what the various cell types do when they encounter various situations. Exactly how these rules are executed is determined by very complex intracellular processes that must be understood if cellular behavior is to be modified, for example, by pharmaceutical intervention. The essential nature of how cells maintain the lung parenchyma, however, can be understood at the tissue level by focusing on *what* cells do rather than on *how* they do it.

The natural vehicle for investigating questions about collective cell behavior computationally is the *agent-based model* (ABM). An ABM represents the world as collections of autonomous agents such as cells, viruses or even molecules operating within an environment that can be as detailed and tortuous as the model builder desires. The agents themselves behave according to logic-based rules that take the nature of the environment and the presence of other agents into account. This allows reality to be modeled more or less exactly as it is perceived, without the constraints of smoothness and differentiability imposed by models based on differential equations. ABMs are also much more conceptually accessible to non-mathematical scientists than models implemented via mathematical equations. Indeed, the majority of the intellectual content in an ABM is embodied in the rules of agent behavior and the details of the environment, things that are the purview of the biologist more than the esoteric mathematician or computer scientist. Of course, actually implementing an ABM requires coding skills. This is becoming progressively less of a barrier, however, due to the development of user-friendly development platforms such as NetLogo (NetLogo Home Page (northwestern.edu)).

An ABM capable of maintaining tissue against the forces that work to degrade it must include agents that continually break down and remove damaged tissue components, as well as agents that continually add new components. Over the long term, this will keep the tissue in near pristine condition in the face of ongoing minor degradation resulting from daily wear and tear. It will also return the tissue to pristine condition following more major damage caused by injury, albeit over a longer time scale. In other words, a system capable of self-maintenance is also capable of self-repair. Figure 6.1 illustrates a simple ABM capable of achieving this [14]. The tissue itself is represented by a square $N \times N$ sheet comprised of individual patches. Equal numbers of catabolic and anabolic agents roam about randomly from patch to patch (the system is closed by a circular boundary such that if any agent moves off one of the four edges of the tissue sheet, it immediately appears on the corresponding patch at the opposite edge).

Each patch in the model is comprised of a mass $M_{m,\,n}$ ($m, n = 1, \ldots, N$) of tissue. The simplest set of agent rules for ongoing tissue maintenance are that, at each time step, each catabolic agent removes a fixed small amount δM_{catab} of tissue from the patch it is sitting on, while each anabolic agent adds a fixed amount δM_{anab} to its patch. The tissue is thus continually being degraded and replaced. However, homeostasis in this scenario requires the total rate of tissue removal to be exactly equal to the total rate at which tissue is added. The slightest imbalance in these rates leads to either progressive atrophy or hypertrophy. Even when anabolism and catabolism are in overall perfect balance, however, this open-loop scheme leads to progressively

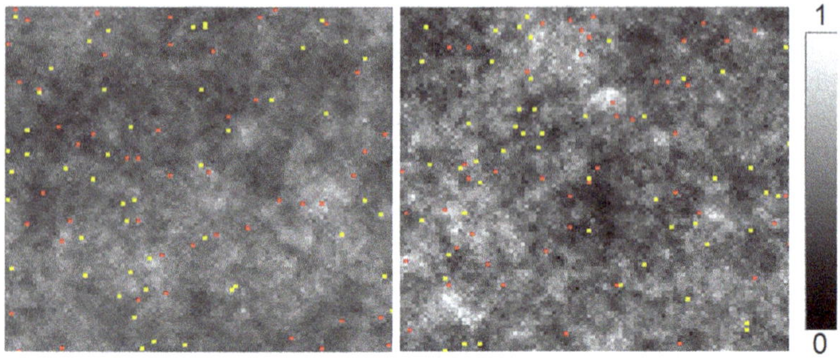

Fig. 6.1 A sheet of tissue comprised of 100 x 100 patches on which 50 catabolic and 50 anabolic cells move about randomly with $\beta = 0.5$ and two different values of α (0.01 on the left and 0.02 on the right). Catabolic cells (red) digest a fixed fraction α of the tissue on each patch at each time step, while anabolic cells (yellow) lay down a fixed amount $\alpha\beta$ of tissue on each patch. White corresponds to a tissue density of 1, while black corresponds to a tissue density of 0. The positions of the anabolic and catabolic cells shown correspond to the end of the simulations continued until the variance of $M_{i,\,j}$ reached steady state. (Adapted with permission from [6])

increasing tissue heterogeneity because the mass of tissue on any given patch changes over time in a similar same way to the distance that a random walker finds itself from its starting point, which has a variance that increases linearly with time.

In order to keep tissue heterogeneity bounded, agent behavior must be influenced in some way by feedback about the current state of the tissue. A simple way to achieve this is to have each catabolic agent remove, from the patch it is currently on, an amount of tissue that is proportional to the total amount already on that patch. At the same time, each anabolic cell continues to operate in open-loop fashion as before. That is, $\delta M_{catab} = \alpha M_{m,\,n}$ and $\delta M_{anab} = \alpha\beta$, where α and β are constants. Figure 6.2 shows that the steady-state behavior of the model with this simple form of feedback is indeed bounded. What is also clear from Fig. 6.2, however, is that the degree of spatial heterogeneity across the tissue at steady state increases with α. In other words, ongoing maintenance by this mechanism never leaves the tissue in its original pristine state.

To see how this happens, consider a single patch of tissue that has probabilities of ν_j and $(1 - \nu_j)$ of being occupied by an anabolic or a catabolic agent, respectively, at each time step j, where $0 \leq \nu_j \leq 1$. The change in tissue mass (representing a discrete conservation of mass) on the patch is then

$$M_{j+1} - M_j = \alpha\beta\nu_j - \alpha M_j\left(1 - \nu_j\right) = \delta_j \tag{6.1}$$

where ν_j is an uncorrelated random variable that takes a value of 0 with probability $0 < q < 1$ and a value of 1 with probability $1 - q$. The expected value of ν_j is

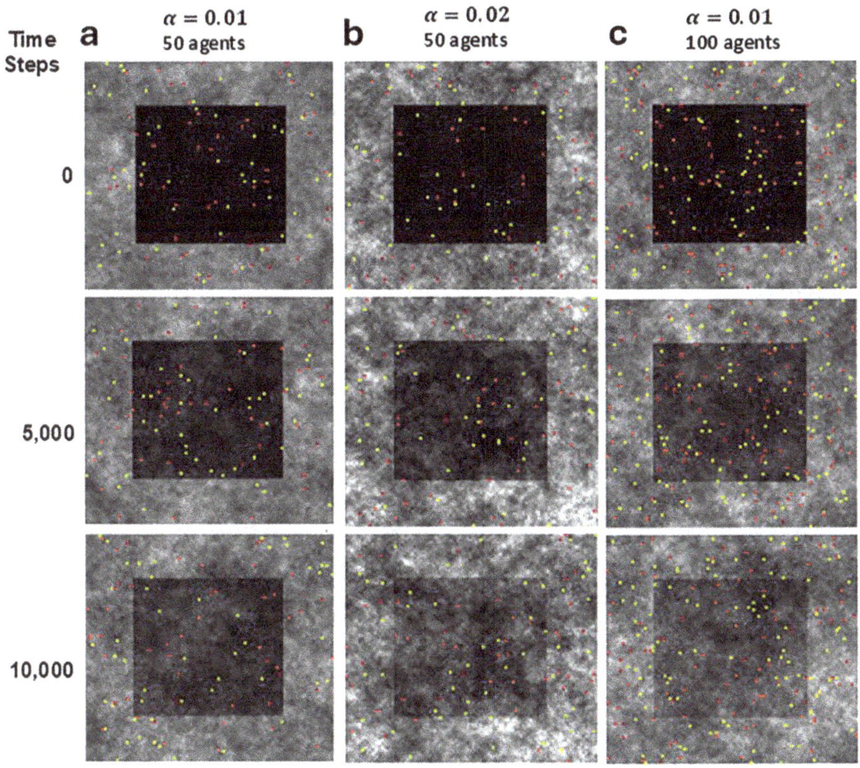

Fig. 6.2 A sheet of tissue comprised of 100 × 100 patches on which 50 catabolic and 50 anabolic cells move about randomly, with $\beta = 0.5$ and two different values of α (0.01 on the left and 0.02 on the right). Catabolic cells (red) digest a fixed fraction α of the tissue on each patch at each time step, while anabolic cells (yellow) lay down a fixed amount $\alpha\beta$ of tissue on each patch. White corresponds to a tissue density of 1, while black corresponds to a tissue density of 0. The positions of the anabolic and catabolic cells shown correspond to the end of the simulations. (Adapted with permission from [14])

$$\mathbb{E}\{\nu_j\} = 0 \times q + 1 \times (1 - q) = 1 - q \tag{6.2}$$

where $\mathbb{E}\{\cdot\}$ denotes the statistical expectation operator. The right-hand side of Eq. 6.2 also applies to the expected value of ν_j^2, so $\mathbb{E}\{\nu_j^2\} = 1 - q$.

Because ν_j and M_j are uncorrelated, the expectation of their product is the product of their expectations, so the expectation of Eq. 6.1 is

$$\mathbb{E}\{\delta_j\} = \mathbb{E}\{\alpha\beta\nu_j - \alpha M_j(1 - \nu_j)\} = \alpha\beta(1 - q) - \alpha q \mathbb{E}\{M_j\}. \tag{6.3}$$

If the sequence $\{M_j\}$ converges on average toward a fixed point M^* for large j then, in the limit as j tends to infinity, $\mathbb{E}\{\delta_j\} = 0$, in which case Eq. 6.3 gives

$$\mathbb{E}\{M_j\} = M^* = \frac{\beta(1-q)}{q}. \tag{6.4}$$

The steady state variance of $\{M_j\}$ is found by first noting that

$$\mathbb{E}\{M_{j+1}{}^2\} = \mathbb{E}\left\{(M_j+\delta_j)^2\right\} = \mathbb{E}\{M_j{}^2\} + 2\mathbb{E}\{M_j\delta_j\} + \mathbb{E}\{\delta_j{}^2\}. \tag{6.5}$$

The second and third terms on the right-hand side of Eq. 6.5 can be expressed respectively as

$$2\mathbb{E}\{M_j\delta_j\} = 2\mathbb{E}\{M_j(\alpha\beta\nu_j - \alpha M_j + \alpha M_j\nu_j)\} = 2\alpha\left[\frac{\beta^2(1-q)^2}{q} - q\mathbb{E}\{M_j{}^2\}\right] \tag{6.6}$$

and

$$\mathbb{E}\{\delta_j{}^2\} = \mathbb{E}\left\{\alpha^2(\beta\nu_j - M_j + M_j\nu_j)^2\right\} = \alpha^2\left[\beta^2(1-q) + q\mathbb{E}\{M_j{}^2\}\right]. \tag{6.7}$$

Substituting Eqs. 6.6 and 6.7 into Eq. 6.5 gives

$$\mathbb{E}\{M_{j+1}{}^2\} = \mathbb{E}\{M_j{}^2\} + \frac{2\alpha\beta^2(1-q)^2 + \alpha^2\beta^2(1-q)q}{q} + q\alpha(\alpha-2)\mathbb{E}\{M_j{}^2\}. \tag{6.8}$$

Because $\mathbb{E}\{M_{j+1}{}^2\} = \mathbb{E}\{M_j{}^2\}$ in the limit as j tends to infinity, Eq. 6.8 reduces to

$$\mathbb{E}\{M_j{}^2\} = \frac{2\beta^2(1-q)^2 + \alpha\beta^2(1-q)q}{q^2(2-\alpha)}. \tag{6.9}$$

The steady-state variance of M_j over time, σ^2, is given by the difference between Eq. 6.9 and the square of Eq. 6.4, which is

$$\sigma^2 = \frac{2\beta^2(1-q)^2 + \alpha\beta^2(1-q)q}{q^2(2-\alpha)} - \left[\frac{\beta(1-q)}{q}\right]^2 = \frac{\alpha\beta^2(1-q)}{q^2(2-\alpha)}. \tag{6.10}$$

(Eq. 6.10 has a singularity at $\alpha = 2$ but this is moot because α is a fraction <1.) Finally, the variance of M_j over time is the same as its variance over space because the patch dynamics are independent of each other, so Eq. 6.10 shows that the steady-state spatial heterogeneity of the tissue depends on both α and β.

Another way of getting feedback to the agents of the ABM so that they modulate their behavior to achieve a bounded degree of tissue heterogeneity is to have the catabolic agents move preferentially toward regions of either low or high density [14], a process that mimics *durotaxis* [15].

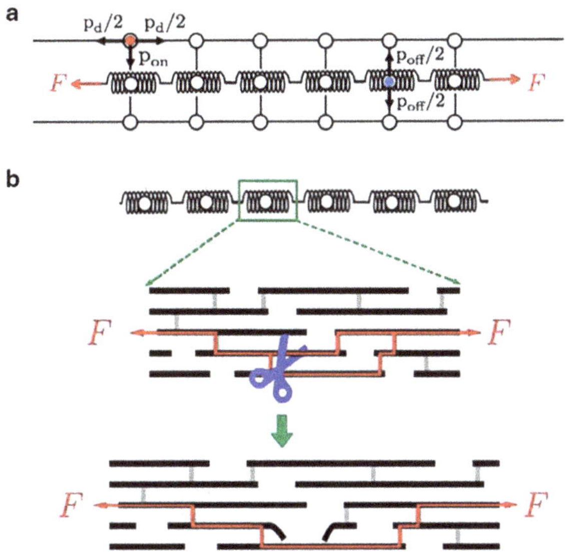

Fig. 6.3 (**a**) The ECM is represented by a chain of springs and binding sites (middle layer of open circles). The two outer layers of circles allow the agents to diffuse along the fiber. Red and blue filled circles corresponding to catabolic and anabolic agents, respectively, that move randomly from site to site. When an agent binds to a spring it either adds collagen (blue anabolic) or removes it (red catabolic). The chain of springs is under constant tension F. (**b**) A zoom into one of the springs comprised of parallel collagen monomers (black) reinforced by cross-links (gray). Force transmission pathways are shown in red. The blue scissors represent a proteolytic enzyme cleaving a collagen monomer. Once the monomer is cleaved, part of the monomer still participates in force transmission via the cross-links. (Adapted with permission from [16])

Having a catabolic agent remove an amount of tissue in proportion to that already present clearly requires some kind of biologic explanation that is not contingent on the presence of intelligent purpose. One possibility is that each cell automatically secretes a fixed amount of proteolytic enzyme, but its effect is dependent on the amount of tissue accessible to the enzyme, which in turn is proportional to the amount of tissue present [8]. A more detailed account of how this might happen in the extracellular matrix (ECM) is depicted in Fig. 6.3 [16]. In this model, agents represent enzymes that walk randomly around collagen fibers. The binding affinity of the anabolic agents depend on the stiffness of the local springs that determine the mechanical properties of the tissue.

The capacity of a self-maintaining tissue to recover from injury is illustrated in Fig. 6.2, which shows the agent-based model at the time of initial injury and then at 5000 and 10,000 time-steps later for two different healing rates (governed by the value of the parameter α). This demonstrates that a fixed program of ongoing breakdown and repair allows a tissue to repair itself from injury so long as some form of feedback about the state of the tissue is provided to the cells involved. It does not guarantee, however, that the rate of healing will be rapid enough to allow an organism to deal successfully with all the other challenges it must meet to remain alive.

One way to increase the rate of healing is to increase the value of α, as demonstrated in Figs. 6.2a, b, but this occurs at the expense of a greater degree of regional tissue heterogeneity that may not be acceptable. Nature has therefore evolved various strategies to increase the healing rate without increased heterogeneity. One is simply to increase the number of agents by attracting more effector cells to the region of tissue requiring repair, as illustrated in Fig. 6.2c. Another strategy is to temporarily increase the activation level of any agent that finds itself on a damaged tissue patch and then return the activation level to baseline once repair is complete. This is illustrated in Fig. 6.2 (where damage is represented by $M_{i,j} < 0.2$) by having activated anabolic agents add an increased amount $\alpha\beta\gamma$ of tissue to each patch at each time-step, instead of the original amount $\alpha\beta$, where $\gamma > 1$. This strategy must be terminated, of course, once repair is at or near completion or else something akin to fibrotic remodeling will develop. In the ABM, termination is achieved by giving each activated agent a very small but fixed probability of becoming deactivated at each time step, so the expected lifetimes of the agents are exponentially distributed with half-life T. This corresponds to apoptosis of activated cells followed by their replacement with normal unactivated cells.

Mechanisms that accelerate the rate of healing can also lead to temporary deposition of excess tissue. This is illustrated by the ABM in Fig. 6.4 which shows that while temporarily increasing α and T can result in substantial increases in the rate of healing, it can also lead to an overshoot in the degree of repair characterized by excess tissue deposition both within and around the site of injury. Until such excess tissue is eventually cleared, it manifests as a scar [17] (Fig. 6.5). While such evolutionary tricks bring clear survival advantages to the endowed

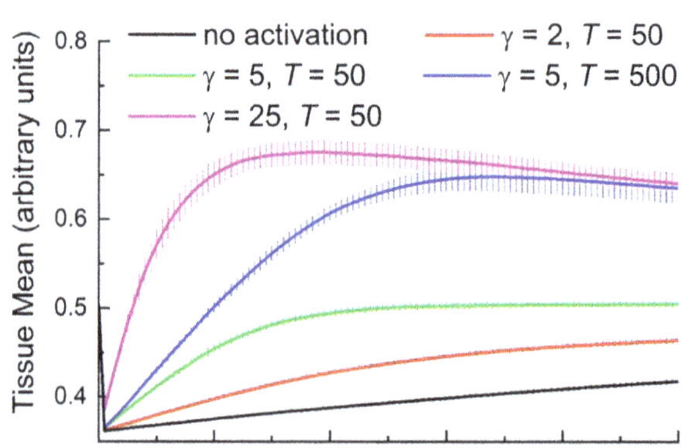

Fig. 6.4 Effects of injury-induced activation of anabolic agents on rate of recovery from injury. The parameter α controls the amount of tissue produced or removed by each agent at each time step and is temporarily increased by the factor γ. Activated cells persist for an average half-life of T time-steps. The lines show the mean of 20 stochastically independent simulations. The error bars show 1 standard deviation. (Adapted with permission from [14])

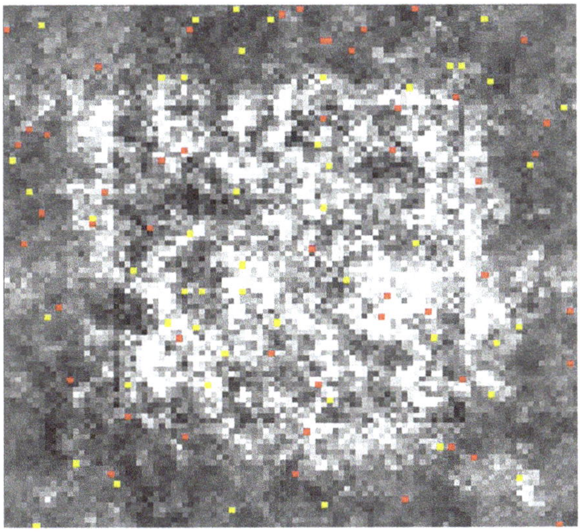

Fig. 6.5 Collateral fibrosis of self-repair. The model tissue after 10,000 time-steps following injury with $\gamma = 20$ and $T = 50$ time steps. This caused the addition of tissue to overshoot the homeostatic set point. (Adapted with permission from [14])

organism, they also bring the specter of disease. For example, dysregulation of the collagen deposition that occurs when anabolic cells become activated can result in progressive fibrotic disease [18, 19]. Fibrosis can also occur as a result of impaired apoptosis of anabolic cells [20].

6.3 Self-Repair

It is thus crucial for the health of an organism that the active processes involved in self-repair eventually return to quiescence once repair is complete. An agent-based model in which agent behavior is a continuous analytic function of tissue properties and time provides insight into the mathematical conditions that lead to self-limited self-repair [21]. The model considers a 2-dimensional sheet of tissue on which sits an agent representing a pro-fibrotic cell. The agent is characterized by a single state variable, a', representing its level of pro-fibrotic activity. (Enzymatic softening, as occurs in emphysema, could also be represented by a model like this.) The local patch of tissue under the agent is characterized by another state variable, k', representing its mechanical stiffness. Both a' and k' are functions of time, t'. Since the mechanical properties of tissue are known to influence the anabolic activities of pro-fibrotic cells [22], the dynamics of the cell and its tissue substrate can be represented by a pair of coupled ordinary differential equations,

$$\frac{da'}{dt'} = f(a', k') \tag{6.11}$$

and

$$\frac{dk'}{dt'} = g(a', k') \tag{6.12}$$

where f and g are continuously differentiable functions.

If the tissue stiffness is at baseline and is not being altered by the agent, the agent activation level decays toward a steady-state baseline level of β according to

$$\frac{da'}{dt'} = \alpha\beta - \alpha a' \tag{6.13}$$

so that

$$a'(t') = \beta + We^{-\alpha t'} \tag{6.14}$$

where W is a constant determined by the initial conditions.

In general, the activation state of an agent is affected by tissue stiffness such that β increases with k'. Activation cannot increase indefinitely, of course, because it requires energy which is in limited supply. Tissue stiffness, on the other hand, being a passive property is less constrained in this regard. This situation can be represented by having the dependence of activation on stiffness exhibit first-order saturation behavior according to

$$\frac{da'}{dt'} = \alpha\frac{Ak'}{Q + k'} - \alpha a' \tag{6.15}$$

where A is the maximum possible value of a', and Q is the value of k'at which a' is half its maximum value (i.e., $A/2$). The value of a' thus depends linearly on k' when k' is small but asymptotes to A as k' becomes large.

The drive toward homeostasis arises when Eq. 6.12 specifies that enzymatic digestion of excess collagen occurs if tissue stiffness is too large, while collagen deposition occurs if stiffness is too low. These conditions can be achieved by having k' decrease or increase when a' is, respectively, above or below a threshold q'. A simple way to specify this is to write $g(a', k') = (a' - q')h(k')$ where $h(k')$ is a monotonic function of k'. For example, if $h(k') = -D'k'$, where D' is a constant, then

$$\frac{dk'}{dt'} = (a' - q')(-D'k') + P' \tag{6.16}$$

where the constant P' is a production term that serves to prevent (0,0) from being a fixed point (i.e., a solution of the equation when the left-hand size is zero) of the system.

By introducing the dimensionless variables $t = \alpha t'$, $a = \frac{a'}{A}$, and $k = \frac{k'}{Q}$, and by normalizing the digestion rate $D = \frac{AD'}{\alpha}$, the production rate $P = \frac{P'}{\alpha Q}$, and the activity threshold $q = \frac{q'}{A}$, Eqs. 6.15 and 6.16 become

$$\frac{da}{dt} = \frac{k}{1+k} - a \tag{6.17}$$

and

$$\frac{dk}{dt} = (a-q)(-Dk) + P. \tag{6.18}$$

The fixed points of the system defined by Eqs. 6.17 and 6.18 correspond to the condition $\frac{da}{dt} = \frac{dk}{dt} = 0$. From Eq. 6.17, this gives $a = \frac{k}{1+k}$, which, when substituted into Eq. 6.18, gives a quadratic equation for k having the two solutions

$$k_{1,2} = \frac{qD + P \pm \sqrt{(qD+P)^2 + 4D(1-q)P}}{2D(1-q)}. \tag{6.19}$$

The model parameters q, D and P are all positive, and $q < 1$ (since $0 < a < 1$), so $qD + P < \sqrt{(qD+P)^2 + 4D(1-q)P}$. This means that the only physically realistic (i.e., positive) stiffness solution for the fixed point is

$$k^* = \frac{qD + P + \sqrt{(qD+P)^2 + 4D(1-q)P}}{2D(1-q)} \tag{6.20}$$

with corresponding activity

$$a^* = \frac{k^*}{1+k^*}. \tag{6.21}$$

The stability of this fixed point is determined by the Jacobian $J = \begin{pmatrix} \frac{df}{da} & \frac{df}{dk} \\ \frac{dg}{dk} & \frac{dg}{da} \end{pmatrix} = \begin{pmatrix} 1 & \frac{1}{(1+k)^2} \\ -Dk & -D(a-q) \end{pmatrix}$, which has trace $\tau = \; -1 \; -D(a-q)$ and determinant $\Delta = D(a-q) + \frac{Dk}{(1+k)^2}$. The system is stable at the fixed point if $\Delta > 0$ and $\tau < 0$ [23], which gives the condition

$$a^* > q - \frac{1}{D} \tag{6.22}$$

and

$$k^* > \frac{qD-1}{D-(qD-1)}. \tag{6.23}$$

The above analytic model can be implemented on a computational model of soft tissue consisting of a hexagonal lattice of identical pre-stressed Hookean springs with fixed external boundaries [21]. At baseline (i.e., healthy), each spring has spring-constant k^*. Injury was simulated by setting the spring-constant of a single central spring to $2k^*$. Self-healing of the network is simulated by placing a number of agents at random spring junctions and then allowing them to migrate to an adjacent junction at each time step. Agent migration can be either random or biased by local tissue stiffness such that the probability of an agent moving to a given adjacent spring junction node is equal to the stiffness of the intervening spring as a fraction of the sum of the stiffnesses of all three springs impinging on the current junction. The initial activities of all agents are set equal to a^*. At each subsequent time step, the activity level of every agent and the stiffness of any spring that an agent had just passed across are simultaneously updated according to Eqs. 6.17 and 6.18, respectively. The elastic equilibrium configuration of the network is computed by finding the positions of the spring junctions that minimized the total strain energy of the system using simulated annealing [24].

Figure 6.6 illustrates that fibrotic injury resolves in this model. A total of 64 agents were allowed to migrate over the hexagonal spring network, which consisted of 252 nodes and 345 springs (time-step 0 in Fig. 6.6). At each subsequent time step, each agent moved randomly to an adjacent node with a bias toward stepping over stiffer rather than softer springs. By time-step 4 in Fig. 6.6, two agents have already visited the injury site where their activation levels become elevated. They then move to neighboring sites and so raise the stiffness of the springs that they step over in the process. As this process continues, a region of stiffer springs forms around the original injury site (time step 8). The stiff region grows initially (time step 32), similar to the development of local fibrosis that often occurs around an injury site as

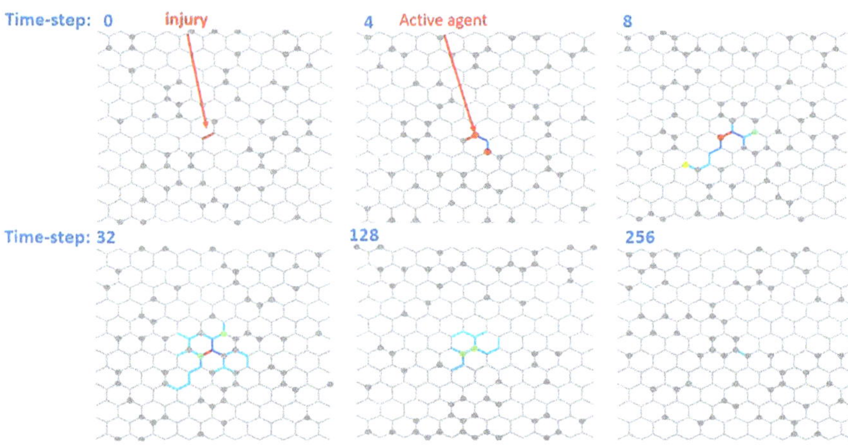

Fig. 6.6 Resolution of injury modeled as a single spring with a high stiffness (red line element) at time-step 0 (indicated in blue). As unactive agents (gray circles) visit the injury site, they become active (red circles at time-step 4) and then proceed to stiffen neighboring sites, creating a halo (shown as shades of blue and green at time-step 8). Subsequently, the stiffened region first increases (time-step 32), then gradually shrinks (time-step 128) so that by time-step 256 the tissue has returned almost entirely to normal. (Adapted with permission from [21])

it is in the process of healing. Eventually, however, the increased stiffness resolves back to baseline. This demonstrates that the reparative response predicted by this model is temporary and spatially limited provided that the conditions 6.22 and 6.23 are met.

6.4 Health and Disease as Network Phenomena

Life is a dynamic process involving an unimaginably large number of components interacting via feedback loops that are both positive (excitatory) and negative (inhibitory). Normally, these interactions keep the state space of the system confined within an appropriately bounded region corresponding to homeostasis. Transient departures from this bounded region are to be expected as a result of sufficiently noxious stimuli, but a healthy system capable of recovery will eventually return itself to the homeostatic region. Persistent departures from the homeostatic region, either instigated by some external stimulus or for unknown reasons, correspond to chronic disease. ABMs such as those discussed above recapitulate these ideas using collections of agents that interact with each other either directly when they come into contact, or indirectly via their effects on the environment. The agents in these models can thus be viewed as nodes in a network, while their modes of interaction correspond to the network links (edges). Such a network of agents is also dynamic because its state, defined by the positions and activities of each of its nodes, is in a constant state of flux as it either explores the region of state space corresponding to healthy homeostasis, or ventures outside this region as a result of injury or disease.

This raises two general questions about the nature of health and disease. First, what are the conditions under which a dynamic network returns to the homeostatic region of its state space when perturbed. Second, what can cause an initially healthy network to lose this ability to return to homeostasis? The answer to the first question amounts to identifying the conditions under which a dynamic network has an attractor state. This is often not hard to do. Systems of coupled ordinary differential equations, for example, are often used to model biological systems in terms of interconnected compartments, and typically converge toward a fixed point when integrated over time.

There appear to be, broadly speaking, three distinct answers to the second question above. The most obvious answer is that the structure of the network becomes altered in some way, that is, its nodes and/or their links become damaged. Genetic defects, for example, correspond to components of a living network being malformed from birth, and if these defects are critical to normal function, then function will be abnormal. Perhaps the classic pulmonary example of this is cystic fibrosis where a genetically malfunctioning chloride channel results in abnormalities in mucus rheology and epithelial ionic milieu, leading to chronic and eventually lethal lung infections [25]. Recent advances in gain-of-function drugs have greatly improved the outlook for many cystic fibrosis patients [26], but actual cure requires that the defective gene be replaced by a normal one. Such gene therapies are still out of reach, but when they arrive, they will represent the structural correction of an abnormal network. Alpha-1-antitrypsin deficiency is another example of a genetically determined network structural defect, in this case leading to accelerating age-related onset of emphysema and its associated decline in lung function (Chap. 9). Exogenous replacement of alpha-1-antitrypsin can slow the progress of disease [27], essentially acting to replace a missing node in the biologic network. Idiopathic pulmonary fibrosis may turn out to be yet another example, given its familial characteristics [28], but much about the pathophysiology of this insidious condition remains unclear (Chap. 8). Altered network structure can also be caused by externally applied insults, such as the pulmonary fibrosis that results from administration of bleomycin to treat cancer [29]. Finally, network breakdown also occurs gradually with increasing age (Chap. 12).

A second way in which dynamic network behavior can become aberrant is through the presence of a persistent insult, that is, a chronic environmental stimulus that pushes the state of the network away from homeostasis so long as the stimulus is around. This is the proximate cause of chronic lung diseases such as silicosis and asbestosis. Here, fine particles of silica or asbestos, respectively, are inhaled into the lungs where they take up residence in the tissues. Phagocytic macrophages recognize these particles as foreign and ingest them [30, 31]. However, silica and asbestos particles are immune to the intracellular chemicals that normally work well on viruses and bacteria, so the phagocytes eventually become overloaded with material they cannot process. This kills the cells, causing them to release their loads, and the cycle begins again. The caustic chemicals released by the macrophages in the process lead to the progressive development of fibrosis. The silica and asbestos particles, once inhaled into the lungs, thus act as permanent activators of the immune system whose defensive actions cause progressive scarring of the lung. Indeed, adding a persistent activation to the ABM-network model in Fig. 6.6 leads to scarring and progressive fibrosis [21].

A third possible way of chronically altering network behavior arises because nonlinear dynamic networks have the potential to have more than one stable attractor, or fixed point. If the main attractor lies at the center of the homeostatic region of state space, the network will remain under its pull while it is healthy. On the other hand, an environmental stimulus that pushes the network into the clutches of another attractor could cause it to converge toward an aberrant stable state corresponding to disease. This presents the intriguing possibility that chronic disease might occur in a network that is structurally perfectly normal, and that the insult that caused the transition to the abnormal attractor could be long gone by the time the disease is diagnosed even though the system remains stuck in the abnormal attractor. The resulting pathology would then not be traceable to any apparent cause or abnormality, making it truly idiopathic. Multiple attractors have been described for gene regulatory networks [32, 33] and immunological networks [34]. Whether their existence explains the pathology of any of the idiopathic systemic diseases is still something of an open question, although there is compelling evidence that it is behind some forms of cardiac arrhythmia [35].

The archetypical nonlinear dynamic network is the Hopfield net consisting of a set of H nodes, each being an artificial neuron. Each of the nodes communicates with itself and all the other nodes over links having assigned weights, $w_{i,\,j}$ corresponding to synaptic strengths in actual neural networks [36]. The Hopfield net is also symmetric, meaning $w_{i,\,j} = w_{j,\,i}$. The state of the network at time point j is the set $\{\psi_{i,\,k}\}$ of values, $\psi_{i,k}$, of each of the H nodes. The network is iterated from time step k to time step $k + 1$ according to

$$\psi_{i,k+1} = f\left\{\sum_{j=1}^{H} w_{i,j}\psi_{j,k}\right\}; i \in [1, H] \tag{6.24}$$

where $f\{\cdot\}$ is a step function that proceeds from -1 to 1 when x reaches a specified threshold value, ψ_t. Specifically,

$$\begin{aligned} \psi_{i,k+1} &= -1; \sum_{j=1}^{H} w_{i,j}\psi_{j,k} < \psi_t \\ &= 1; \sum_{j=1}^{H} w_{i,j}\psi_{j,k} \geq \psi_t. \end{aligned} \tag{6.25}$$

Equation 6.25 is an idealized representation of how a real neuron sends a signal down its axon to other neurons when the summed input on all its dendrites exceeds an activation threshold.

It can be shown that the Hopfield net will always converge toward a stable state $\{\psi_i\}_0$ when initiated from an arbitrary starting state $\{\psi_{i,\,0}\}$ [37]. Furthermore, in general, the net has multiple stable states $\{\psi_i\}_m$; $m = 1, \ldots, M$, where the maximum value of M increases with H. The Hopfield net can have a predesignated set of $\{\psi_i\}$ through appropriate choice of the $w_{i,\,j}$. One of the key uses of the Hopfield net arises from its ability to exhibit associative memory. That is, it will converge to the $\{\psi_i\}_m$ that is closest to $\{\psi_{i,\,0}\}$, thereby associating its initial state with the nearest attractor. The Hopfield net exhibits this remarkable ability despite it being essentially nothing more than a network of nonlinear summing junctions connected by links of varying strength, where the nonlinear activation function, $f(\cdot)$, associated with each node is a step function.

In order for the Hopfield net to exhibit multiple attractors, the precise form of $f(\cdot)$ is not particularly important so long as it exhibits a strong saturation-type nonlinearity. For example, a nonlinearity of the form exhibited by the Michaelis-Menten equation

$$f(x) = \frac{x}{x+1} \tag{6.26}$$

works perfectly well in this regard [38], where x is any independent variable. The significance of this is that many biochemical and enzymatic reactions in the body occur as the result of random associations between a large number of ligands and a fixed number of receptors, the kinetics of which are governed by an equation of the form of Eq. 6.26. Reactions of this sort frequently involve large numbers of players, such as those taking place within in the second messenger system in cells, or among different cell types communicating via cytokines, or between organs communicating via hormones, essentially forming a multiscale communication network. It therefore seems entirely plausible that dynamic networks within living systems might exhibit multiple attractors. If all the attractors represent beneficial modes of biological behavior, then their existence could be viewed as advantageous, perhaps allowing an organism to adapt its behavior to changes in its surrounding conditions. On the other hand, if some of the attractors are not advantageous then the possibility arises that an organism might become trapped in an unhealthy stable state.

It is not clear if any of the common chronic diseases of the lung (or, indeed, any other organ) are examples of this kind of dynamic entrapment. If diseases of this nature do not exist, then it begs an explanation for why not, given that all the necessary machinery seems to be in place for there to be multiple attractors. One possibility is that biological networks have evolved structures that do not permit the existence of more than a single attractor. The multiple attractors of the Hopfield net owe their profligacy to the fact that the network is fully connected—every neuron talks to every other neuron. Biological networks, on the other hand, are typically rather sparsely connected [39]. Interestingly, if the synaptic connections in a Hopfield net are pruned sufficiently, a range of connectivity can be found over which only a single attractor can be supported (Fig. 6.7) [40]. It could thus be that natural selection has exploited the dangers of multiple attractors in living systems and has taken care to evolve networks in which this does not arise.

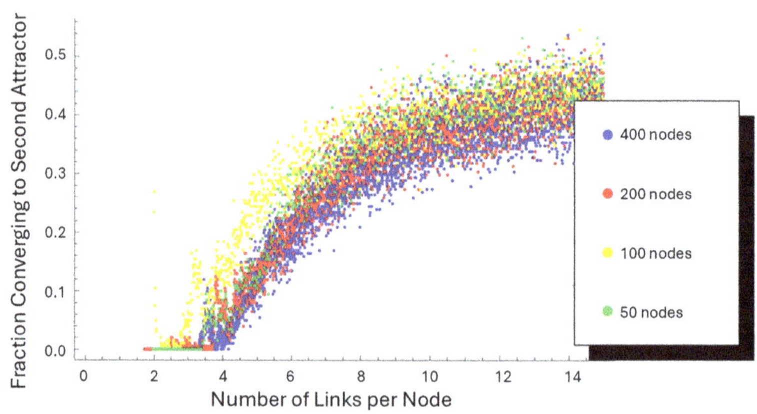

Fig. 6.7 Results of a Monte-Carlo simulation showing the fraction of times that sparse Hopfield-type networks converge to a second orthogonal attractor versus the number of links per node in the network, for nets of N = 50, 100, 200 and 400 nodes. This reveals a connectivity sweet spot between about 2–4 links per node where the net supports only a single attractor. Reproduced with permission from [40]

6.5 Tissue Remodeling and Mixture Theory

The above discussion might seem to suggest a dichotomy in the behavior of nonlinear dynamic networks—either they adhere to a homeostatic attractor representing health, or they succumb to an aberrant one corresponding to chronic disease. Reality is not so clear cut. An intermediate possibility is the phenomenon known as tissue remodeling whereby tissue structure adapts to chronic stress through a structural change that is typically long-term or permanent. Remodeled tissue is clearly not normal in the strict sense of the word, but it may be relatively inconsequential depending on its extent and severity. Indeed, transient remodeling during the course of repair from injury is a mark of healthy tissue able to repair itself [41].

Tissue remodeling is a common feature of lung disease. For example, remodeling of the airway wall characterized by excess collagen deposition is a common finding in asthma [42]. This may be an adaptation to the stress of frequent bronchospasm rather than a fundamental component of asthma pathophysiology [43]. On the other hand, when the remodeled airway wall becomes thickened sufficiently, it may impact the geometry of the lumen and alter the mechanical opposition to shortening of the airway smooth muscle sufficiently to become a significant contributor to airways hyperresponsiveness [44]. Tissue remodeling also constitutes the fundamental structural derangement in pulmonary fibrosis, which itself exists on a continuum from mild enough to be symptomless to severe enough to be incompatible with life [45].

Tissue remodeling, when it involves the incorporation of new material into an existing structure, is similar in many ways to normal growth or to the development of a tumor. This makes the tissue structure bulkier, causing it to displace neighboring tissues. The displaced tissues, being elastic yet incompressible, develop internal stresses as a result. The remodeled tissue also develops internal stresses itself because new material must be incorporated alongside the existing tissue that is already present, while at the same time retaining material continuity. A simple illustration of this point is shown in Fig. 6.8 which shows what happens when new tissue is deposited in the middle of an initially unstressed airway wall

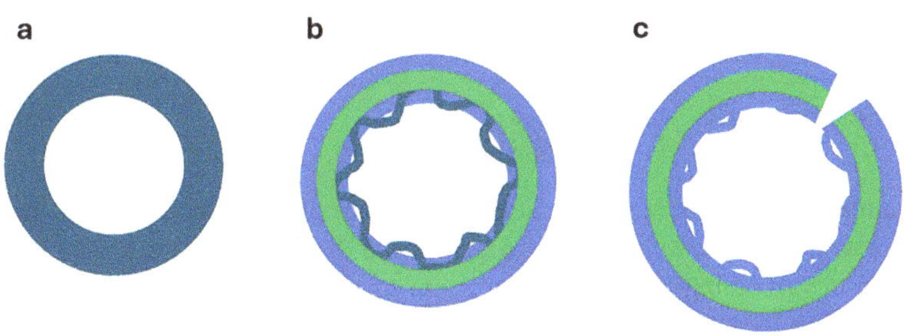

Fig. 6.8 (**a**) Cross-section of an unstressed airway wall. (**b**) Compression of inside wall tissue and stretch of outside tissue due to central deposition of new tissue (green). (**c**) Radial cut through the wall causing it to spring out due to the outer wall tension

(Fig. 6.8b). The part of the wall inside the new tissue becomes compressed while the part outside becomes stretched (Fig. 6.8b). If the remodeled wall is cut through radially, its outer portion will retract to leave a wedge opening caused by the residual stress (Fig. 6.8c).

A continuum mechanics approach known as mixture theory has been developed to model such processes [46, 47]. Suppose that new collagen having a constant mass density ρ equal to that of the existing wall material is being added uniformly throughout the airway wall tissue (i.e., not just in a central band as in Fig. 6.8b). The mass (μ) rate of growth in a unit volume V of wall tissue is

$$\frac{d\mu}{dt} = \frac{d(\rho V)}{dt} = \rho \frac{d(V)}{dt}. \tag{6.27}$$

Conservation of mass requires that the rate of change of μ equals the rate at which mass is being generated within V plus the rate at which mass is being pushed into V due to it being displaced by remodeling taking place in adjacent regions of tissue. If $\boldsymbol{v}$ is the velocity vector describing the bulk movement of the tissue, then

$$\frac{d\mu}{dt} = \frac{\partial \rho}{\partial t} + \mathrm{div}(\rho \boldsymbol{v}) = \rho \mathrm{div}(\boldsymbol{v}) = \mathrm{tr}\boldsymbol{D}_g \tag{6.28}$$

where div(·) denotes the divergence of the bracketed vector quantity and $\mathrm{tr}\boldsymbol{D}_g$ denotes the trace of the growth tensor. The growth deformation gradient $\boldsymbol{F}$ (see Chap. 4) maps the original un-remodeled tissue into its final remodeled form complete with internal stresses. This is equal to the product of the deformation gradient describing the expansion of the tissue due to growth, with no constraint on continuity, and the deformation gradient describing how the remodeled tissue must be deformed to be continuous [47]. This theory can be applied to multiple components of a tissue that simultaneously occupy the same physical space and so contribute fractionally to the total tissue volume. The mixture theory has been applied successfully, for example, to the remodeling of vascular tissue [47] and bones [46], but has yet to be applied to airway wall remodeling such as occurs in asthma or COPD.

6.6 Conclusions

The maintenance of a homeostatic state throughout life, once growth and development are complete, and an ability to self-repair from injury are among the key characteristics of living organisms that seem to place them apart from non-living systems. Being thermodynamically open is a crucial requirement for a homeostatic self-repairing system, but this alone does not explain how these crucial functions have emerged and are performed. The prime effectors of these functions are anabolic and catabolic cells that operate on the tissues they maintain and repair. Agent-based modeling is a convenient and powerful tool for investigating the possible strategies used by communities of cells to regulate homeostasis and repair, and to explore ways in which these vital functions can potentially fail in chronic disease. These strategies are represented by the behavioral rules that the agents obey as they interact with their environment and with each other. The agent rules may be represented as Boolean logic statements or as systems of ordinary or partial differential equations. Computational models based on these principles provide insight into the ways in which living systems can give the appearance of violating the second law of thermodynamics, opening the possibility of exploiting these strategies in artificial self-healing systems.

6.7 Problems

1. Equation 6.10 predicts the temporal variance of tissue density of a single patch when the patch is randomly visited by either an anabolic or a catabolic cell at each time step, as specified by Eq. 6.1. Figure 6.1 illustrates how the spatial variance of tissue density varies in a similar manner. What argument can be made that the temporal density variation of a single patch is the same as the spatial variation of multiple patches at a single point in time (at steady state).
2. Code an agent-based model similar to that described in Sect. 6.2. Assign the agents rules that correspond to durotaxis.
   (a) Does the spatial variance of tissue patch density in your model asymptote to a finite value in the limit of a large number of iterations?
   (b) How does spatial density variance change with the strength of the durotactic signal?
   (c) Is there any other form of feedback that the anabolic and/or catabolic agents could receive from the tissue to cause the model to maintain homeostasis?
3. Equations 6.17 and 6.18 describe a dynamical system that has a single fixed point given by Eqs. 6.20 and 6.21.
   (a) How can this system of equations be extended so that it has two distinct fixed points?
   (b) Can this new system be put in the context of a Hopfield net as in Sect. 6.4?

References

1. Peterson J. Understanding the thermodynamics of biological order. Am Biol Teach. 2012;
2. Macklem PT. Emergent phenomena and the secrets of life. J Appl Physiol. 1985;2008(104): 1844–6.
3. Prigogine I, Stengers I. Order out of chaos. New York: Bantam; 1984.
4. Schrodinger E. What is life? Cambridge: Cambridge University Press; 1944.

5. Cannon WB. In the wisdom of the body. New York: W.W. Norton; 1932. p. 177–201.
6. Iizuka H. Epidermal turnover time. J Dermatol Sci. 1994;8:215–7.
7. Molina PE. Endocrine physiology. 4th ed. New York: McGraw-Hill; 2013.
8. Jesudason R, Sato S, Parameswaran H, Araujo AD, Majumdar A, Allen PG, Bartolak-Suki E, Suki B. Mechanical forces regulate elastase activity and binding site availability in lung elastin. Biophys J. 2010;99:3076–83.
9. Wellman TJ, Mondonedo JR, Davis GS, Bates JHT, Suki B. Topographic distribution of idiopathic pulmonary fibrosis: a hybrid physics- and agent-based model. Physiol Meas. 2018;39:064007.
10. Strange C. Alpha-1 antitrypsin deficiency associated COPD. Clin Chest Med. 2020;41:339–45.
11. Cox LA Jr. A mathematical model of protease-antiprotease homeostasis failure in chronic obstructive pulmonary disease (COPD). Risk Anal. 2009;29:576–86.
12. Warheit-Niemi HI, Hult EM, Moore BB. A pathologic two-way street: how innate immunity impacts lung fibrosis and fibrosis impacts lung immunity. Clin Transl Immunol. 2019;8:e1065.
13. Suki BS. Structure and function of the extracellular matrix. London: Academic; 2022.
14. Bates JHT, Herrmann J, Casey DT, Suki B. An agent-based model of tissue maintenance and self-repair. Am J Physiol Cell Physiol. 2023;324:C941–50.
15. Plotnikov SV, Pasapera AM, Sabass B, Waterman CM. Force fluctuations within focal adhesions mediate ECM-rigidity sensing to guide directed cell migration. Cell. 2012;151:1513–27.
16. Alves C, Araujo AD, Oliveira CL, Imsirovic J, Bartolak-Suki E, Andrade JS, Suki B. Homeostatic maintenance via degradation and repair of elastic fibers under tension. Sci Rep. 2016;6:27474.
17. Rouillard AD, Holmes JW. Mechanical regulation of fibroblast migration and collagen remodelling in healing myocardial infarcts. J Physiol. 2012;590:4585–602.
18. Jones MG, Andriotis OG, Roberts JJ, Lunn K, Tear VJ, Cao L, Ask K, Smart DE, Bonfanti A, Johnson P, et al. Nanoscale dysregulation of collagen structure-function disrupts mechano-homeostasis and mediates pulmonary fibrosis. elife. 2018:7.
19. Burnham EL, Janssen WJ, Riches DW, Moss M, Downey GP. The fibroproliferative response in acute respiratory distress syndrome: mechanisms and clinical significance. Eur Respir J. 2014;43:276–85.
20. Chang W, Wei K, Jacobs SS, Upadhyay D, Weill D, Rosen GD. SPARC suppresses apoptosis of idiopathic pulmonary fibrosis fibroblasts through constitutive activation of beta-catenin. J Biol Chem. 2010;285:8196–206.
21. Suki B, Herrmann J, Bates JHT. An analytic model of tissue self-healing and its network implementation: application to fibrosis and aging. Front Physiol. 2020;11:583024.
22. Younesi FS, Miller AE, Barker TH, Rossi FMV, Hinz B. Fibroblast and myofibroblast activation in normal tissue repair and fibrosis. Nat Rev Mol Cell Biol. 2024;25:617–38.
23. Strogatz SH. Nonlinear dynamics and chaos. 3rd ed. New York: Chapman and Hall/CRC; 2024.
24. Suki B, Parameswaran H, Alves C, Araujo AD, Bartolák-Suki E. Multi-scale extracellular matrix mechanics and mechanobiology. In: Zhang Y, editor. Studies in mechanobiology, tissue engineering and biomaterials, vol. 23. Cham: Springer; 2020. p. 83–121.
25. Bergeron C, Cantin AM. Cystic fibrosis: pathophysiology of lung disease. Semin Respir Crit Care Med. 2019;40:715–26.
26. Roda J, Pinto-Silva C, Silva IAI, Maia C, Almeida S, Ferreira R, Oliveira G. New drugs in cystic fibrosis: what has changed in the last decade? Ther Adv Chronic Dis. 2022;13: 20406223221098136.
27. Barjaktarevic I, Campos M. Management of lung disease in alpha-1 antitrypsin deficiency: what we do and what we do not know. Ther Adv Chronic Dis. 2021;12(suppl):20406223211010172.
28. Zhang D, Newton CA. Familial pulmonary fibrosis: genetic features and clinical implications. Chest. 2021;160:1764–73.
29. Mohammed SM, Al-Saedi HFS, Mohammed AQ, Amir AA, Radi UK, Sattar R, Ahmad I, Ramadan MF, Alshahrani MY, Balasim HM, Alawadi A. Mechanisms of bleomycin-induced lung fibrosis: a review of therapeutic targets and approaches. Cell Biochem Biophys. 2024;

30. Adamcakova J, Mokra D. New insights into pathomechanisms and treatment possibilities for lung silicosis. Int J Mol Sci. 2021;22
31. Warheit DB, Hill LH, George G, Brody AR. Time course of chemotactic factor generation and the corresponding macrophage response to asbestos inhalation. Am Rev Respir Dis. 1986;134: 128–33.
32. Ribeiro AS, Kauffman SA. Noisy attractors and ergodic sets in models of gene regulatory networks. J Theor Biol. 2007;247:743–55.
33. Taherian Fard A, Ragan MA. Modeling the attractor landscape of disease progression: a network-based approach. Front Genet. 2017;8:48.
34. Segel LA. Multiple attractors in immunology: theory and experiment. Biophys Chem. 1998;72: 223–30.
35. Bates OR, Suki B, Spector PS, Bates JH. Structural defects lead to dynamic entrapment in cardiac electrophysiology. PLoS One. 2015;10:e0119535.
36. Hopfield JJ. Neurons with graded response have collective computational properties like those of two-state neurons. Proc Natl Acad Sci USA. 1984;81:3088–92.
37. Haykin S. Neural networks and learning machines. Upper Saddle River: Pearson Education.
38. Bates JHT. Nonlinear network theory of complex disease. In: Bar-Yam Y, editor. International conference on complex systems; June 25–30, 2006. Boston: New England Complex Systems Institute; 2006.
39. Ravasz E, Somera AL, Mongru DA, Oltvai ZN, Barabasi AL. Hierarchical organization of modularity in metabolic networks. Science. 2002;297:1551–5.
40. Anafi RC, Bates JH. Balancing robustness against the dangers of multiple attractors in a Hopfield-type model of biological attractors. PLoS One. 2010;5:e14413.
41. Fehrenbach H, Wagner C, Wegmann M. Airway remodeling in asthma: what really matters. Cell Tissue Res. 2017;367:551–69.
42. Hough KP, Curtiss ML, Blain TJ, Liu RM, Trevor J, Deshane JS, Thannickal VJ. Airway remodeling in asthma. Front Med (Lausanne). 2020;7:191.
43. O'Sullivan MJ, Phung TN, Park JA. Bronchoconstriction: a potential missing link in airway remodelling. Open Biol. 2020;10:200254.
44. Moreno RH, Hogg JC, Pare PD. Mechanics of airway narrowing. Am Rev Respir Dis. 1986;133:1171–80.
45. Knudsen L, Ruppert C, Ochs M. Tissue remodelling in pulmonary fibrosis. Cell Tissue Res. 2017;367:607–26.
46. Ambrosi D, Ben Amar M, Cyron CJ, DeSimone A, Goriely A, Humphrey JD, Kuhl E. Growth and remodelling of living tissues: perspectives, challenges and opportunities. J R Soc Interface. 2019;16:20190233.
47. Rodriguez EK, Hoger A, McCulloch AD. Stress-dependent finite growth in soft elastic tissues. J Biomech. 1994;27:455–67.

Chapter 7
Asthma

Abstract Asthma is a common affliction that is often mild and intermittent but may be debilitating or even fatal. A key component of the asthma phenotype is airways hyperresponsiveness (AHR) characterized by excessive derangements in mechanical lung function in response to standardized doses of airway smooth muscle agonist. Understanding AHR is a multi-scale problem. At the microscopic scale, dynamic interactions between myosin cross-bridges and actin filaments within the ASM cell are ultimately responsible for force generation. The Huxley model of skeletal muscle contraction has been used to account for these interactions, but a coarse-grained version of the model can be derived that acknowledges the much more haphazard arrangement of the contractile proteins in ASM compared to skeletal muscle. At the mesoscopic scale, differential equation models describe airway narrowing due to force generation by the ASM when the pulmonary airways are embedded in visco-elastic parenchymal tissue. Such models recapitulate the strong dependence of narrowing on lung volume due to the forces of airway-parenchymal interdependence. At the macroscopic scale, models explain how distributions of constricting airways in the pulmonary airway tree collectively give rise to the changes in lung mechanical impedance seen, for example, following airway challenge with methacholine aerosol.

Keywords Airway smooth muscle · Actin fiber · Myosin cross-bridge · Huxley model · Differential equation model · Airway narrowing · Lung impedance

7.1 Introduction

Virtually everyone has heard of *asthma*, a common affliction that affects roughly 10% of the population to a degree that is often mild and intermittent, but it may also be debilitating or even fatal. Asthma was recognized as such by Hippocrates in Ancient Greece, but its history dates back to about 5000 years ago in China [1]. Nevertheless, the precise pathogenesis of asthma remains a subject of intense research even today. Indeed, asthma is more appropriately defined as a syndrome rather than a disease, since it is defined on the basis of its symptoms rather than a specific pathophysiologic mechanism [2]. Chief among the symptoms is reversible

B. Suki, J. H. T. Bates, *Mathematical Modeling of the Healthy and Diseased Lung*,
https://doi.org/10.1007/978-3-031-95322-4_7

airway constriction, demonstrated by the fact that asthmatic individuals derive significant relief from drugs that relax the *airway smooth muscle* (ASM). This allows the airways to expand so that breathing becomes easier. Accordingly, inappropriate and/or excessive ASM contraction, commonly termed *bronchospasm*, has received a great deal of attention as a potential root cause of asthma and as a target for therapy [3]. Whether ASM contraction is the ultimate culprit behind asthma remains controversial, however, due to limited and conflicting evidence that asthmatic ASM is itself abnormal [4, 5].

The alternative is that bronchospasm in asthma is simply an effector acting on a lung that has become pathologically altered [6]. Even Hippocrates realized that asthma could be linked to environmental triggers [1], and it is now known that allergic inflammation of the airways is a key component of the main asthma *phenotypes* [7]. This inflammation involves thickening of the bronchial walls and epithelial release of mucus into the airway lumen, all of which can serve to either partially or completely occlude the airways and lead to the breathing difficulties for which asthma is infamous. The immune system thus plays a central role in asthma pathogenesis, particularly when involved in a Th2-type inflammation against which corticosteroids are frequently effective [8]. The role of allergy has also been invoked in the so-called "hygiene hypothesis" to explain why asthma has become more prevalent in recent decades; the cleanliness of modern first-world living has relieved our immune systems from having to battle parasites via Th1-type inflammation, leaving it free to focus its energy on developing allergies associated with Th2-type inflammation [9].

The story behind asthma is more complex than a simple Th1-Th2 dichotomy, however. Asthma is now recognized to exist in a range of different *endotypes* [9], some of which are associated with Th1-type immune bias. It typically develops in childhood and frequently resolves post-puberty, but it may begin in adulthood when it typically has a poorer prognosis. The modern obesity epidemic has also given rise to an obesity-specific form of asthma, the underlying cause of which remains unclear [10]. Asthma is thus perhaps best viewed as the common clinical endpoint for a number of different pathological processes.

Whether or not there is something abnormal about the ASM in individuals with asthma, there is little doubt that bronchospasm is a key component of the acute asthma attack. ASM activation results in circumferential forces being applied around the pulmonary airways, which in turn distorts the nearby viscoelastic parenchyma in which the airways are embedded. The dynamic balance between these forces and their consequences for lung function are amenable to mathematical and computational modeling, which can lead to important insights into what may distinguish individuals who have asthma from those who do not. Understanding the consequences of ASM contraction for the lung as a whole is a non-trivial problem, however, because it involves phenomena that manifest over a wide range of length scale. At the microscopic level of scale, it begins with activation of the ASM and the resultant generation of ASM contractile force. How this affects airway caliber then involves mesoscopic force-balance events at the tissue level. Finally, at the macroscopic organ level, overall lung function is strongly affected by the regionally heterogeneous manner in which these microscopic and mesoscopic events manifest. Accordingly, this chapter will pursue the modeling of these phenomena in order of increasing length scale.

7.2 The Dynamics of Airway Smooth Muscle Contraction

ASM, the engine of bronchospasm, is composed of individual smooth muscle cells aligned roughly circumferentially around the walls of the pulmonary airways from the mainstem bronchi all the way down to the alveolar ducts (except for the trachea, which in humans has a band of ASM tissue along its posterior wall). The cells are embedded in a matrix of connective tissue that transmits mechanical forces between the cells such that when they contract, the airway cross-section is reduced. ASM shares some important similarities with skeletal muscle, but there are also some important differences [11]. Like *skeletal muscle*, ASM force generation occurs through the cyclic binding of *myosin cross-bridges* to adjacent *actin filaments* within the ASM cell. The coupling of this process to the signals that activate the cells to contract is quite different, however. In skeletal muscle, activation-contraction coupling is mediated by the intracellular release of calcium ions from the *sarcoplasmic reticulum*. This leads to a configurational change in *tropomyosin*, a protein that normally hides the actin binding sites but leaves them exposed in the presence of calcium so that cross-bridge cycling can begin. Cycling comes to an end, and the muscle relaxes, when the calcium ions are actively re-sequestered back into the sarcoplasmic reticulum. In ASM, calcium influx into the cell instigates cross-bridge cycling by starting a process that leads to a configurational change in the enzyme *myosin light-chain kinase*, which then causes myosin cross-bridge cycling. Cessation of cycling is brought about by a second enzyme, *myosin light-chain phosphatase*.

There are also structural differences between ASM and skeletal muscle that have an important bearing on their differences in function. The actin and myosin filaments responsible for force generation in skeletal muscle cells are arranged in an orderly parallel array aligned in the direction of muscle shortening. They are also arranged serially in a sequence of contractile units called *sarcomeres* that span the muscle cell from one end to the other. Each sarcomere is composed of a set of interdigitated actin and myosin filaments, the former being anchored to structural proteins that transmit the active force to the embedding tissue (Fig. 7.1a). In contrast, the actin and myosin filaments in ASM are arranged much more haphazardly in a variety of orientations (Fig. 7.1b) [12]. Furthermore, it is believed that a significant portion of the actin in an unstimulated ASM cell may exist as disconnected monomers that polymerize into filaments when the cell is activated [13]. This imbues the ASM cell with a degree of structural plasticity, allowing it to adapt its force generating capacity to the configuration of the cell at the time of stimulation. This is advantageous for the purposes of generating a constricting force around a conduit that varies its circumference throughout the breathing cycle.

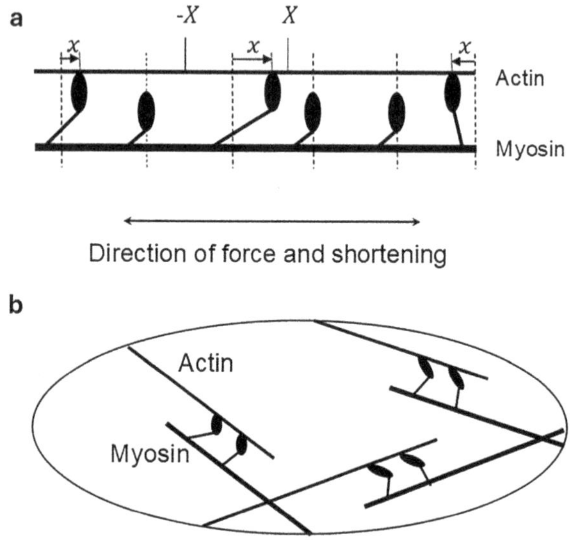

Fig. 7.1 (**a**) The contractile proteins actin and myosin in skeletal muscle are arranged in an orderly fashion such that cross-bridge shortening occurs in the direction of overall muscle shortening and force generation. The figure shows 3 unbound cross-bridges and 3 bound cross-bridges. The extension of each of the bound cross-bridges past its unstressed position (shown by the vertical dashed lines) is represented by x. X and $-X$ denote the limits of x. (**b**) In an ASM cell, the actin and myosin filaments are arranged at a variety of angles to the direction of overall muscle shortening. (Reproduced with permission from [21])

Because of its sarcomeric structure, skeletal muscle exhibits a distinct and fixed dependence of maximal force on its length when stimulated. Maximal force is produced when the interdigitated actin and myosin fibers within each sarcomere are fully overlapped so that all myosin cross-bridges can engage in force production. At sufficiently short muscle lengths, the myosin fibers bump into the perpendicular borders of the sarcomere (the so-called z-lines), resulting in reduced force. As length increases, a gap appears between the opposing sets of actin filaments so that the middle myosin cross-bridges can no longer bind to them, again resulting in a progressively decreasing force. The less organized arrangement of the contractile proteins in ASM, on the other hand, means that although total force production is less than in skeletal muscle, it is also less constrained by muscle length. Evolution has thus crafted ASM so that effective force production occurs over a wider range of lengths than in skeletal muscle, in addition to being adaptable to the length of the muscle at the time of stimulation [14].

The velocity with which a muscle shortens during contraction is related to the contractile force it develops, resulting in a characteristic force-velocity relationship. Maximally stimulated muscle contracts more rapidly when it is unloaded (i.e., not pulling on anything) compared to when it is connected to a load. This makes intuitive physical sense because the mechanical power applied by a muscle to a load is given by the product of the reactive force of the load and the velocity of the load in the

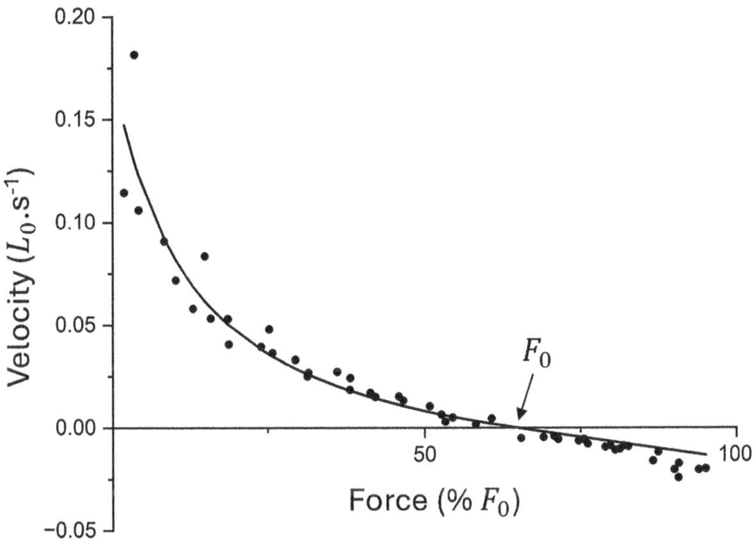

Fig. 7.2 Measurements of the force-velocity behavior of ASM (circles) showing a hyperbolic relationship over the concentric contraction range between maximum shortening velocity and zero at the isometric force F_0 (Eq. 7.1), and a linear relationship for force $>F_0$ when contraction is eccentric (Eq. 7.2). Also shown (solid line) is the fit provided by Eqs. 7.1 and 7.2. L_0 is the length of the muscle when it is stretched to the point just before passive strain begins to develop. (Force and velocity data taken from [15])

direction of force. Since muscle power is ultimately traced to the biochemical energy of its molecular reactions, this obviously has a maximum value when a muscle is maximally stimulated, which implies that velocity and load should be inversely related. Indeed, the force-velocity behaviors of both skeletal muscle and ASM are hyperbolic between the limits of maximum unloaded velocity and the isometric force achieved when the muscle is unable to move the load at all. Muscle contraction over this range of velocities is referred to *concentric*. Even larger loads will overcome the contractile ability of a maximally activated muscle, causing it to lengthen in what is referred to as *eccentric* contraction. Muscle force during eccentric contraction increases approximately linearly with the rate of extension. These force-velocity characteristics of ASM are illustrated in Fig. 7.2 [15].

The hyperbolic portion of the force-velocity relationship shown in Fig. 7.2 pertains when active muscle force, $F(t)$, is less than isometric force F_0 [16]. This relationship is described by the *Hill equation* that was originally developed for skeletal muscle but has been applied to ASM [17]. The Hill equation is

$$\dot{L}(t) = \frac{b[F_0 - F(t)]}{a + F(t)} \tag{7.1}$$

where a and b are constants, $L(t)$ is muscle length at time t, and velocity is defined as positive during muscle shortening. The relationship for $F > F_0$ can be represented by a linear continuation of the relationship at $F = F_0$, namely

$$\dot{L}(t) = \frac{bF_0}{a + F_0} - \frac{bF(t)}{a + F_0}. \tag{7.2}$$

A simple representation of the contraction dynamics of a strip of ASM can be constructed by connecting a force generator obeying Eqs. 7.1 and 7.2 in series with a spring representing the connective tissue that transmits the force of cross-bridge cycling to the muscle as a whole. Dynamic force-length loops simulated with this model are reminiscent of experimental observations, particularly if the spring becomes stiffer as it lengthens [18]. However, this model fails to recapitulate the rapid transient decreases in force that occur when an activated strip of ASM initially held at fixed length is suddenly subjected to rapid length oscillations (Fig. 7.3).

A potential source of the force transients observed in Fig. 7.3a is stress adaptation due to the viscoelastic properties of the structural components of the ASM cell and the extracellular matrix in which it is embedded. Figure 7.4 shows a model incorporating force generation according to Eqs. 7.1 and 7.2 along with nonlinear Kelvin bodies (see Sect. 5.2.2) representing the series and parallel viscoelastic elements in a

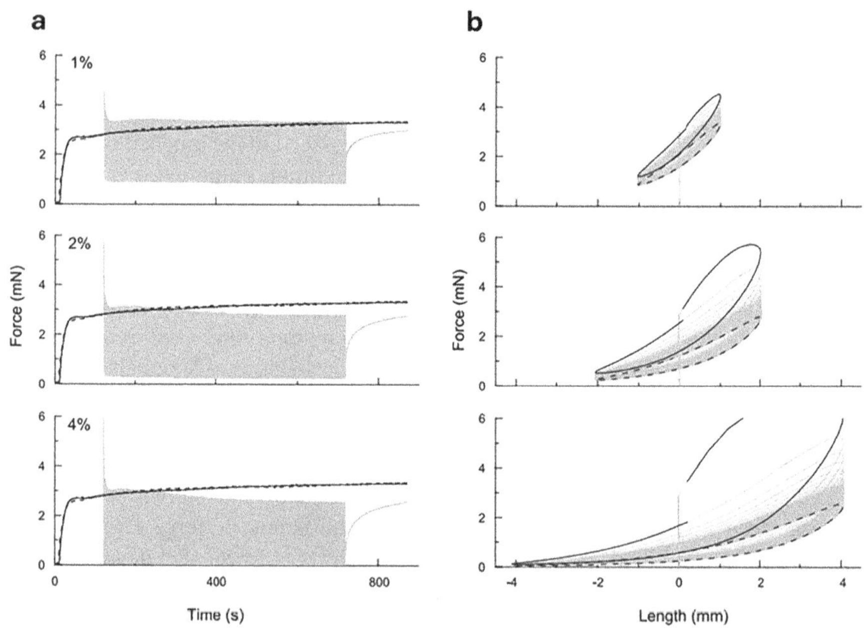

Fig. 7.3 (**a**) Force-time and force-length traces obtained in a strip of rat trachealis muscle. The gray curves show experimental data obtained when 2 Hz length oscillations of ±1%, ± 2%, and ± 4% were instigated. Isometric force-time curves are shown in black (solid lines before oscillations, dashed lines after oscillations). (**b**) Force-length loops during length oscillation (first loop shown as solid black line, last loop as dashed black line). (Reproduced with permission from [19])

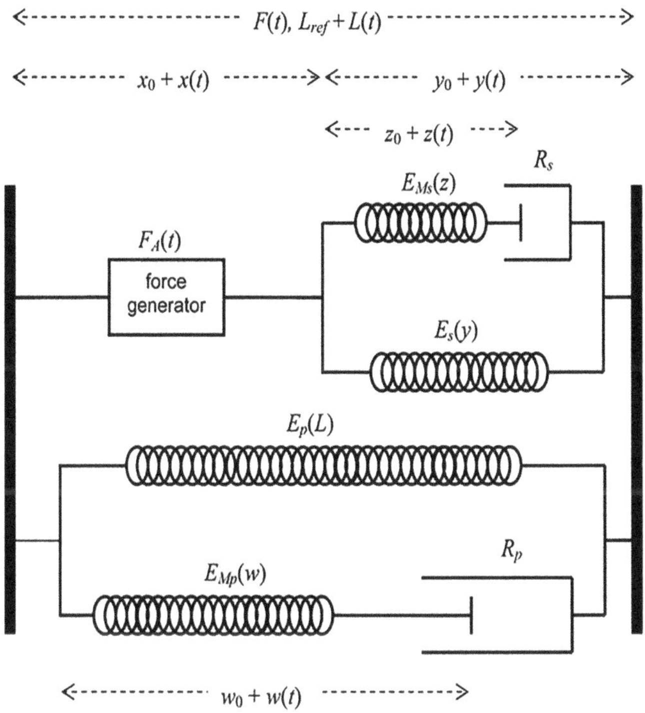

Fig. 7.4 Model of ASM consisting of a nonlinear Kelvin body in series with an active force generator, together with a second nonlinear Kelvin body connected in parallel. The 2 Kelvin bodies account for the passive properties of the strip. The five mechanical degrees of freedom $L(t)$, $x(t)$, $y(t)$, $w(t)$ and $z(t)$ are indicated by arrows. L_0, x_0, y_0, w_0 and z_0 are the unstressed lengths of the springs. $F_A(t)$ is active force. Spring elastances and dashpot resistances are shown above their corresponding model elements. (Reproduced with permission from [19])

strip of ASM tissue [19]. The model is driven with an imposed length perturbation $L(t)$. When the force generator is not active, the stiffness of the force generator is assumed to be negligible, so the model dynamics are determined only by the parallel Kelvin body. In this case, the force, $F(t)$, across the model is

$$F(t) = E_p(L)L(t) + E_{MP}(w)w(t) \tag{7.3}$$

where $L(t)$ is the extension beyond the unstressed length (L_{ref}) of the main spring in the parallel Kelvin body, and $w(t)$ is the extension beyond w_0 of the spring connected in series with the dashpot. The stiffnesses of the two springs, E_p and E_{Mp}, are nonlinear functions of their respective lengths according to

$$E_p(L) = \alpha_p|L|^3 \tag{7.4}$$

and

$$E_{Mp}(L) = \alpha_{Mp}|w|^3 \tag{7.5}$$

where α_p and α_{Mp} are constants, and the absolute values of L and w allow for strain stiffening during both extension and compression of the springs. In addition, the force across E_{Mp} is equal to the force across the resistive dashpot R_p, expressed as

$$\alpha_{Mp}|w|^3 w(t) = R_p\left[\dot{L}(t) - \dot{w}(t)\right]. \tag{7.6}$$

In order to have the Kelvin body conform to quasi-linear viscoelasticity (Sect. 5.4) in keeping known tissue properties, R_p is given the same functional dependence on w as E_{Mp} (i.e., that defined by Eq. 7.5).

When the force generator is activated, it produces an equivalent force across the series Kelvin body given by

$$F_A(t) = E_s(y)y(t) + E_{Ms}(z)z(t) \tag{7.7}$$

where the spring constants E_s and E_{Ms} are again cubic functions of the spring extensions y and z, respectively, similar to Eqs. 7.4 and 7.5. As with the series Kelvin body above, the forces across the spring E_{Ms} and dashpot R_s are equal, and are expressed as

$$\alpha_{Ms}|z(t)|^3 z(t) = \alpha_{Ms}|z(t)|^3 R_s[\dot{y}(t) - \dot{z}(t)] \tag{7.8}$$

where the stiffnesses of the springs in the series viscoelastic element are assumed to have a cubic dependence on stretch, as used above for the parallel element. The force across the entire model is then

$$F(t) = E_p(L)L(t) + E_{MP}(w)w(t) + F_A(t). \tag{7.9}$$

The length of the force generator changes with time according to

$$\dot{L}(t) = \dot{x}(t) + \dot{y}(t). \tag{7.10}$$

The experimental data in Fig. 7.3 were simulated using the above model by starting the model at its relaxed length and activating the force generator. Sinusoidal variations in L were initiated at $t = 60$ s, which resulted in immediate increases in peak force that subsequently decreased, although not rapidly or pronounced enough to match experimental observation (Fig. 7.3). The only way to recapitulate the experimental data with this model is to postulate that the maximum active isometric force, F_{0max}, that is established prior to length oscillation is very rapidly reduced once oscillations begin. In fact, this is known to occur and reflects the forced detachment of bound cross-bridges by the externally imposed length oscillations. These detached cross-bridges then reattach when oscillations cease. When the cross-bridges are detached, both the force generating capacity and the mechanical stiffness

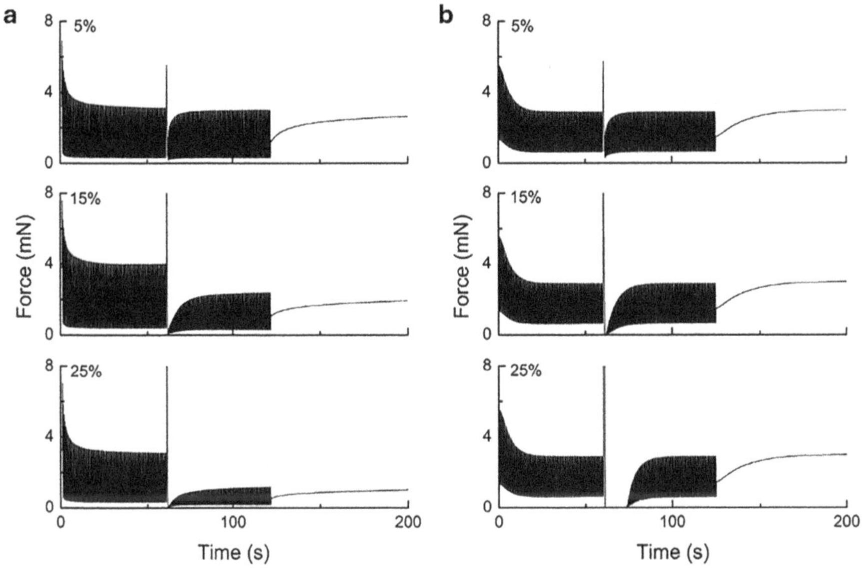

Fig. 7.5 (**a**) Experimental measurements of maximally activated ASM force versus time when ±2% length oscillations are initiated at $t = 0$ and a sudden brief stretch of 5%, 15% and 25% is applied at $t = 60$ s. (**b**) Corresponding simulations by the model based on Eqs. 11.1, 11.2, 11.3, 11.4, 11.5, 11.6, 11.7, 11.8, 11.9 and 11.10. (Reproduced with permission from [19])

of the ASM are temporarily decreased, a phenomenon that has been referred to as fluidization [20].

Assuming first-order kinetics for both detachment (rate-constant k_{off}) and reattachment (rate-constant k_{on}) gives the following equation for the dynamics of F_0 [21]:

$$\dot{F}_0(t) = k_{on}[F_{0max} - F_0(t)] - k_{off}|\dot{x}(t)|F_0(t). \tag{7.11}$$

Incorporating Eq. 7.11 into the model produces transient force behavior that is similar in a number of key ways to that seen experimentally in strips of ASM following instigation of ±2% length oscillations that are then interrupted by a sudden brief length increase of 5%, 15% and 25% (Fig. 7.5). Nevertheless, this model still fails to recapitulate the depressions in recovery of force that occur once the oscillations stop, as seen most dramatically in the lowest panel of Fig. 7.5a.

7.3 The Huxley Model of Muscle Contraction

The descriptions of ASM considered above are based on lumped-parameter models described by ordinary differential equations. These models can be made to account for the key qualitative features of ASM dynamic force-length behavior, but their

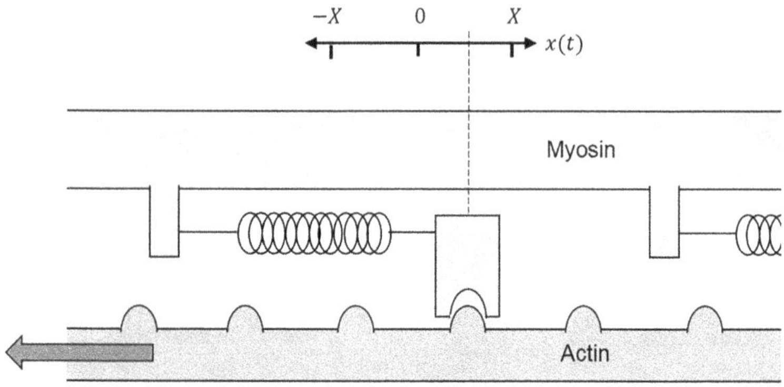

Fig. 7.6 The Huxley model of skeletal shortening dynamics shown at the level of a single bound cross-bridge. When attached to a specific actin binding site, the elastic myosin head is stretched by length $x(t)$, which lies between the extremes of $\pm X$. The direction of movement of the actin filament relative to the myosin filament is shown by the grey arrow

representations of active force generation and tissue viscoelasticity are purely phenomenological. A mechanistically based model of skeletal muscle contraction dynamics actually predates all these efforts. In 1957, Andrew Huxley linked force production and shortening velocity to the cycling of myosin cross-bridges [22]. He began by assuming that any cross-section in a contracting muscle contains a large population of parallel pairs of interacting actin and myosin fibers, all aligned in the direction of shortening. Each myosin cross-bridge in this ensemble is either unbound, and thus not contributing to active force, or is bound to an actin site. When a cross-bridge is bound, it is stretched to some degree and thus generates a recoil force. It is this recoil force that comprises the contribution of the cross-bridge to the total active force of the muscle. As with any spring, the recoil force of a cross-bridge depends on the length by which it is stretched past its unstressed position, represented by x in Fig. 7.6. x itself is a function of t because as the myosin head cycles, it is continually in the process of attaching and detaching from one or another actin binding site, and it is continually changing length while bound because of the relative motion of the actin and myosin fibers.

If a macroscopic piece of muscle contains a very large number of cross-bridges acting in parallel over its cross-section, the cross-bridge lengths comprise a continuous distribution $n(x, t)$, as illustrated in Fig. 7.7. The total number of attached cross-bridges with lengths between $x = a$ and $x = b$ is then

$$N((a,b),t) = \int_a^b n(x,t)dx. \tag{7.12}$$

$N((a, b), t)$ changes with time due to the rate, $G(x, t)$, that cross-bridges attach and the rate, $H(x, t)$, at which they detach within the position window $[a, b]$. But these are not the only factors that cause $N((a, b), t)$ to change. As the muscle shortens, bound

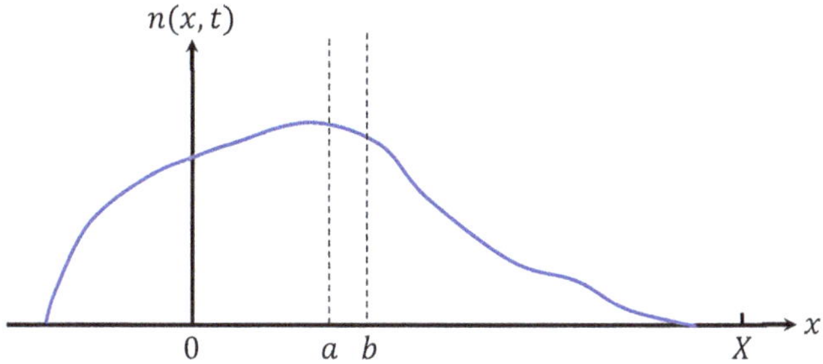

Fig. 7.7 Equation 11.12 gives the number of cross-bridges stretched at lengths between a and b at time t

cross-bridges that are stretched just slightly more than length b will decrease their degree of stretch and move into the $[a, b]$ window. Conversely, those stretched just above a length of a move out of the $[a, b]$ window. Differentiating Eq. 7.12 and setting the result equal to the sum of these various rates of changes gives

$$\begin{aligned}\frac{\partial N((a,b),t)}{\partial t} &= \int_a^b \frac{\partial n(x,t)}{\partial t} dx \\ &= \int_a^b [G(x,t) - H(x,t)]dx + v(t)[n(b,t) - n(a,t)] \\ &= \int_a^b [G(x,t) - H(x,t)]dx + \int_a^b v(t)\frac{\partial n(x,t)}{\partial x} dx\end{aligned} \tag{7.13}$$

where $v(t)$ is the velocity of muscle shortening (i.e., the relative velocity of actin and myosin). Since the above equation is true for arbitrary x and t, and the integrals are all determined with respect to x between the same limits a and b, the integrands must also equate. That is,

$$\frac{\partial n(x,t)}{\partial t} = [G(x,t) - H(x,t)] + v(t)\frac{\partial n(x,t)}{\partial x}. \tag{7.14}$$

The total active muscle force generated at time t is then the sum of the forces in all the parallel cross-bridges, which is

$$F(t) = \int_{-X}^{X} f(x)n(x,t)dx \tag{7.15}$$

where $f(x)$ is the force-length relationship of a single cross-bridge.

Equation 7.14 is a partial differential equation in x and t and is known as the *Huxley model* of skeletal muscle. This equation can be solved using the methods of characteristics, but to obtain a specific solution, the functions $G(x, t)$ and $H(x, t)$ must be known. Huxley assumed that bound cross-bridges become detached and unbound cross-bridges become re-attached according to first-order kinetics, with rate constants that are functions only of x. In other words,

$$G(x,t) = [n_{max}(x) - n(x,t)]g(x) \tag{7.16}$$

and

$$H(x,t) = n(x,t)h(x) \tag{7.17}$$

where $n_{max}(x)$ is the maximum possible number of bound cross-bridges at length x. Huxley further assumed the rate-constant functions $g(x)$ and $h(x)$ to be linear with x. Substituting Eqs. 7.16 and 7.17 into Eq. 7.14 allows analytic solutions to the Huxley model to be obtained. Notably, because of the influence of shortening velocity on cross-bridge attachment and detachment, the Huxley model predicts force transients of the type seen in Fig. 7.5a.

The Huxley model has been extended to incorporate a mechanism thought to be specific to smooth muscle, namely the formation of latch-bridges [23, 24]. The latch-bridge purports to explain how smooth muscle can maintain force production without undue expenditure of energy. In skeletal muscle, force generation is the result of active cross-bridge cycling that requires dephosphorylation of a molecule of ATP each time a cross-bridge releases its hold on an actin binding site. This occurs continually in the presence of cytosolic calcium. Active force in skeletal muscle is thus either turned on or turned off. In smooth muscle, due to its rather different excitation-contraction coupling mechanism, actin and myosin can interact in four distinct ways: (1) unphosphorylated and thus unbound to actin (no force generation), (2) phosphorylated but still unbound to actin (no force generation), (3) phosphorylated and bound to actin (active cross-bridge cycling that generates force in the conventional manner), and (4) unphosphorylated but nevertheless still bound to actin (the latch state in which force is maintained by cross-bridges that do not actively cycle). A kinetic scheme has been proposed that defines how myosin transitions between these four states [25]. The Huxley model thus not only applies to skeletal muscle but can also be adapted to account for the excitation-contraction coupling peculiarities of smooth muscle. Nevertheless, the latch-bridge hypothesis remains controversial [26], so its incorporation into the Huxley model may still be questioned.

A limitation of the Huxley model is that it does not account for any dependence of muscle force on length. This is less of an issue for ASM compared to skeletal muscle because of the labile nature of the ASM contractile apparatus that allows force to adapt to changes in muscle length [27]. Nevertheless, models of how the contractile apparatus dynamically rearranges in ASM have been proposed [28–30]. There is another conceptual problem, however, with direct application of the Huxley model to ASM. Whereas the highly regular parallel arrangement of the contractile fibers in skeletal muscle (Fig. 7.1a) motivates precise specification of $n(x, t)$ in Eq. 7.14, this is more difficult to justify in ASM because the contractile proteins are aligned at various angles with respect to the direction of shortening (Fig. 7.1b). Furthermore, these angles change as the ASM cell shortens. One way of potentially dealing with this limitation is to specifically include the orientation distribution of actin-myosin pairs and its evolution in a computational model [31], but this significantly increases the computational complexity of the model.

An alternative approach, which still retains the key operational features of the Huxley model, is to avoid requiring a specific functional form for $n(x, t)$ altogether by eliminating its dependence on x through spatial averaging. This begins with the assumption that n_{max} is a constant and thus independent of x. The rate of change of active muscle force is then, from Eqs. 7.15, 7.16 and 7.17,

$$\begin{aligned}\dot{F}(t) &= \int_{-X}^{X} f(x)\frac{\partial n(x,t)}{\partial t}dx \\ &= \int_{-X}^{X} f(x)[(n_{max}-n(x,t))g(x)-n(x,t)h(x)]dx + \int_{-X}^{X} f(x)v(t)\frac{\partial n(x,t)}{\partial x}dx.\end{aligned} \tag{7.18}$$

Applying integration by parts to the final term above gives

$$\int_{-X}^{X} f(x)v(t)\frac{\partial n(x,t)}{\partial x}dx = v(t)[f(x)n(x,t)]_{-X}^{X} - v(t)\int_{-X}^{X}\frac{\partial f(x,t)}{\partial x}n(x,t)dx. \tag{7.19}$$

Choosing X to extend beyond all physically possible extensions of the cross-bridge spring means that $n(-X, t) = n(X, t) = 0$. Assuming that the cross-bridges behave as Hookean springs such that $f(x) = Ex$ where E is a spring constant, Eq. 7.19 then becomes

$$-v(t)\int_{-X}^{X}\frac{\partial f(x,t)}{\partial x}n(x,t)dx = -Ev(t)\int_{-X}^{X} n(x,t)dx = -Ev(t)N(t) \tag{7.20}$$

where $N(t)$ is the total number of cross-bridges bound at time t.

In order to deal with the first integral on the final line of Eq. 7.18 it is necessary to know the functions $g(x)$ and $h(x)$. These functions remain a matter of debate even for skeletal muscle [32], so there is little to guide their choice for ASM. Assume for simplicity, therefore, that

$$g(x) = k_{on}; x \in [0, X] \tag{7.21}$$

where k_{on} is a constant, and the lower limit of 0 for x ensures that cross-bridge binding acts only to shorten the muscle and not lengthen it. Cross-bridge detachment, on the other hand, is increased by the relative movement of interacting actin and myosin fibers regardless of direction, so detachment occurs during both lengthening and shortening. There is experimental evidence that detachment rate depends linearly on the magnitude of velocity [33], implying

$$h(x) = k_{off} \frac{|v(t)|}{v_{max}}; x \in [-X, X] \tag{7.22}$$

where k_{off} is another constant, and v_{max} is the maximum (unloaded) shortening velocity of the muscle. (Normalizing $v(t)$ by v_{max} ensures that k_{off} has the same units as k_{on}.)

Employing these expressions for $g(x)$ and $h(x)$ in the first integral term in Eq. 7.18 gives

$$\begin{aligned} &\int_{-X}^{X} f(x)[(n_{max} - n(x,t))g(x) - n(x,t)h(x)]dx \\ &= Ek_{on}n_{max} \int_0^X xdx - Ek_{on} \int_0^X xn(x,t)dx - Ek_{off} \frac{|v(t)|}{v_{max}} \int_0^X xn(x,t)dx \\ &= Ek_{on}n_{max} \frac{X^2}{2} - Ek_{on} \int_0^X xn(x,t)dx - k_{off} \frac{|v(t)|}{v_{max}} F(t). \end{aligned} \tag{7.23}$$

The middle term in Eq. 7.23 is proportional to the component of $F(t)$ arising from cross-bridges bound at $x \geq 0$, which act in the direction of muscle shortening. These are opposed by the cross-bridges bound at $x < 0$. The greater the velocity of shortening, the greater will be the fraction of cross-bridges bound at $x < 0$, until the force opposing contraction eventually equals that producing it and the muscle reaches v_{max} and generates zero force. There is also experimental evidence from skeletal muscle that the total number of attached cross-bridges decreases as shortening velocity increases [33]. Assuming these effects cause $F(t)$ to decrease linearly with shortening velocity gives

$$Ek_{on} \int_0^X xn(x,t)dx = k_{on}F(t)\left[1 - \frac{v(t)}{v_{max}}\right]. \tag{7.24}$$

Substituting Eqs. 7.20, 7.23 and 7.24 into Eq. 7.18 gives

$$\dot{F}(t) = Ek_{on}n_{max}\frac{X^2}{2} - k_{on}F(t)\left[1 - \frac{v(t)}{v_{max}}\right] - k_{off}\frac{|v(t)|}{v_{max}}F(t) - Ev(t)N(t). \quad (7.25)$$

The first term in Eq. 7.25 can be expressed as $k_{on}(n_{max}X)EX/2$. The factor $n_{max}X$ is the maximum number, N_{max}, of cross-bridges that can be bound across the entire muscle cross-section. The factor $EX/2$ is the mean force, F_{br}, in each individual bound cross-bridge based on the assumptions that the cross-bridges behave as Hookean springs and that they bind in uniform concentration over x. Furthermore, the product $N_{max}F_{br}$ is the maximum force, F_{max}, that the muscle can generate. Finally, an increase in the relative motion of the actin and myosin fibers (i.e., increasing $v(t)$) causes increasing numbers of cross-bridges to become detached, thereby decreasing $N(t)$. Assuming an inverse relationship between these two variables reduces the last term in Eq. 9.25 to a constant, α. With these observations, Eq. 7.25 becomes

$$\begin{aligned} \dot{F}(t) &= k_{on}F_{br}N_{max} - k_{on}F(t)\left[1 - \frac{v(t)}{v_{max}}\right] - k_{off}\frac{|v(t)|}{v_{max}}F(t) - \alpha \\ &= (k_{on}F_{br}N_{max} - \alpha) - k_{on}F(t) + \frac{k_{on}}{v_{max}}v(t)F(t) - \frac{k_{off}}{v_{max}}|v(t)|F(t). \end{aligned} \quad (7.26)$$

N_{max} is the maximum number of cross-bridges that can be bound, which in turn determines F_{max}. Experimental evidence suggests that suddenly stretching activated ASM by a sufficient amount temporarily damages its contractile machinery (Fig. 7.5a). This reduces N_{max}. Suppose there is a threshold tissue length, l_{crit}, such that if activated ASM is stretched above this length, its contractile ability becomes ablated to a degree proportional to the amount by which tissue strain, $l(t)$, exceeds l_{crit}. Further assume that recovery from damage occurs exponentially with time (i.e., the muscle eventually regains its full force-generating capacity as the disrupted contractile components re-polymerize). N_{max} is then governed by

$$\begin{aligned} \dot{N}_{max}(t) &= k_{fix}[N_{max,0} - N_{max}(t)] - k_{break}[l(t) - l_{crit}]N_{max}(t); l(t) \geq l_{crit} \\ &= k_{fix}[N_{max,0} - N_{max}(t)]; l(t) < l_{crit} \end{aligned} \quad (7.27)$$

where k_{break} and k_{fix} are rate-constants, and $N_{max,0}$ is the maximum value of N_{max} achieved when the contractile machinery is fully intact. Equations 7.26 and 7.27 together predict the oscillatory force-velocity relationships shown in Fig. 7.8, which recapitulate the post stretch decreases in recovery force seem experimentally in Fig. 7.5a [21].

Equations 7.26 and 7.27 also explain the essential features of the ASM force-velocity relationships. When the muscle contracts at a constant rate so that $|v| = v = dx(t)/dt$, Eq. 7.26 becomes

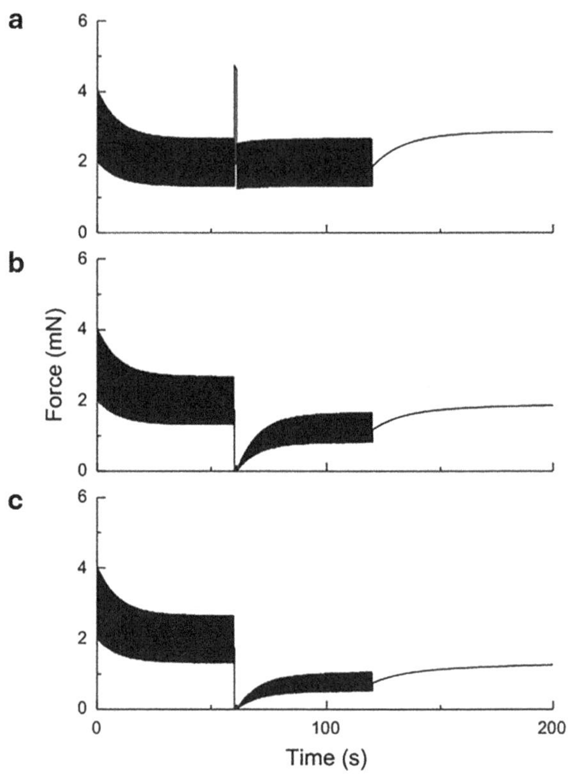

Fig. 7.8 Effect of brief stretches, applied at 60 s on the oscillatory force-length behavior of ASM as determined by Eq. 11.26, with N_{max} being modulated according to Eq. 7.27. The stretch amplitudes are (**a**) 5% l_{ref}, (**b**) 15% l_{ref}, and (**c**) 25% l_{ref}. This recapitulates the stretch induced decrements in plateau recovery pressure seen in the experimental data in Fig. 7.5a and is not evident in the model simulations shown in Fig. 7.5b. (Reproduced with permission from [21])

$$\dot{F}(t) = A - (B + Cv)F(t) \tag{7.28}$$

where $A = P_{br}N_{max}k_{on}$, $B = k_{on}$ and $C = \frac{(k_{off} - k_{on})}{v_{max}} + \frac{E}{P_{br}}$. The solution to Eq. 7.28 is

$$F(t) = \frac{A}{B + Cv} + Me^{-(Cv+B)t} \tag{7.29}$$

where M is a constant determined by the initial conditions. The steady state solution to this equation is

$$F = \frac{A}{B + Cv} = F_i + F_e \tag{7.30}$$

where F_e is the external load and F_i is the internal load due to the structural components of ASM itself. Equation 7.30 rearranges to give

$$v = \frac{\frac{B}{C}\left[\left(\frac{A}{B} - F_i\right) - F_e\right]}{F_i + F_e} = \frac{a(F_0 - F_e)}{a + F_e} \tag{7.31}$$

where a and b are constants, and F_0 is isometric force. Equation 7.31 has precisely the form of the classic Hill equation (Eq. 7.1), which gives an underlying mechanistic explanation to the parameters in this equation.

The influence of muscle shortening velocity on the dynamics of cross-bridge attachment and detachment is thus an inherent consequence of the Huxley model (Eq. 7.14) for skeletal muscle, or its coarse-grained counterpart in the case of the less organized contractile proteins in ASM (Eq. 7.26). ASM also involves additional dynamics concerning the transient lability of its contractile machinery, as embodied in Eq. 7.27.

7.4 Muscle-Airway Interactions

ASM is aligned roughly circumferentially around the airway wall and thus acts like a ligature that, when activated, narrows the airway lumen. In order for narrowing to occur, however, the ASM must generate sufficient force to overcome the mechanical loads that oppose its shortening. These loads arise from a variety of sources (Fig. 7.9), which can be grouped into those that contribute to the compressive stiffness of the airway wall and those that determine the outward tethering forces

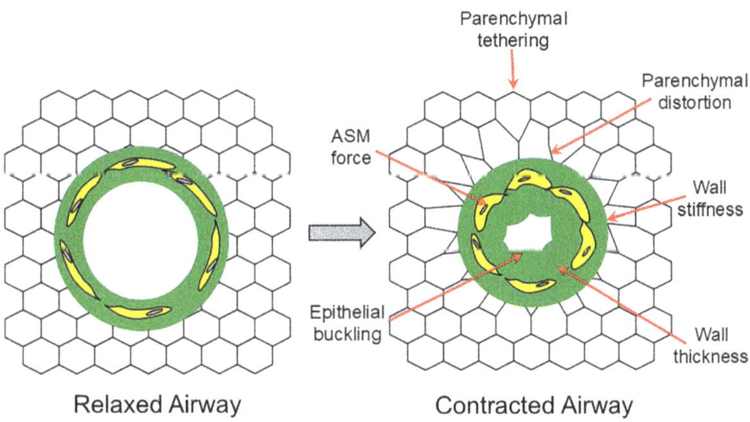

Fig. 7.9 Key factors influencing how much airway narrowing can occur when ASM is stimulated to contract. The compressive stiffness of the airway wall (green) and the outward tethering forces exerted by the parenchyma, exacerbated by local distortion, act to oppose the contractile forces generated by the ASM cells. Thickening of the submucosa geometrically amplifies the narrowing of the airway lumen, although the forces required to buckle the inner border have been hypothesized as protective against bronchoconstriction. Thickening of the adventitia may decouple the airway wall from parenchymal tethering and enhance airway responsiveness

exerted on the airway wall by its alveolar attachments. The airway wall itself is divided into the submucosa and the adventitia, which are located on the luminal and parenchymal sides of the ASM, respectively. The alveolar attachments impinge on the outer boundary of the adventitia.

The compressive stiffness of the airway wall as a whole arises from the mechanical integrity of its connective tissue components (including those in the ASM) that, under baseline conditions, are configured to be at equilibrium with the outer tethering forces exerted by the parenchyma (discussed below). Wall stiffness varies with airway generation since the relative amount of stiff cartilage in the wall decreases as the airways become smaller and more distal. A model that has been used to express the relationship between transmural pressure (*P*) and airway area (*A*) is based on two hyperbolic equations matched for their values at zero pressure thus [34]:

$$\begin{aligned} a_n &= a_{n0}\left(1-\frac{P}{P_1}\right)^{-b_1} ;P\leq 0 \\ a_n &= 1-(1-a_{n0})\left(1-\frac{P}{P_2}\right)^{-b_2} ;P\geq 0 \end{aligned} \tag{7.32}$$

where $a_n = A/A_{max}$ is the normalized area, a_{n0} is the value of a_n when $P = 0$, and A_{max} is the maximum possible airway area. Eqs. 7.31 and 7.32 determine the values of P_1 and P_2, respectively, in terms of the other parameters. The value of $da_n/dP = a'_{n0}$ at $P = 0$ must also be specified and must be identical for both expressions above. Based on limited experimental data, the resulting five free parameters for this model (a_{n0}, a'_{n0}, b_1, b_2 and A_{max}) have been evaluated for the first 17 generations of the airway tree in a computational model of maximal expiratory flow [34]. These airway pressure-area relationships have also been used in other models of lung mechanics [35, 36], but they are not the only options. For example, a more detailed model with finite wall thickness has been used to analyze the effects of the mechanical environment on cells resident within the airway walls [37].

The lung parenchyma mediates the pressure between the pleural space and the airway lumen, which under static conditions equals transpulmonary pressure. The alveolar walls that attach to the outsides of the intraparenchymal airways exert an outward tethering force on the airway wall that approximates transpulmonary pressure, at least to the extent that the parenchyma is uniformly inflated. Accordingly, the parenchymal load opposing ASM shortening increases with increasing transpulmonary pressure. This gives rise to a profound dependence of airways responsiveness on lung volume [38–40]. Indeed, taking a deep breath (or imposing one with an externally applied airway pressure) is one of the most effective ways to reverse bronchoconstriction, at least in normal lungs [41].

Equating the parenchymal tethering forces on the airway to transpulmonary pressure is only exact, however, if the parenchyma expands isotropically with lung inflation (and any flow-related pressure drops along the airways are neglected). Airway radius then increases with the 1/3 power of lung volume thus:

$$\frac{r_{iso}}{r_{isoTLC}} = \left(\frac{P_{tp}}{P_{tpTLC}}\right)^{\frac{1}{3}} \tag{7.33}$$

where r_{iso} is the radius of the airway under conditions of isotropic parenchymal expansion, r_{isoTLC} is the value of r_{iso} at total lung capacity (TLC), and P_{tpTLC} is transpulmonary pressure at TLC. In general, however, airway radius (r) does not equal r_{iso} because the airways and parenchyma have different intrinsic elastic properties and thus expand at different relative rates. (The parenchyma also expands anisotropically, but this will be neglected here.) This causes the parenchyma in the vicinity of the airway to become distorted, as illustrated in Fig. 7.10, giving rise to an additional component to the transmural pressure (ΔP_{tm}) across the airway wall. Lai-Fook [42] showed that $\Delta P_{tm}/2\mu_s$ is a single-valued curvilinear function of r/r_{iso}, where μ_s is the parenchymal shear modulus. Using a value for $\mu_s = 0.7P_{tp}$ [42] gives

$$\frac{\Delta P_{tm}}{P_{tp}} = \frac{1}{0.7}\left(\frac{r_{iso}}{r} - 1\right). \tag{7.34}$$

Combining Eqs. 7.33 and 7.34 leads to

$$P_{tm} = P_{tp} + \Delta P_{tm} = \left[\frac{r_{isoTLC}}{r}\left(\frac{P_{tp}}{P_{tpTLC}}\right)^{\frac{1}{3}} - 0.3\right]\frac{P_{tp}}{0.7}. \tag{7.35}$$

The outward pull of the parenchyma on the airway wall is balanced by the inward recoil generated by the elastic properties of the airway wall, the latter producing a passive hoop stress F_P (Fig. 7.10a). In the case where the elastic properties of the airway and parenchyma are perfectly matched and the parenchyma expands isotropically, F_P is estimated from the Laplace law for a cylinder to be

$$P_{tm} = \frac{F_P}{r}. \tag{7.36}$$

Strictly speaking, Eq. 7.36 only applies in the case of a cylindrical airway with a very thin wall that has no bending stiffness. However, it has been shown numerically that the Laplace law provides a good approximation to the pressure-radius relationship for a thick-walled cylinder [43].

Combining Eqs. 7.33 and 7.36 gives the following expression for the hoop stress when the lung parenchyma expands isotropically:

$$F_{isoP} = r_{iso}P_{tm} = r_{isoTLC}P_{tp}\left(\frac{P_{tp}}{P_{tpTLC}}\right)^{\frac{1}{3}}. \tag{7.37}$$

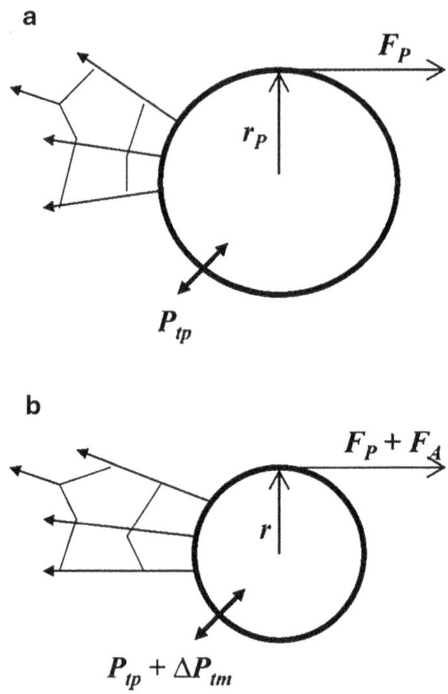

Fig. 7.10 (**a**) When ASM is relaxed, a circular airway with radius r develops an inward recoil pressure determined by the passive force in the airway wall (F_P) through the Laplace law (Eq. 7.36). When the parenchyma expands isotropically, this inward pressure is balanced by the outward pull of the parenchymal attachments that give rise to transpulmonary pressure (P_{tp}). (**b**) When the ASM develops an active force (F_A), the airway narrows until the increased peribronchial pressure (ΔP_{tm}) due to local parenchymal distortion balances the inward recoil of the airway. (Adapted with permission from [44])

The hoop stress increases when the ASM is stimulated to contract through the addition of an active component, F_A, causing Eq. 7.36 to become

$$P_{tm} = \frac{F_P + F_A}{r}. \tag{7.38}$$

Combining Eqs. 7.35, 7.37 and 7.38 then gives F_A for the case of isotropic parenchymal expansion to be

$$F_{isoA} = rP_{tm} - F_{isoP} = \left[r_{isoTLC} \left(\frac{P_{tp}}{P_{tpTLC}} \right)^{\frac{1}{3}} - r \right] \frac{0.3P_{tp}}{0.7}. \tag{7.39}$$

In general, however, the relatively stiffer airway expands more slowly than the parenchyma as the lung inflates. A simple way to model the expansion of the airway in this anisotropic parenchyma case [36] is

$$\frac{r}{r_{anisoTLC}} = \sigma + (1-\sigma)\left(\frac{P_{tm}}{P_{tpTLC}} \right)^{\frac{1}{3}} = \sigma + (1-\sigma)\left(\frac{F_{anisoP}}{rP_{tpTLC}} \right)^{\frac{1}{3}} \tag{7.40}$$

where $0 < \sigma < 1$ is a constant and $r_{anisoTLC}$ is the airway radius achieved at TLC. This model assumes that the airway wall is composed of two components, one being completely inextensible and comprising a fraction σ of the total circumference, and the other expanding as described by Eq. 7.33.

Even though the airway and the parenchyma expand with P_{tp} at different rates, there exists some value of $P_{tp} = P_0$ for which the state of inflation of the airway and parenchyma are matched. At this pressure, there is no local parenchymal distortion and $\Delta P_{tm} = 0$, in which case r_{iso} in Eq. 7.33 and r in Eq. 7.40 are equal, yielding

$$r_{anisoTLC} = \frac{r_{isoTLC}\left(\frac{P_0}{P_{tpTLC}}\right)^{\frac{1}{3}}}{\sigma + (1-\sigma)\left(\frac{P_0}{P_{tpTLC}}\right)^{\frac{1}{3}}}. \tag{7.41}$$

Combining Eqs. 7.40 and 7.41 gives the hoop stress under conditions of anisotropic parenchymal expansion:

$$F_{anisoP} = rP_{tpTLC}\left[\frac{r\sigma}{(1-\sigma)r_{isoTLC}\left(\frac{P_0}{P_{tpTLC}}\right)^{\frac{1}{3}}} + \frac{r}{r_{isoTLC}} - \frac{\sigma}{1-\sigma}\right]^3. \tag{7.42}$$

In analogy with Eq. 7.39, this gives the component of the hoop stress due to the active ASM force as

$$\begin{aligned} F_{anisoA} = &\left[r_{isoTLC}\left(\frac{P_{tp}}{P_{tpTLC}}\right)^{\frac{1}{3}} - 0.3r\right]\frac{P_{tp}}{0.7} \\ &- rP_{tpTLC}\left[\frac{r\sigma}{(1-\sigma)r_{isoTLC}\left(\frac{P_0}{P_{tpTLC}}\right)^{\frac{1}{3}}} + \frac{r}{r_{isoTLC}} - \frac{\sigma}{1-\sigma}\right]^3. \end{aligned} \tag{7.43}$$

If F_{anisoA} from Eq. 7.43 is equated to $F(t)$ in Eqs. 7.1 and 7.2, the dynamics of airway narrowing can be calculated (assuming active muscle force is independent of its length). Specifically, Eq. 7.43 is integrated using a prescribed P_{tp} profile to determine F_{anisoA}. At each time step in this process, the result is substituted into either Eq. 7.1 or 7.2 (depending on whether the ASM is shortening or lengthening, respectively) to provide dr/dt, which is then integrated to give $r(t)$. Finally, $r(t)$ is converted to the flow resistance of the airway (R_{aw}) assuming that airway length remains fixed and that flow is laminar, in which case

$$R_{aw} \propto \left(\frac{r_{isoTLC}}{r}\right)^4. \tag{7.44}$$

The complete model has 5 independent free parameters—F_0, a, and b from Eqs. 7.1 and 7.2, and σ and P_0 from Eq. 7.43. Five parameters are few enough for inverse modeling (Sect. 1.4). That is, the values of the parameters can be adjusted to provide a least-square fit of the model predictions of R_{aw} to experimental data [44]. This model accurately accounts for the strong volume dependencies of R_{aw} in rats ventilated at different levels of positive-end expiratory pressure (Fig. 7.11a) and rabbits ventilated with different tidal volumes (Fig. 7.11b) following intravenous bolus injections of methacholine.

The fits in Fig. 7.11 show no obvious systematic deviations from the experimental data suggestive of the need for additions to the model to improve its predictive accuracy. Nevertheless, this model of dynamic airway constriction is arguably the simplest that is at all physiologically reasonable. It certainly cannot mimic every type of behavior that ASM exhibits. In particular, it takes no account of the transient post-stretch decrements in active force that are clearly evident at the level of the ASM strip (Fig. 7.5a). This is likely because such decrements are not relevant to the data shown in Fig. 7.11 for which the stretches applied to the ASM were not large and/or vigorous enough to disrupt the contractile machinery as represented in Eq. 7.27. In fact, evidence of such disruption in vivo requires large and very rapid inflations of the lung during the period when bronchoconstriction is developing [38]. For example, combining Eq. 7.45 with Eqs. 7.26 and 7.27 provides accurate predictions of the time-course of R_{aw} in mice subjected to sudden very large lung inflations at various time points following the bolus intravenous injection of methacholine (Fig. 7.12) [45]. These data cannot be recapitulated without representing the transient decrements in ASM contractility that come with a very large and sudden stretch. Nevertheless, these are conditions that are unlikely to ever pertain clinically, demonstrating that phenomena evident at one level of length scale do not always translate up to higher levels of scale with the same degree of importance.

The model considered above ignores numerous other phenomena that potentially have a significant influence on the link between ASM activation and narrowing of the airway lumen. Important among these are geometric effects related to the thickness of the submucosa [46]. When an airway constricts, the submucosal tissue is constrained to remain toward the luminal side of the ASM, so its cross-sectional area is conserved. This causes the radius of the lumen to narrow more rapidly than that of the ring of ASM itself, a discrepancy that accelerates as the thickness of the submucosa increases. On the other hand, it has been proposed that the forces required to buckle the inner border of the submucosa may be protective against bronchoconstriction [47], while having implications for bronchial perfusion and remodeling [48]. Expansion of the submucosa, as occurs in allergic inflammation, is thus a potent mechanism of airway hyperresponsiveness that manifests even when the degree of ASM shortening is perfectly normal. Another geometric factor is the orientation of the ASM cells relative to the airway wall. These cells are arranged at a variety of pitch angles relative to the circumferential direction, with a mean angle of about 30° [49]. Computational modeling supports the possibility that variations in this angle could have a significant effect on the degree of airway narrowing caused by a given degree of ASM shortening [50]. Expansion of the airway adventitia, also a consequence of airway inflammation, has the effect of partially decoupling the

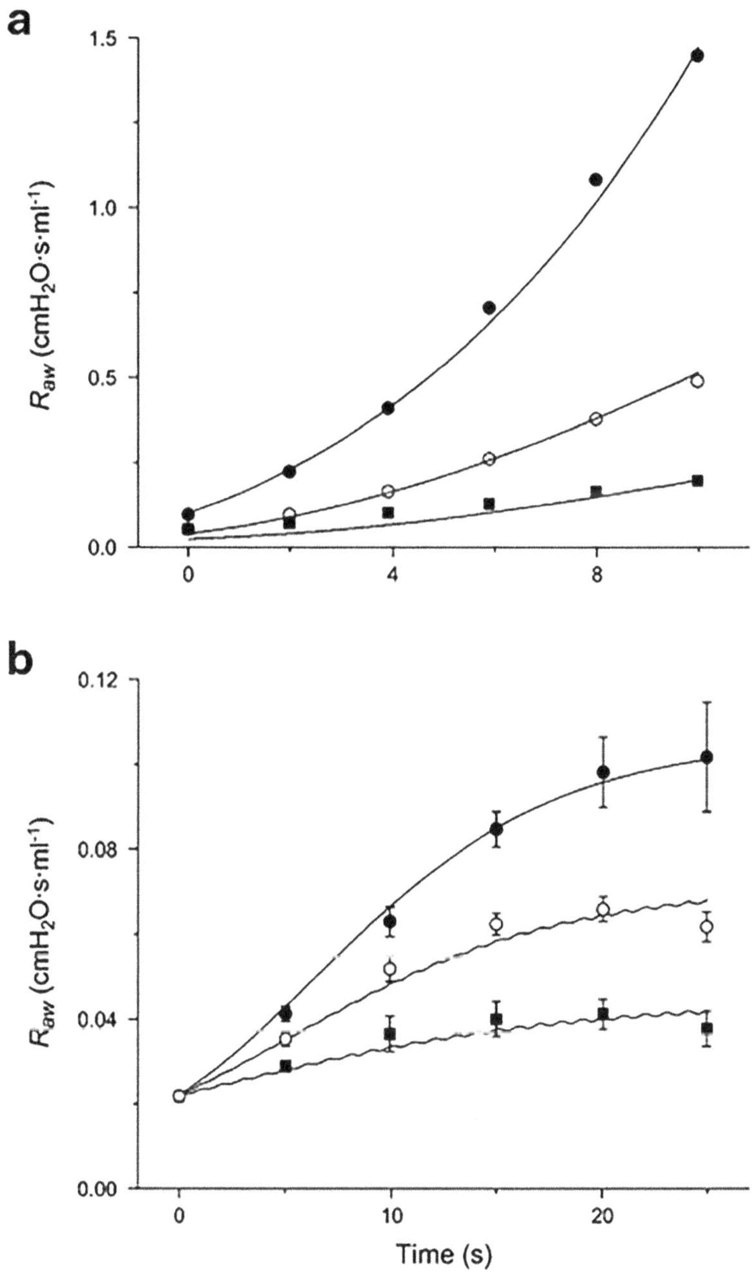

Fig. 7.11 Experimentally measured time courses of airway resistance (R_{aw}) together with simulated curves obtained using Eqs. 7.1 and 7.2 together with Eqs. 7.43 and 7.44. (**a**) Data from rats at PEEP levels of positive end-expiratory pressure (PEEP). (**b**) Data from rabbits ventilated with different tidal volumes (V_T). (Reproduced with permission from [44])

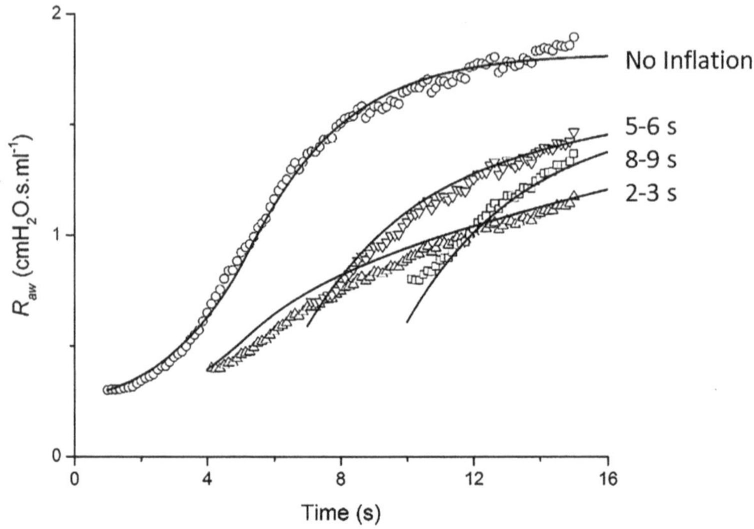

Fig. 7.12 Effects of a sudden large brief lung inflation given at various times after methacholine injection. Symbols show mean experimental time courses of R_{aw} without inflation (○), with an inflation at 2–3 s (Δ), 5–6 s (∇), and 8–9 s (□) post-methacholine injection. The fit from the model to the non-inflation data is shown as the upper solid line. The three lower solid lines are model predictions for the 3 differently timed inflations. (Reproduced with permission from [45])

airway wall from its alveolar attachments, thereby reducing the load on the ASM similar to what happens when lung volume is reduced [51].

Another interesting phenomenon related to airway narrowing arises from the nature of the Laplace law (Eq. 7.38), which shows that for a fixed F_A, the pressure P_{tm} maintaining airway patency decreases as the lung inflates (r increases) if F_P remains unchanged. Of course, the mechanical properties of the airway wall are such that F_P typically does increase with lung volume, as do the outward tethering forces of the parenchymal attachments. However, if the Laplace law effect is stronger than the mitigating actions of airway wall stiffness and parenchymal tethering, it is theoretically possible for an instability to develop in which there are two values of r compatible with a given value of F_A. (A similar instability is possible for the alveolus—see Sect. 4.6.) The situation is further complicated by the axial pressure gradient due to gas flow along the airway and the peribronchial pressure that develops as the alveoli inflate around it, raising the possibility that an airway might suddenly transition between two different values of r when the ASM becomes activated, rather than there being a smooth decrease in r with time as implied by the data and model predictions in Figs. 7.11 and 7.12. It remains unclear whether such a bistability manifests in practice, but the concept has been formulated in a computational model that shows how such rapid transitions in airway radius might actually happen [52].

7.5 Airway Responsiveness at the Level of the Whole Lung

The foregoing makes it clear that bronchoconstriction is a multi-scale event involving phenomena taking place at the molecular, tissue, airway and whole lung scales. This makes bronchoconstriction a prime candidate for multiscale modeling [53]. Indeed, this has already been exemplified in the previous section with respect to the modeling of ASM and its interactions with the airway and parenchymal tissues. Experimental data recorded during length cycling of activated strips of ASM demand that cross-bridge population dynamics be modeled explicitly, as exemplified in the partial differential equation of the Huxley model (Eq. 7.14) or its coarse-grained ordinary differential equation counterpart (Eqs. 7.26 and 7.27). Without these details it is impossible to account for the temporary ablation of ASM contractile ability caused by a sudden large stretch (Figs. 7.5 and 7.8). At the level of the whole lung, however, unless an extreme and non-physiologic maneuver is applied (Fig. 7.12), the much simpler description of ASM contraction dynamics given by the Hill equation (Eqs. 7.1 and 7.2) suffices to accurately account for the time-course of airway narrowing (as shown in Fig. 7.11). In other words, not all the phenomena modeled at the micro scale must necessarily be carried up explicitly to models at the macro scale. Macro scale behavior in a complex system is almost always emergent, meaning it arises from the ensemble behavior of the system components operating at the micro scale. This emergent behavior can invariably be accounted for in ways that are empirical and much simpler than would be the case if all activity at the micro scale were specifically included. In fact, doing the latter may mitigate against insight by obfuscating what is most important at the macro scale.

On the other hand, moving up in scale can introduce completely new phenomena that are not apparent at lower levels of scale. An example of this is the appearance of regional mechanical heterogeneity in the lung. The airway tree contains very many individual branches that can potentially narrow by different amounts during bronchoconstriction, causing regional heterogeneities in lung ventilation. Such heterogeneities are considered to be an important part of the pathophysiology of asthma [54], yet obviously cannot be represented in a model of a single constricting airway or of the crossbridges within a single ASM cell. The first attempts to represent ventilation heterogeneities used models of two compartments (see Chap. 3). These models recapitulate the essential mechanical consequences of ventilation heterogeneity and thus have been useful conceptually, but they have little to say about the precise nature and location of the events causing the heterogeneity. The need for a better link between structure and function eventually led to the advent of airway tree models bearing some degree of correspondence to anatomic reality [55]. Early models incorporated the Weibel symmetrically bifurcating airway tree [56] in which the branches of the tree are labeled according to their generation number, beginning with the trachea as generation 0 (Fig. 2.1a). This has the computational advantage that each generation can be represented by a single equivalent airway that has a resistance equal to that of a single branch divided by the number of branches in that generation. Such models have been used to investigate the biomechanical and geometric factors that potentially influence R_{aw} in asthma [57, 58].

A key limitation of the Weibel model, however, is that it does not represent the airway branching asymmetry that is present to different degrees in different animal species, including humans. This asymmetry is conveniently approximated by the self-similar asymmetric branching scheme introduced by Horsfield [59] in which the individual airway branches are assigned an order number starting with the terminal bronchioles as order 1 (Fig. 2.1b). The two daughter airways branching off from any parent in the tree differ in their own order numbers by the integer Δ. The great advantage of the Horsfield scheme from a computational perspective is that it allows the mechanical input impedance of the airways, Z_{aw} (a generalization of R_{aw} that includes the effects of airway inertance and elastance), to be calculated with very efficient software code that uses a recursive subroutine (a subroutine that can call itself) as follows. Starting with the trachea (order N), Z_L is the sum of the impedance of the trachea (Z_N) and the parallel combination of the total impedances of the down-steam sub-trees that are subtended by an order $N - 1$ airway ($Z_{N-1,tot}$) and an order $N - 1 - \Delta_N$ airway ($Z_{N-1-\Delta_N,tot}$). That is,

$$Z_L = Z_N + \left(\frac{1}{Z_{N-1,tot}} + \frac{1}{Z_{N-1-\Delta_N,tot}}\right)^{-1} \tag{7.45}$$

where

$$Z_{N-1,tot} = Z_{N-1} + \left(\frac{1}{Z_{N-2,tot}} + \frac{1}{Z_{N-2-\Delta_{N-1},tot}}\right)^{-1} \tag{7.46}$$

and

$$Z_{N-1-\Delta_N,tot} = Z_{N-1} + \left(\frac{1}{Z_{N-2-\Delta_{N-1},tot}} + \frac{1}{Z_{N-2-2\Delta_{N-1},tot}}\right)^{-1} \tag{7.47}$$

and so on. The impedance of the entire lung, Z_L (Sect. 3.6) can thus be determined using a subroutine that performs Eqs. 7.46 and 7.47 repeatedly for each branch of the airway tree, each time calling itself to calculate the two down-stream impedances. The subroutine proceeds in this way until it reaches a terminal airway (order 1) at which point it appends the impedance of a terminal tissue unit, Z_{tis}, and then stops [36, 60, 61].

So long as the wavelength of sound is large compared to the dimensions of the structures being studied, $Z_L(f)$ can be represented by a lumped-parameter model in which the resistive, inertive and elastic properties of an airway branch are encapsulated in single parameters (Sect. 3.6). This condition pertains in most studies of respiratory system impedance, for which f is typically <40 Hz. If an airway tree branch of order j is represented by a cylindrical conduit of radius r_j and length l_j, and assuming Poiseuille flow, the resistance of the branch, R_j, is given by (Sect. 3.2.2)

$$R_j = \frac{8\mu l_j}{\pi r_j^4} \tag{7.48}$$

where μ is the viscosity of air. The inertance, I_j, of the airway (due to the mass of air it contains) is (Sect. 3.2.5)

$$I_j = \frac{\rho l_j}{\pi r_j^2} \tag{7.49}$$

where ρ is the density of air. The compressibility of the gas in the airway (inversely proportional to airway volume, $\pi r^2 l$) together with the stiffness of the airway wall (e.g., from Eq. 7.40) give rise to a shunt elastance for the airway branch of E_j (Sect. 3.2.4). The contributions of R_j, I_j and E_j to Z_j are, respectively [62],

$$Z_{R_j}(f) = R_j, \tag{7.50}$$

$$Z_{I_j}(f) = i2\pi f I_j, \tag{7.51}$$

and

$$Z_{E_j}(f) = \frac{E_j}{i2\pi f} \tag{7.52}$$

where $i = \sqrt{-1}$.

Assuming that the shunt impedance due to E_j branches off halfway along the airway to create the T-network shown in Fig. 7.13a, Z_j is given by

$$Z_j(f) = \frac{Z_{R_j}(f) + i\pi f I_j}{2} + \left(\frac{1}{\frac{Z_{R_j}(f) + i\pi f I_j}{2}} + \frac{i2\pi f}{E_j} \right)^{-1}. \tag{7.53}$$

The viscoelastic properties of the alveolar parenchyma mean that the impedance of the lung tissues, Z_{tis}, must account for both its elastic and dissipative (resistive) properties. In contrast to the predictions of lumped parameter models such as the Kelvin body (see Sect. 5.2.2), experimental data indicate that the ratio of tissue resistance to reactance is essentially independent of f [63]. This results in Z_{tis} having a phase that is also independent of frequency [64, 65], expressed as

$$Z_{tis}(f) = \frac{(G - iH)}{(2\pi f)^{\alpha}} \tag{7.54}$$

where

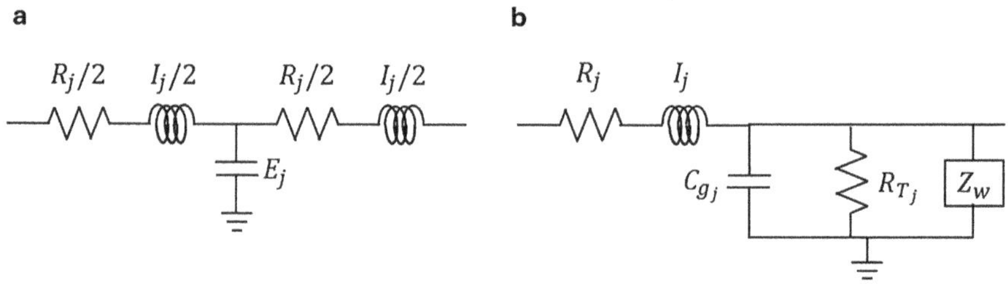

Fig. 7.13 (**a**) A T-network configuration for the impedance of a single airway branch using the lumped-parameter equations. R and I correspond to R_j and I_j in Eqs. 7.48 and 7.49, respectively [36]. (**b**) The unit configuration for the transmission line model; a series of these networks may be used to represent a single airway branch that is long compared to the wavelength of sound. The parameters in this model correspond to those in Eqs. 7.56, 7.57, 7.58, 7.59, 7.60, 7.61, 7.62, 7.63 and 7.64 [67]. The wall impedance Z_w could be of the form of the constant-phase model (Eqs. 7.54 and 7.55)

$$\alpha = \frac{2}{\pi}\tan^{-1}\frac{H}{G} \tag{7.55}$$

and the parameters G and H are referred to as tissue damping and elastance, respectively (Sect. 5.3). A branching airway model of the above form has been used to mimic experimental observations of the dynamics of bronchoconstriction in the dog lung by forcing the calibers of the airways and lung elastance to follow prescribed time courses [36].

When f increases into the acoustic range (up to several hundred Hz or more), the long wavelength assumption for sound no longer applies. In this case, all the individual airway impedances can no longer be expressed as lumped parameter models (particularly for the longer branches such as the trachea and mainstem bronchi). Now it is necessary to model each branch as an acoustic transmission line as illustrated in Fig. 7.13b [66]. Defining c as the velocity of sound, C_P and C_V as the specific heats of air at constant pressure and volume, $\psi = C_P/C_V$, and J_0 and J_1 as the zeroth- and first-order Bessel functions, the lumped-parameter equations above are replaced by [67].

$$R_j = \left(\frac{2\pi f\rho}{\pi r^2}\right)\left(\frac{F_v \sin\phi_v}{D^2}\right), \tag{7.56}$$

$$I_j = \left(\frac{\rho}{\pi r^2}\right)\left(\frac{1 - F_v \sin\phi_v}{D^2}\right), \tag{7.57}$$

$$R_T = \frac{-\rho c^2}{2\pi^2 f r^2 (\psi - 1) F_t \sin\phi_t}, \tag{7.58}$$

and

$$C_g = \frac{\pi r^2[1 + (\psi - 1)F_t \cos \phi_t]}{\rho c^2} \tag{7.59}$$

where

$$F_v e^{i\phi_v} = \left(\frac{2}{r_v\sqrt{-i}}\right)\frac{J_1\left(r_v\sqrt{-i}\right)}{J_0\left(r_v\sqrt{-i}\right)}, \tag{7.60}$$

$$r_v = r\sqrt{\frac{2\pi f\rho}{\mu}}, \tag{7.61}$$

$$F_t e^{i\phi_t} = \left(\frac{2}{r_v\sqrt{-i}}\right)\frac{J_1\left(r_v\sqrt{-i}\right)}{J_0\left(r_v\sqrt{-i}\right)}, \tag{7.62}$$

$$r_t = r\sqrt{\frac{2\pi f C_P}{\kappa}}, \tag{7.63}$$

and

$$D^2 = (1 - F_v \cos \phi_v)^2 + (F_v \sin \phi_v)^2. \tag{7.64}$$

where r is tube radius, c is sound velocity in air, and κ as thermal conductivity.

Equations 7.56, 7.57, 7.58, 7.59, 7.60, 7.61, 7.62, 7.63 and 7.64 are used to describe propagation of flow along one segment of the airway (Fig. 7.13b). Multiple segments are used to model an entire airway branch when the branch is long compared to the wavelength of sound at the oscillation frequency f [67–69].

The deterministic versions of the branching airway models given above assume that the dimensions of all airways of a given order are identical. The unique utility of branching airway models is not realized, however, until the radii of the individual airways within a given generation are allowed to vary to create the kinds of regional ventilation heterogeneities that typify bronchoconstriction. Monte-Carlo methods may then be used, for example, to determine how statistical variations in airway radii translate into the statistics of R_{aw} [70], or to investigate putative mechanisms underlying the behavior observed experimentally during induced bronchoconstriction [36]. Variations in airway dimensions within a given order can still be implemented with a recursive subroutine by assigning dimensions to each individual airway in the tree, with each call of the subroutine, based on appropriate probability distribution functions [61], thereby retaining the computational efficiency of this approach.

The most anatomically faithful way of representing the pulmonary airway tree is to have it reproduce the precise branching pattern in a given lung. This personalized approach to lung modeling requires that the branching pattern be accurately known, which is possible for the higher airway orders via CT imaging. The lower order airways cannot be resolved by CT, however, so must be guessed at, for example, by using an algorithm that grows the airway tree into underserved regions of the lung so as to optimally ventilate those regions [71–73]. Models of this nature are computationally expensive, but they provide the most precise description of a particular lung at a fixed point in time [74].

There are now a large number of published studies utilizing anatomically based computational models of the lung to investigate factors potentially responsible for the airways hyperresponsiveness that characterizes asthma. Key among these factors is regional ventilation heterogeneity caused by variations in the amounts of narrowing of the different branches of the airway tree. Simulating lung impedance using stochastic airway narrowing based on various statistical distributions demonstrates the profound effect that heterogeneity has on the $Z(f)$ profile (Fig. 7.14) [75]. Most notably, increasing heterogeneity causes a marked and increasing negative dependence of the apparent resistance of the lung on frequency (Fig. 7.14a). Heterogeneity also has a dramatic effect on the apparent elastance of the lung (Fig. 7.14b), largely due to the effects of central airway shunting, which occurs

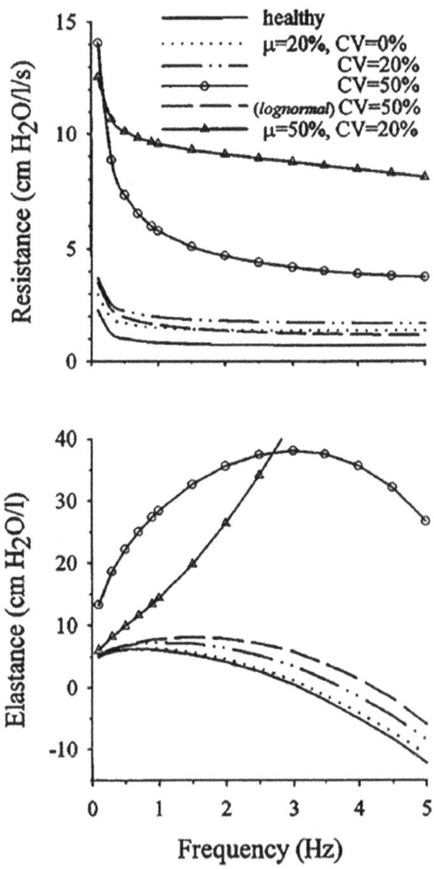

Fig. 7.14 Simulated lung resistance (**a**) and elastance (**b**) for the healthy human lung (solid line) and a constricted lung showing various degrees of % airway narrowing and heterogeneity (Gaussian coefficient of variation, CV). (Reproduced with permission from [75])

when high frequency oscillations in flow are unable to progress far into a constricted lung and so remain confined to the relatively stiff upper airways.

Heterogeneity of airway narrowing during bronchoconstriction, however, is a dynamic process that develops over time. If it develops while an aerosolized bronchial agonist is being delivered to the lung, the deposition pattern of the agonist throughout the airway tree can be affected. Incorporating ASM contraction kinetics in an airway tree model indicates that this may be protective against further constriction because airways that are already partly constricted limit further access of agonist [76].

The development of regional ventilation heterogeneity in the lung during bronchospasm is not, however, necessarily contingent upon the existence of regional differences in lung mechanics or ASM contractility. It is conceivable that a lung that is almost perfectly homogeneous under normal conditions can develop heterogeneities when its symmetry is broken by feedback amplification of very slight regional differences in lung mechanics, even when ASM activation is spatially uniform. Computational modeling has been used to show that this can lead to the formation of clusters of poorly ventilated lung regions that are difficult to predict a priori, but which have been observed by positron emission tomography [77]. Figure 7.15 shows a section of such a model after ventilation defects have formed.

Lung models can also be used to infer likely mechanisms of bronchospasm among possible contenders. For example, respiratory system impedance in mice up to 20 Hz is accurately described by the constant-phase model of lung tissue (Eq. 7.54) with the addition of a Newtonian resistance, R_N, that approximates R_{aw} [78] thus:

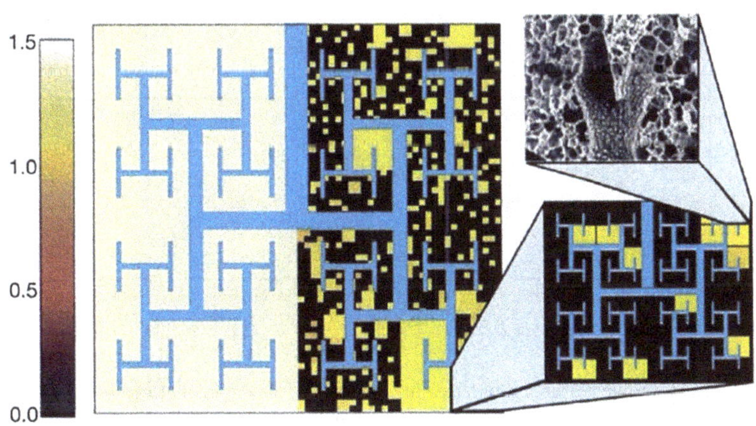

Fig. 7.15 A fractal model of the airway tree showing clusters of poorly ventilated regions on the right side of the lung (black areas). The upper right inset shows a scanning electron micrograph of a terminal lung unit. The bottom right inset shows the final seven generations of the model. (Adapted with permission from [77])

$$Z(f) = R_N + \frac{(G - iH)}{(2\pi f)^{\alpha}}. \tag{7.65}$$

In mice administered an aerosol of methacholine, the parameters R_N, G and H of this model have been shown to follow characteristic time profiles that can be quite accurately predicted by an anatomically based lung model provided that the airways in the model are assigned a critical closure radius. Any airway that narrows to this radius is suddenly closed completely, simulating the formation of a mucus plug across the airway lumen [61]. (Experimentally, such plugs remain in place until the lungs are given a deep inflation.) Importantly, the model representing the normal mice cannot be made to behave like mice with allergically inflamed lungs simply by increasing the degree of ASM shortening induced by a given dose of agonist. Rather, the inflamed mice behave as if their ASM shortening is normal, but they have substantially enhanced degrees of small airway closure, likely due to enhanced formation of mucus plugs (Fig. 7.16). This model thus supports the hypothesis that the ASM in inflamed mice behaves normally, but it operates within an abnormal lung environment.

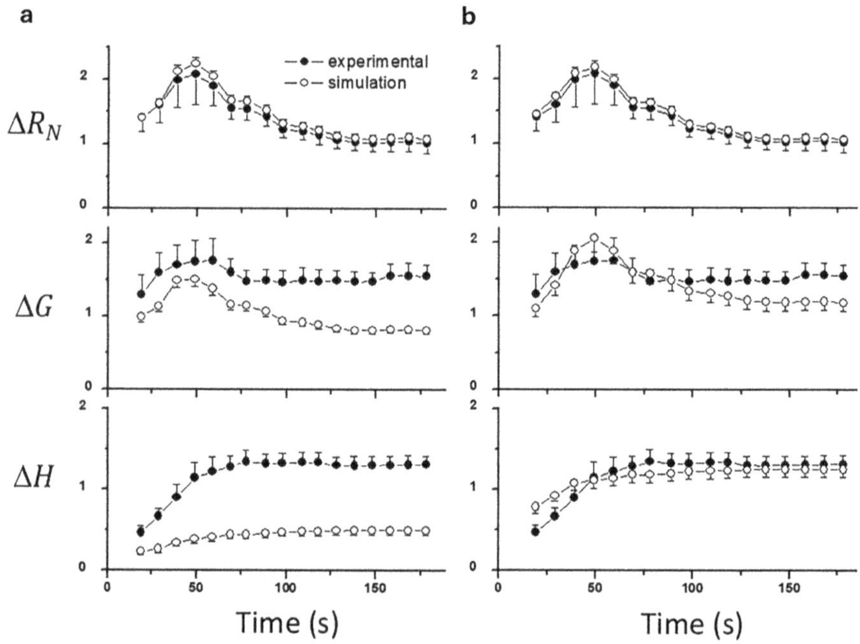

Fig. 7.16 (**a**) Experimental and simulated time courses of the fractional changes in impedance parameters in allergically inflamed mice following inhalation of methacholine aerosol. The simulated change in R_N from baseline (ΔR_N) was forced to match the experimental values by having the airway wall radii in the model following an exaggerated (compared to normal) narrowing profile, but this resulted in predictions of the changes in G and H from baseline (ΔG and ΔH, respectively) that differed dramatically from experiment. (**b**) When the degree of wall narrowing was reduced to normal, but a lining was imposed inside all the airways along with an increased propensity for liquid plugs to occlude the lumen, a substantially improved match between prediction and data was achieved. (Adapted with permission from [61])

Interestingly, when the same modeling approach is applied to mice that have been pre-treated with a lung instillation of the cationic protein poly-L-lysine, the opposite conclusion is reached. Specifically, the ASM appears more contractile than in non-treated mice, apparently because poly-L-lysine compromises the epithelial barrier. This makes the underlying ASM more accessible to any bronchial agonist that is inhaled into the airways [79]. In other words, the ASM is still normal, but the dose of agonist it receives has been enhanced. This is a completely different mechanism of airways hyperresponsiveness than that due to allergic inflammation. In fact, when both mechanisms are present at the same time, they synergize to produce an extreme level of airways hyperresponsiveness [80].

The anatomically based computational models considered above have been used in a forward fashion to simulate lung impedance by assigning values to their rather large number of parameters based on knowledge of lung morphology. The complexity of such models might seem to mitigate against their use as inverse models in which parameter values are estimated through the process of model fitting. Nevertheless, an anatomically based inverse model is possible with the appropriate simplifying assumptions [81]. This task is aided by the self-similar nature of the airway tree (section 2.2), which allows the dimensions of each airway order to be related to those of the next higher order by a fixed ratio $0 < \gamma < 1$ such that $r_j = r_N \square \gamma^{N-j}$, where r_N is the radius of the trachea. A reasonable value of γ for the human lung is 0.883 [56]. The ratio between the length, l, of an airway and r can also reasonably be assumed to be fixed such that $l = \lambda r$, except for the relatively longer trachea for which $l = \beta \lambda r$. Reasonable values for β and λ are 4 and 3, respectively. The entire airway tree is thus determined by the four parameters γ, λ, β and r_N.

Because of the regular scaling of the airway dimensions as a function of order, the Poiseuille flow resistance, R_j, of an airway of order j is

$$R_j = \frac{8\mu\lambda}{\pi r_N{}^3 \gamma^{3(N-j)}}. \tag{7.66}$$

Because of its relatively greater length compared to the other branches of the airway tree, the resistance of the trachea, R_N, is increased over that given by Eq. 7.66 by the factor β. Note that R_j has units of $cmH_2O.s.mL^{-1}$ when r_N is expressed in cm and μ is expressed in units of $g.s^{-1}.cm^{-1}$.

The elastance, E_j, of an airway of order j has two components: (1) the elastic stiffness of the airway wall, E_{wj}, and (2) the elastic compressibility of the gas contained within its lumen, E_{gj} [81]. If the airway wall tissue has Young's modulus E_{tis}, and the wall thickness is assumed to be proportional to radius (with constant of proportionality α), the increase in circumferential wall tension during radial expansion. Together with the Laplace law (Eq. 7.36) result in E_{wj} being given by Suki et al. [82]:

$$E_{wj} = \frac{E_{tis}\alpha}{2\pi\lambda r_N{}^3 \gamma^{3(N-j)}}. \tag{7.67}$$

E_{gj} is also proportional to airway volume and, in units of cmH$_2$O.mL^{-1}, is given by

$$E_{gj} = \frac{1000}{\pi r_j{}^2 l_j} = \frac{1000}{\pi\lambda r_N{}^3 \gamma^{3(N-j)}} \tag{7.68}$$

where, again because of its relatively greater length, the elastance of the trachea is given by Eq. 7.68 divided by the factor β. E_{gj} and E_{wj} act in parallel to give

$$E_j = \frac{E_{wj}E_{gj}}{E_{wj} + E_{gj}}. \tag{7.69}$$

The inertance, I_j, of each airway branch is proportional to the length of the branch and inversely proportional to the square of branch radius [62] thus:

$$I_j = \frac{\rho l_j}{\pi r_j{}^2} = \frac{\rho\lambda}{\pi r_N \gamma^{(N-j)}} \tag{7.70}$$

with I_N for the relatively longer trachea being given by Eq. 7.70 multiplied by the factor β.

The terminal airways of order 1 each connect to a parenchymal tissue unit having an impedance, Z_{ti}, representing the mechanical properties of the lung and chest wall tissues according to the constant-phase model [64]

$$Z_{tis}(f) = \frac{(G_u - iH_u)}{(2\pi f)^\alpha} \tag{7.71}$$

where G_u and H_u are tissue damping and elastance, respectively, for a single parenchymal unit. Setting the values of G_u and H_u and their ratio, η, the same for each of the n_t terminal units yields the stiffness of the entire respiratory system, H_{rs}, as

$$H_{rs} = \frac{H_u}{n_t}. \tag{7.72}$$

A shunt impedance

$$Z_{shunt} = H_{shunt} \frac{(\eta - i)}{(2\pi f)^\alpha} \tag{7.73}$$

representing the resistive and elastic properties of the upper airway structures proximal to the trachea [60], is also included in the model. The parameter H_{shunt} defines the stiffness of the shunt compartment.

The total impedance of the model is determined using the recursive approach described above. This impedance is defined by only 5 free parameters γ, r_N, E_{tis}, H_{rs} and H_{shunt}. This would seem to present the possibility that it could be used as an inverse model, allowing the 5 parameters to be estimated by fitting the model to measured impedance data. It turns out, however, that separately estimating wall tissue elastance and shunt elastance makes the fitting problem too ill-conditioned because these two quantities have very similar effects on the predicted shape of the impedance spectrum [83]. The conditioning difficulties are greatly reduced if E_j is subsumed into H_{shunt} [83], leaving only 4 free parameters to be estimated via the fitting process. The remaining model parameters (λ, β, η, ρ and μ) are either known physical constants or can be reasonably assigned a priori based on lung anatomy.

The above model is entirely deterministic, but a stochastic version can be created by allowing the individual radii of each of the branches in the airway tree to vary randomly about their order-assigned mean values. This allows the model to simulate the effects of small airway closure by identifying a critical airway closure radius, r_{crit}, below which airways transition to complete closure. This model has been used to investigate the nature of airway responsiveness in the asthma of obesity [83]. Figure 7.17 shows that it provides accurate fits to respiratory system impedance measurements in both lean and obese control individuals and individuals with asthma. The best-fit model parameters indicate that the major derangement in lung function in the asthma of obesity is elevated lung stiffness reflected in the value of the parameter H_{rs}, and that this is likely explained by enhanced closure of small peripheral airways [83].

7.6 Conclusions

Asthma is a complex syndrome displaying a number of different endotypes that reflect differing etiologies. Common among these various endotypes is airways hyperresponsiveness characterized by an excessive decrement in lung mechanical function following exposure to a standardized dose of ASM agonist. There are a number of disparate but physiologically plausible explanations for airways hyperresponsiveness that invoke mechanisms manifesting over a wide range of length scales from the molecular to the whole organ. Incorporation of these mechanisms in computational models leads to recapitulation of the dynamic processes involved in bronchospasm, significantly advancing our understanding of why individuals with asthma exhibit airway hyperresponsiveness.

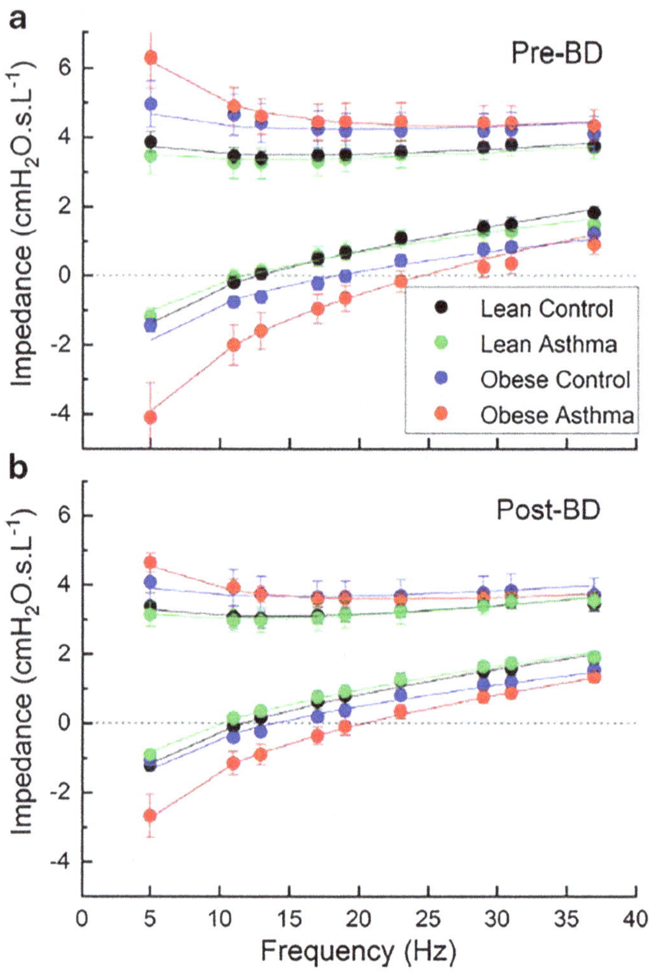

Fig. 7.17 Mean (± SE) of the real and imaginary parts of Z_{rs} (the upper and lower four groups of symbols in each graph) together with the fits provided by the deterministic inverse model with the four free parameters γ, r_N, H_{rs}, and H_{shunt} (solid lines) (**a**) before receiving a bronchodilator (BD), and (**b**) afterwards. (Reproduced with permission from [83])

7.7 Problems

1. Simulate the model shown in Fig. 7.4 (Eqs. 7.1, 7.2, 7.3, 7.4, 7.5, 7.6, 7.7, 7.8, 7.9, 7.10 and 7.11), driving it with a length signal L(t) corresponding to a deep inspiration followed by an expiration, with the rates of inspiration and expiration similar to what you would expect to see if a person was sighing. Do you see any evidence of force reduction due to temporary cross-bridge detachment as seen in Fig. 7.5b after the sudden length change at 60 s? Would you conclude that cross-bridge detachment is a significant feature of normal breathing?

2. Substitute Eqs. 7.16 and 7.17 into Eq. 7.14. Try to solve this analytically.
3. Simulate the differential equation model of smooth muscle contraction given by Eqs. 7.26 and 7.27, driving the model with a length signal L(t) corresponding to a sigh as in Problem 1 above.
    (a) Do you see any evidence of force reduction due to temporary cross-bridge detachment, such as is seen in Fig. 7.5b after the sudden length change at 60 s?
    (b) How does this simulation compare to that from Problem 1 above?
4. Equation 7.38 states the transmural pressure, P_{tm}, across an airway will increase with an increase in active muscle force (F_A) and decrease with increase in airway radius (r). However, when F_A increases, r will normally decrease because the airway wall is elastic. Derive an equation for the relationship between F_A and r that will ensure that airway narrowing is always stable (i.e., there is no bi-stable situation in r will suddenly collapse from one value to a lower one without smoothly transitioning between these two values).

References

1. Kapri A, Pant S, Gupta N, Paliwal S, Nain S. Asthma history, current situation, an overview of its control history, challenges, and ongoing management programs: an updated review. Proc Natl Acad Sci India Sect B Biol Sci. 2022;93:1–13.
2. Expert Panel Working Group of the National Heart, Lung, and Blood Institute (NHLBI) administered and coordinated National Asthma Education and Prevention Program Coordinating Committee (NAEPPCC), Cloutier MM, Baptist AP, Blake KV, Brooks EG, Bryant-Stephens T, DiMango E, et al. Focused updates to the asthma management guidelines: a report from the National Asthma Education and Prevention Program Coordinating Committee Expert Panel Working Group. J Allergy Clin Immunol. 2020;2020(146):1217–70.
3. Doeing DC, Solway J. Airway smooth muscle in the pathophysiology and treatment of asthma. J Appl Physiol (1985). 2013;114:834–43.
4. Ijpma G, Kachmar L, Matusovsky OS, Bates JH, Benedetti A, Martin JG, Lauzon AM. Human trachealis and main bronchi smooth muscle are normoresponsive in asthma. Am J Respir Crit Care Med. 2015;191:884–93.
5. James AL, Elliot JG, Jones RL, Carroll ML, Mauad T, Bai TR, Abramson MJ, McKay KO, Green FH. Airway smooth muscle hypertrophy and hyperplasia in asthma. Am J Respir Crit Care Med. 2012;185:1058–64.
6. Lauzon AM, Martin JG. Airway hyperresponsiveness; smooth muscle as the principal actor. F1000Res. 2016;5:F1000 Faculty Rev-306.
7. Gillissen A, Paparoupa M. Inflammation and infections in asthma. Clin Respir J. 2015;9:257–69.
8. Harker JA, Lloyd CM. T helper 2 cells in asthma. J Exp Med. 2023;220:e20221094.
9. Stiemsma LT, Reynolds LA, Turvey SE, Finlay BB. The hygiene hypothesis: current perspectives and future therapies. Immunotargets Ther. 2015;4:143–57.
10. Peters U, Dixon AE, Forno E. Obesity and asthma. J Allergy Clin Immunol. 2018;141:1169–79.
11. Seow CY. Introduction to smooth muscle mechanics. Victoria: FriesenPress; 2016. p. 1–15.
12. Gunst SJ, Zhang W. Actin cytoskeletal dynamics in smooth muscle: a new paradigm for the regulation of smooth muscle contraction. Am J Physiol Cell Physiol. 2008;295:C576–87.

13. Seow CY. Myosin filament assembly in an ever-changing myofilament lattice of smooth muscle. Am J Physiol Cell Physiol. 2005;289:C1363–8.
14. Seow CY, Solway J. Mechanical and structural plasticity. Compr Physiol. 2011;1:283–93.
15. Hanks BS, Stephens NL. Mechanics and energetics of lengthening of active airway smooth muscle. Am J Phys. 1981;241:C42–6.
16. Hill AV. The heat of shortening and the dynamic constants of muscle. Proc R Soc B. 1938;126: 136–95.
17. Stephens NL, Li W, Jiang H, Unruh H, Ma X. The biophysics of asthmatic airway smooth muscle. Respir Physiol Neurobiol. 2003;137:125–40.
18. Bates JH, Lauzon AM. Modeling the oscillation dynamics of activated airway smooth muscle strips. Am J Physiol Lung Cell Mol Physiol. 2005;289:L849–55.
19. Bates JH, Bullimore SR, Politi AZ, Sneyd J, Anafi RC, Lauzon AM. Transient oscillatory force-length behavior of activated airway smooth muscle. Am J Physiol Lung Cell Mol Physiol. 2009;297:L362–72.
20. Krishnan R, Trepat X, Nguyen TT, Lenormand G, Oliver M, Fredberg JJ. Airway smooth muscle and bronchospasm: fluctuating, fluidizing, freezing. Respir Physiol Neurobiol. 2008;163:17–24.
21. Bates JH. Modeling the impairment of airway smooth muscle force by stretch. J Appl Physiol (1985). 2015;118:684–91.
22. Huxley AF. Muscle structure and theories of contraction. Prog Biophys Biophys Chem. 1957;7: 255–318.
23. Hai CM, Murphy RA. Regulation of shortening velocity by cross-bridge phosphorylation in smooth muscle. Am J Phys. 1988;255:C86–94.
24. Hai CM, Murphy RA. Cross-bridge phosphorylation and regulation of latch state in smooth muscle. Am J Phys. 1988;254:C99–106.
25. Mijailovich SM, Butler JP, Fredberg JJ. Perturbed equilibria of myosin binding in airway smooth muscle: bond-length distributions, mechanics, and ATP metabolism. Biophys J. 2000;79:2667–81.
26. Delmotte P, Han YS, Sieck GC. Cytoskeletal remodeling slows cross-bridge cycling and ATP hydrolysis rates in airway smooth muscle. Physiol Rep. 2020;8:e14561.
27. Qi D, Mitchell RW, Burdyga T, Ford LE, Kuo KH, Seow CY. Myosin light chain phosphorylation facilitates in vivo myosin filament reassembly after mechanical perturbation. Am J Physiol Cell Physiol. 2002;282:C1298–305.
28. Donovan GM. Modelling airway smooth muscle passive length adaptation via thick filament length distributions. J Theor Biol. 2013;333:102–8.
29. Brook BS. Emergence of airway smooth muscle mechanical behavior through dynamic reorganization of contractile units and force transmission pathways. J Appl Physiol (1985). 2014;116:980–97.
30. Silveira PS, Butler JP, Fredberg JJ. Length adaptation of airway smooth muscle: a stochastic model of cytoskeletal dynamics. J Appl Physiol (1985). 2005;99:2087–98.
31. Kroon M. Influence of dispersion in myosin filament orientation and anisotropic filament contractions in smooth muscle. J Theor Biol. 2011;272:72–82.
32. Williams WO. Huxley's model of muscle contraction with compliance. J Elast. 2011;105:365–80.
33. Piazzesi G, Reconditi M, Linari M, Lucii L, Bianco P, Brunello E, Decostre V, Stewart A, Gore DB, Irving TC, et al. Skeletal muscle performance determined by modulation of number of myosin motors rather than motor force or stroke size. Cell. 2007;131:784–95.
34. Lambert RK, Wilson TA, Hyatt RE, Rodarte JR. A computational model for expiratory flow. J Appl Physiol Respir Environ Exerc Physiol. 1982;52:44–56.
35. Lambert RK, Pare PD. Lung parenchymal shear modulus, airway wall remodeling, and bronchial hyperresponsiveness. J Appl Physiol (1985). 1997;83:140–7.
36. Thorpe CW, Bates JH. Effect of stochastic heterogeneity on lung impedance during acute bronchoconstriction: a model analysis. J Appl Physiol (1985). 1997;82:1616–25.

37. LaPrad AS, Lutchen KR, Suki B. A mechanical design principle for tissue structure and function in the airway tree. PLoS Comput Biol. 2013;9:e1003083.
38. Bates JH, Cojocaru A, Lundblad LK. Bronchodilatory effect of deep inspiration on the dynamics of bronchoconstriction in mice. J Appl Physiol (1985). 2007;103:1696–705.
39. Ding DJ, Martin JG, Macklem PT. Effects of lung volume on maximal methacholine-induced bronchoconstriction in normal humans. J Appl Physiol (1985). 1987;62:1324–30.
40. Shen X, Gunst SJ, Tepper RS. Effect of tidal volume and frequency on airway responsiveness in mechanically ventilated rabbits. J Appl Physiol (1985). 1997;83:1202–8.
41. Lavoie TL, Krishnan R, Siegel HR, Maston ED, Fredberg JJ, Solway J, Dowell ML. Dilatation of the constricted human airway by tidal expansion of lung parenchyma. Am J Respir Crit Care Med. 2012;186:225–32.
42. Lai-Fook SJ. A continuum mechanics analysis of pulmonary vascular interdependence in isolated dog lobes. J Appl Physiol Respir Environ Exerc Physiol. 1979;46:419–29.
43. Brook BS, Peel SE, Hall IP, Politi AZ, Sneyd J, Bai Y, Sanderson MJ, Jensen OE. A biomechanical model of agonist-initiated contraction in the asthmatic airway. Respir Physiol Neurobiol. 2010;170:44–58.
44. Bates JH, Lauzon AM. Parenchymal tethering, airway wall stiffness, and the dynamics of bronchoconstriction. J Appl Physiol (1985). 2007;102:1912–20.
45. Bates JHT, Rajendran V. Mitigation of airways responsiveness by deep inflation of the lung. J Appl Physiol (1985). 2018;124:1447–55.
46. Moreno RH, Hogg JC, Pare PD. Mechanics of airway narrowing. Am Rev Respir Dis. 1986;133:1171–80.
47. Lambert RK, Codd SL, Alley MR, Pack RJ. Physical determinants of bronchial mucosal folding. J Appl Physiol (1985). 1994;77:1206–16.
48. Wiggs BR, Hrousis CA, Drazen JM, Kamm RD. On the mechanism of mucosal folding in normal and asthmatic airways. J Appl Physiol (1985). 1997;83:1814–21.
49. Ebina M, Takahashi T. The muscular architeture in the airway wall and its changes in bronchial asthma. Am Rev Respir Dis. 1989;139:A469.
50. Bates JH, Martin JG. A theoretical study of the effect of airway smooth muscle orientation on bronchoconstriction. J Appl Physiol (1985). 1990;69:995–1001.
51. Macklem PT. A theoretical analysis of the effect of airway smooth muscle load on airway narrowing. Am J Respir Crit Care Med. 1996;153:83–9.
52. Anafi RC, Wilson TA. Airway stability and heterogeneity in the constricted lung. J Appl Physiol (1985). 2001;91:1185–92.
53. Politi AZ, Donovan GM, Tawhai MH, Sanderson MJ, Lauzon AM, Bates JH, Sneyd J. A multiscale, spatially distributed model of asthmatic airway hyper-responsiveness. J Theor Biol. 2010;266:614–24.
54. Lui JK, Lutchen KR. The role of heterogeneity in asthma: a structure-to-function perspective. Clin Transl Med. 2017;6:29.
55. Fredberg JJ. Mechanical response of the lungs at high frequencies. J Biomech Eng. 1978;100: 57–66.
56. Weibel ER. Morphometry of the human lung. Berlin: Springer; 1963.
57. Wiggs BR, Bosken C, Pare PD, James A, Hogg JC. A model of airway narrowing in asthma and in chronic obstructive pulmonary disease. Am Rev Respir Dis. 1992;145:1251–8.
58. Wiggs BR, Moreno R, Hogg JC, Hilliam C, Pare PD. A model of the mechanics of airway narrowing. J Appl Physiol (1985). 1990;69:849–60.
59. Horsfield K, Kemp W, Phillips S. An asymmetrical model of the airways of the dog lung. J Appl Physiol Respir Environ Exerc Physiol. 1982;52:21–6.
60. Lutchen KR, Gillis H. Relationship between heterogeneous changes in airway morphometry and lung resistance and elastance. J Appl Physiol (1985). 1997;83:1192–201.
61. Wagers S, Lundblad LK, Ekman M, Irvin CG, Bates JH. The allergic mouse model of asthma: normal smooth muscle in an abnormal lung? J Appl Physiol (1985). 2004;96:2019–27.

62. Bates JHT. Lung mechanics. An inverse modeling approach. Cambridge: Cambridge University Press; 2009.
63. Fredberg JJ, Stamenovic D. On the imperfect elasticity of lung tissue. J Appl Physiol (1985). 1989;67:2408–19.
64. Hantos Z, Daroczy B, Suki B, Nagy S, Fredberg JJ. Input impedance and peripheral inhomogeneity of dog lungs. J Appl Physiol (1985). 1992;72:168–78.
65. Hantos Z, Suki B, Csendes T, Daroczy B. Constant phase modelling of pulmonary tissue impedance. Bull Eur Pathophysiol Respir. 1987;12:326s.
66. Benade AH. On the propagation of sound waves in a cylindrical conduit. J Acoust Soc Am. 1950;22:563–4.
67. Lutchen KR, Giurdanella CA, Jackson AC. Inability to separate airway from tissue properties by use of human respiratory input impedance. J Appl Physiol (1985). 1990;68:2403–12.
68. Suki B, Habib RH, Jackson AC. Wave propagation, input impedance, and wall mechanics of the calf trachea from 16 to 1,600 Hz. J Appl Physiol (1985). 1993;75:2755–66.
69. Habib RH, Suki B, Bates JH, Jackson AC. Serial distribution of airway mechanical properties in dogs: effects of histamine. J Appl Physiol (1985). 1994;77:554–66.
70. Bates JH. Stochastic model of the pulmonary airway tree and its implications for bronchial responsiveness. J Appl Physiol (1985). 1993;75:2493–9.
71. Tawhai MH, Nash MP, Hoffman EA. An imaging-based computational approach to model ventilation distribution and soft-tissue deformation in the ovine lung. Acad Radiol. 2006;13: 113–20.
72. Tawhai MH, Hunter P, Tschirren J, Reinhardt J, McLennan G, Hoffman EA. CT-based geometry analysis and finite element models of the human and ovine bronchial tree. J Appl Physiol (1985). 2004;97:2310–21.
73. Kitaoka H, Takaki R, Suki B. A three-dimensional model of the human airway tree. J Appl Physiol (1985). 1999;87:2207–17.
74. Tawhai MH, Lin CL. Image-based modeling of lung structure and function. J Magn Reson Imaging. 2010;32:1421–31.
75. Gillis HL, Lutchen KR. How heterogeneous bronchoconstriction affects ventilation distribution in human lungs: a morphometric model. Ann Biomed Eng. 1999;27:14–22.
76. Amin SD, Majumdar A, Frey U, Suki B. Modeling the dynamics of airway constriction: effects of agonist transport and binding. J Appl Physiol (1985). 2010;109:553–63.
77. Venegas JG, Winkler T, Musch G, Vidal Melo MF, Layfield D, Tgavalekos N, Fischman AJ, Callahan RJ, Bellani G, Harris RS. Self-organized patchiness in asthma as a prelude to catastrophic shifts. Nature. 2005;434:777–82.
78. Tomioka S, Bates JH, Irvin CG. Airway and tissue mechanics in a murine model of asthma: alveolar capsule vs. forced oscillations. J Appl Physiol (1985). 2002;93:263–70.
79. Bates JH, Wagers SS, Norton RJ, Rinaldi LM, Irvin CG. Exaggerated airway narrowing in mice treated with intratracheal cationic protein. J Appl Physiol (1985). 2006;100:500–6.
80. Bates JH, Cojocaru A, Haverkamp HC, Rinaldi LM, Irvin CG. The synergistic interactions of allergic lung inflammation and intratracheal cationic protein. Am J Respir Crit Care Med. 2008;177:261–8.
81. Bates JHT. The role of airway shunt elastance on the compartmentalization of respiratory system impedance. J Eng Sci Med Diagn Ther. 2019;2:011001-1–8.
82. Suki B, Habib RH & Jackson AC. Wave propagation, input impedance, and wall mechanics of the calf trachea from 16 to 1,600 Hz. J Appl Physiol. 1993;75(6):2755–66. PubMed PMID: 8125900
83. Bates JHT, Kaminsky DA, Garrow OJ, Martin FK, Peters U, Tharp WG, Dixon AE. Lung de-recruitment in the allergic asthma of obesity: evidence from an anatomically based inverse model. J Appl Physiol (1985). 2023;134:356–64.

Chapter 8
Pulmonary Fibrosis

Abstract Pulmonary fibrosis (PF) is an interstitial lung disease characterized by scar tissue formation in the lung parenchyma resulting from aberrant wound healing. Since collagen is the stiffest material in the lung, an increase in either its amount or its degree of cross-linking can elevate both regional and global lung stiffness. Elevated stiffness is not just the end-result of PF; it also drives disease progression via an aberrant mechanotransduction process, the conversion of mechanical cues to cellular signals, which leads to further deposition and/or cross-linking of collagen in a positive feedback loop. Computational models that link multiscale structure to function can advance the understanding of the clinical manifestations of PF. The biologic processes involved in PF signaling and progression can be modeled via systems of coupled differential equations. However, a disconnect exists between the microscale fibrotic process and the development of clinical abnormalities, which can be explained by the concept of percolation. The most complete models of PF to date combine physics-based network representations of the alveolar parenchyma with agent-based representations of fibrotic cell biology. These models offer an explanation for the emergence of characteristic tissue structure in PF such as subpleural honeycombing.

Keywords Stiffness · Collagen · Compliance · Percolation · Network · Agent-based model · Stretch · Mechanotransduction

8.1 Introduction

Pulmonary fibrosis (PF) is an interstitial lung disease characterized by the appearance of collagenous scar tissue in the lung parenchyma. Chronic PF may involve a progressively fibrosing phenotype in which permanent scarring of the lung worsens over time, leading to difficulties in breathing, worsening lung function with exacerbations and complications, and even eventual death [1, 2]. While PF often accompanies a variety of other lung diseases [3], a common and particularly insidious form is idiopathic PF (IPF), which progresses rapidly and has a median survival of only 27.4 months from time of diagnosis in severe cases [4]. Globally, 0.5–12 people per 100,000 die from IPF each year [5], including nearly 40,000 in the US. The factors

B. Suki, J. H. T. Bates, *Mathematical Modeling of the Healthy and Diseased Lung*,
https://doi.org/10.1007/978-3-031-95322-4_8

that initiate the pathogenesis of IPF remain unknown (hence the term "idiopathic"), but several possibilities have been proposed, including the herpes viruses [6] and chronic micro-aspiration from gastroesophageal reflux disease [7]. Environmental factors may also play a role. Inorganic particles such as silicon and aluminum have been reported to appear in higher concentrations in the lymph nodes of patients with IPF compared to healthy controls [8]. Disease progression, defined by worsening lung function and radiology, has been associated with air pollutants including carbon monoxide, nitrogen oxides, and ozone [9]. There is also a genetic component to the risk for IPF as evidenced by familial aggregation and, in rare cases, the presence of genetic disorders such as mutations in surfactant proteins [10].

Regardless of its varied associations and presentations, PF is believed overall to be the end result of an aberrant wound healing process [11]. Normal wound healing takes place in a cascade of four phases: (i) clotting/coagulation, (ii) inflammatory cell migration to the wound site, (iii) fibroblast migration/proliferation/activation, and (iv) extracellular matrix (ECM) deposition leading to tissue repair followed by return to homeostasis [12]. These stages involve many different cell types including both recruited and lung-resident cells that influence each other's activities through the secretion and detection of a variety of signaling molecules (particularly cytokines and chemokines). Fibroblasts and myofibroblasts are especially crucial for wound healing because they produce the structural molecules of the ECM, such as the collagens, that are required to close a wound. They also produce matrix metalloproteases (MMPs) that are necessary for the degradation of debris so it can be eliminated by phagocytic cells of the immune system. When external (e.g., bacterial) or internal (e.g., genetic predisposition) factors interfere with the wound repair cascade, healing becomes abnormal [13]. For example, a bacterial infection may prolong inflammation, or a stimulus such as aspirated stomach acid may be persistently present. In genetically predisposed and/or aging individuals, this can lead to uncontrolled ECM remodeling that progresses to PF.

The functional deficits arising in PF, which invariably manifest as a loss of lung compliance, are not always linked in an obvious way to alterations in parenchymal structure. Something as obvious as an excess of collagen in the tissue can occur, for example, in the specific form of PF known as cryptogenic fibrosing alveolitis [14]. On the other hand, the increased tissue stiffness seen in IPF has been found to result primarily from dysregulated cross-linking of existing collagen fibers [15]. Since collagen is the stiffest material in the lung [16], an increase in either its amount or its degree of cross-linking, or a combination of both, can elevate both regional and global lung stiffness (inverse of compliance). Furthermore, increased lung stiffness may not just be the end-result of a fibrotic process. It can also drive disease progression. Indeed, mechanotransduction, the conversion of mechanical cues to cellular signals, is a major signaling mechanism in fibroblasts [17] that allows these cells to be sensitive to the stiffness of the underlying ECM. This can lead to activation of various signaling pathways [18, 19] that may result in further deposition and/or cross-linking of collagen and other ECM molecules in a positive feedback loop. In fact, progressive disease driven by tissue stiffening has been implicated in the pathogenesis of IPF [20]. Such complex multi-factorial processes involving intricate mechanotransduction feedback loops are difficult to handle in analytic models, but properly formulated computational models allow experimentally inaccessible mechanisms to be studied in a quantitative fashion.

This chapter covers computational models that attempt to link structure to function in PF. As with virtually all lung diseases, this can be pursued at a variety of different levels of length scale. At the macroscopic level of the patient, models can be used to explain the link between global parameters of lung function, such as forced vital capacity (FVC), and the physical processes taking place in the lung during a pulmonary function test [21]. Such parameters are used clinically to describe IPF severity and to predict its progression [22]. At a lower level of scale, the biologic processes involved in IPF signaling and progression can be modeled via systems of coupled differential equations [23]. It is also important to use modeling to forge links between scales, something that often raises the very important and interesting subject of emergent properties. This is exemplified by the disconnect that can occur between underlying fibrotic disease progression and the development of clinical abnormalities in lung function, which can be explained by the concept of percolation [24–27]. Finally, the most complete models of PF pathogenesis to date combine physics-based network representations of the alveolar parenchyma with agent-based representations of fibrotic cell biology [28–30]. These models try to describe relevant events in a way that mirrors how they actually take place in PF.

8.2 Modeling Maximal Expiratory Flow-Volume Curves

Since the middle of the twentieth century, pulmonary function tests have played a central role in the diagnosis of lung disease. The most widely used test of pulmonary function involves having a patient inspire to total lung capacity (TLC) and then immediately exhale as forcibly as possible down to residual volume while the flow leaving the mouth is recorded. Provided expiratory effort exceeds a certain level that is readily achievable by people with normal respiratory muscle function, the flow measured throughout most of this maneuver is maximal in the sense that further increases in effort will not result in further increases in flow. This phenomenon is called expiratory flow limitation and results from the fact that the airways are compliant and will thus narrow to cause a flow choke point if transmural pressure is sufficiently negative [31, 32]. A plot of flow on the vertical axis versus volume on the horizontal axis then produces the maximal expiratory flow-volume (MEFV) curve.

The diagnostic importance of the MEFV curve lies in the fact that it is a reflection of the mechanical properties of the lung rather than patient effort, and thus reflects abnormalities in lung function. Maximum expiratory flow depends on the degree of lung inflation [33], so the MEFV curve has a characteristic almost linear trajectory in normal lungs that changes shape in characteristic ways in both obstructive and restrictive lung disease. The most important indexes derived from the MEFV curve are the forced vital capacity (FVC—equal to the total volume exhaled over the entire maneuver), the volume exhaled during the first second of the maneuver (FEV1), and the peak expiratory flow (PEF). TLC, FEV1 and PEF have all been shown to be shown to be statistically significantly different between obstructive and restrictive lung diseases [34], while FVC correlates with quality of life, exercise capacity and survival in patients with PF [35–37]. It is therefore important to understand the physical mechanisms that determine the MEFV curve in health and disease.

A mathematical model describing pressure-flow relationships and maximum expiratory flows in the lung based on differential equations was introduced as early as 1958 [38]. The next significant modeling advance in the field was the introduction of wave speed theory to explain maximal flow through a compliant tube [39–41]. The idea here is that the flow driven by an increasing upstream pressure becomes limited to a maximum when the air velocity reaches the speed at which elastic waves travel in the tube [42, 43]. Implementing this mechanism in a symmetric airway tree model imbued with appropriate pressure-area relationships for the various airway branches produces good agreement with experimental data [41]. This model suggests that maximal flow is due to the wave speed mechanism at high to medium lung volumes, while at lower volumes it is more a function of viscous pressure losses and the compliance of the airway wall. This model also shows maximal flow to be sensitive to transpulmonary pressure and airway areas but, somewhat surprisingly, relatively insensitive to lung volume, airway compliance and airway length [39]. Introducing mild degrees of airway constriction into the model along with regional variations in airway and parenchymal properties yields results in agreement with experimental data [40].

The MEFV curve has also been simulated using an asymmetrical airway tree model [21]. The proximal airways in this model are based on the self-similar branching morphology of Horsfield [44], while the distal airways are assumed to branch symmetrically to reduce computational burden. Except for the trachea, external airway walls are assumed to be exposed to pleural pressure so that, in analogy with electrical resistor networks, the MEFV curve may be interpreted as resulting from a "bifurcating net of nonlinear, pressure and lung volume dependent resistors, driven by lung regional pressures and 'grounded' through the upper airways to the atmosphere" [21]. This model demonstrates that flow limitation is distributed throughout the airway tree and is critically influenced by central airway heterogeneity in a way that cannot be recreated using symmetrically bifurcating models.

Personalized airway tree models can now be created based on high-resolution 3D CT images (see Fig. 4.11) and used to investigate the relative contributions of various flow limitation mechanisms. In one study, volumetric CT images were obtained from 3 female and 3 male subjects and utilized to create models of the lung tissue in which segmented central airways from generations 6 through 10 were embedded [45]. The subtended distal airways down to the terminal bronchioles were generated using a volume-filling algorithm [46]. The remaining aspects of the model, including the use of wave speed as the flow limiting mechanism, followed earlier

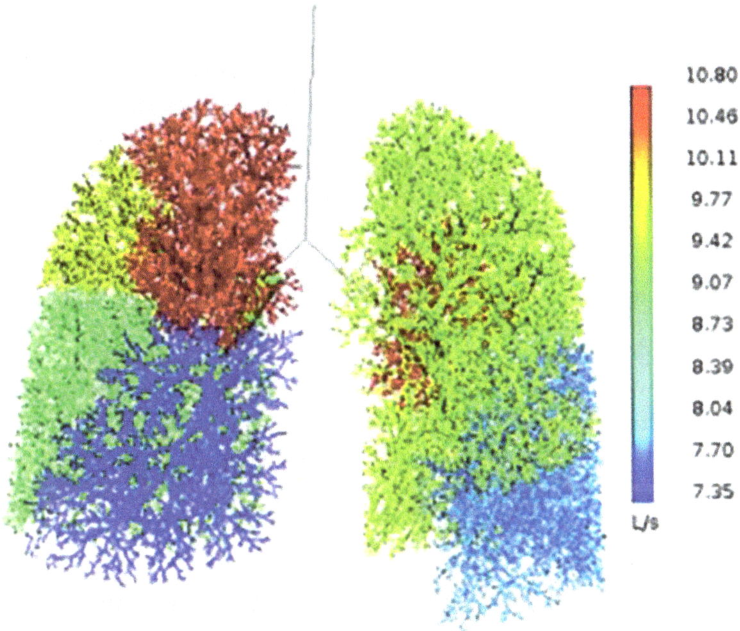

Fig. 8.1 Sites of flow limitation in a personalized model of the airway tree at PEF. Clusters of same-colored segments correspond to regions that are distal to a flow limited airway. The color scale indicates the trachea (whole lung) flow at which the limited airway reached its maximum flow. (Reproduced with permission from Ref. [45])

models [41]. The lung parenchyma was treated as a compressible nonlinearly elastic material with a strain energy function similar to the exponential strain energy discussed in Chap. 4. Using a finite element approach, the driving pressure for flow in each airway was calculated from the elastic recoil of the lung tissue and the pressure generated by the respiratory muscles. This complex model was solved at PEF and yielded lung function metrics such as FEV1 that compared reasonably well with patient measurements. The model also predicts the time evolution of the locations of flow limited airway segments, as well as how FEV1 varies with various patterns of airway constriction. Figure 8.1 shows the map of flow limitation from one subject at the moment when PEF was reached. Interestingly, the subjects exhibited distinct flow limitation patterns without any common trends. The main conclusions from this study were similar to those of previous much simpler computational studies. A complex model such as this one is thus useful for exploring, via simulation, factors that influence the MEFV curve. On the other hand, it cannot function as an inverse model and be fit to measured MEFV curves in the way that a much simpler lumped parameter model can [47]. Thus, despite the vast complexity of the lung, its myriad components and mechanisms somehow conspire to produce relatively low order behavior. In other words, only a limited amount of information about the mechanisms of flow limitation is contained in a measured MEFV curve.

Computational investigation has yet to be performed concerning how the structural changes associated with PF affect MEFV curves. Nevertheless, it is clear from the pressure-volume curve (Fig. 4.4a) that the PF lung is much stiffer than the normal lung. This reflects stiffer lung tissue and reduced lung volumes (both FRC and TLC), all of which contribute to reductions in FEV1 and FVC. A computational model of the MEFV curve in the PF lung must therefore recapitulate these phenomena along with the structural features characteristic of PF. Particularly notable among these features are regions of peripheral honeycombing consisting of distal airspaces surrounded by thick-walled fibrotic tissue that correlate with reduced FVC [48]. The mechanisms responsible for peripheral honeycombing are still not entirely clear, but it can be reproduced computationally by assuming that fibrosis is driven by increased tissue stiffness starting at the pleura [29]. It is likely that such structural changes in PF are directly linked to the characteristic alterations seen in lung function, and that these could be predicted by an anatomically based computational model of expiratory flow limitation incorporating appropriate abnormalities in tissue mechanics.

8.3 The Differential Equation Approach to Modeling Signaling Pathways in PF

Differential equations are often used to describe how biological variables change with time, t, and spatial coordinates. For example, cell positions change continually as cells move around within a tissue. Cell states change in time as a result of proliferation, phenotypic alterations, and apoptosis. In the limit of a large number of cells at a scale much larger than the cell dimension, the representation of cells as discrete entities (i.e., as agents) can be replaced by cell density as a continuous function of position and time. Denoting the density of a particular cell type as X_i ($i = 1, \ldots, M$), the dynamics of the cell can then be expressed as

$$\frac{\partial X_i}{\partial t} = D_{X_i}\Delta X_i + r_{p_i}X_i - r_{a_i}X_i + \sum_{\substack{j=1 \\ j\neq i}}^{M} r_{ji}X_j - X_i \sum_{\substack{j=1 \\ j\neq i}}^{M} r_{ij} \tag{8.1}$$

where Δ is the Laplacian (divergence of the gradient) operator, D_{X_i} is a diffusion coefficient for cell type X_i within the ECM, r_{p_i} is the proliferation rate, r_{a_i} is the apoptosis rate, r_{ij} is the rate at which cell type X_i transitions into cell type X_j, and r_{ji} is the rate at which X_j transitions into X_i. Notice that in general $r_{ij} \neq r_{ji}$. In 3D, all cell densities are functions of the spatial coordinates x, y and z.

Equation 8.1 does not include the effects of chemical mediators such as cytokines that can stimulate signaling events in a variety of cell types. For example, transforming growth factor-beta1 (TGF-β1) is a potent regulatory and inflammatory cytokine [49] that is secreted by many cell types, including macrophages that target and activate other cells including alveolar epithelial cells [50]. TGF-β1 is strongly associated with the pathogenesis and progression of fibrosis [51] because it promotes fibroblast-to-myofibroblast transition accompanied by upregulation of collagen types I, III and IV [52]. Deposition of newly synthetized collagen in turn elevates tissue stiffness and causes normal lung function to deteriorate. The influence of a concentration C_{kj} of cytokine k ($k = 1, \ldots, N$) on the transition of cell type X_j to type X_i can generally be taken into account through the addition of a term to Eq. 8.1 that accounts for receptor-ligand kinetics as follows:

$$\frac{\partial X_i}{\partial t} = D_{X_i}\Delta X_i + r_{p_i}X_i - r_{a_i}X_i + \sum_{\substack{j=1 \\ j\neq i}}^{M} r_{ji}X_j - X_i\sum_{\substack{j=1 \\ j\neq i}}^{M} r_{ij} + \sum_{\substack{j=1 \\ j\neq i}}^{M}\sum_{k=1}^{N} \frac{a_{kj}C_{kj}^{\beta}}{b_{kj}^{\beta} + C_{kj}^{\beta}}X_j. \tag{8.2}$$

The last term in Eq. 8.2 is a sigmoidal Hill function with asymptotic value a_{kj} in the limit of high cytokine concentration, b_{kj} is a constant chosen so that $\sqrt[\beta]{b_{kj}}$ determines the value of X_i at the point where the sigmoidal function reaches half its maximum value, and β is a small integer chosen usually between 1 and 3 that determines the sharpness of the transition from 0 to a_{kj}.

The concentration C_{kj} in the ECM conforms to a similar equation thus:

$$\frac{\partial C_{kj}}{\partial t} = D_{C_{kj}}\Delta C_{kj} + \sum_{j=1}^{M} p_{C_{kj}}X_j - \sum_{j=1}^{M} u_{C_{kj}}X_j - r_{C_{kj}}C_{kj} - \sum_{l}^{L} \frac{e_{kjl}E_{kjl}^{\gamma}}{f_{kjl}^{\gamma} + E_{kjl}^{\gamma}}C_{kj} \tag{8.3}$$

where D_{C_i} is the diffusion coefficient of cytokine i, $p_{C_{kj}}$ is the rate of cytokine production by cell type X_j, $u_{C_{kj}}$ is the rate of uptake by cell type X_j, $r_{C_{kj}}$ is the rate of spontaneous degradation, and the last term is another Hill function describing the elimination of the cytokine by a set of enzymes having concentration E_{kjl} ($l = 1, \ldots, L$) with Hill coefficients of e_{kjl} and f_{kjl} and exponent γ. For more complex (and realistic) behavior, the secretion and uptake terms can be replaced by appropriate Hill functions. There may also be crosstalk between different entities such as the ECM and an enzyme that synergistically affects the production of a cytokine. This can be accounted by adding to Eq. 8.3 a term of the form $C_{kj}E_{kjl}$ in both the nominator and denominator of the Hill function. Lastly, each enzyme must obey an equation similar to Eq. 8.3.

The same modeling approach can be scaled down from the inter-cellular to the intra-cellular level to describe, for example, the production of gene G and protein P when each affects the other's expression. A simple form of this gene-protein interaction can be written as a system of two coupled equations thus:

$$\frac{\partial G}{\partial t} = \frac{r_{GP}}{b_{GP} + P^{\delta}} - r_G G \tag{8.4}$$

$$\frac{\partial P}{\partial t} = r_{PG}G(t-\tau)^{\nu} - r_{P}P. \tag{8.5}$$

The rate constants r_G, r_P, r_{GP} and r_{PG} are defined as above. The first and second terms on the right-hand side of Eq. 8.4 represent a Hill function-like inhibition of the gene by the protein and spontaneous gene degradation, respectively. The first term on the right-hand side of Eq. 8.5 describes protein production delayed by the time, τ, required for gene transcription, translation and secretion. The second term again represents spontaneous degradation, this time of P. Equation 8.5 is an example of a delay differential equation, which almost never admits an analytic solution, so Eqs. 8.4 and 8.5 must be solved numerically. These two equations were developed to model an intracellular circadian rhythm generator in which negative feedback by protein P on the transcription rate of its gene G produces cyclic variations in both [53]. With appropriate choice of the model parameters, this system of equations predicts oscillatory behavior with a period close to 24 h. If P in Eq. 8.5 corresponds to C or E in Eq. 8.3, Eqs. 8.1, 8.2, 8.3, 8.4 and 8.5 can then be combined into a multiscale model of fibrosis development. Since the Laplacian operator describes how each component diffuses in space, these equations fully describe both the diffusion and reactions within a spatial domain, and hence are called diffusion-reaction equations.

The model described above can be simplified by being expressed as a collection of interconnected compartments. Cellular diffusion within each compartment is considered to be negligible (i.e., the compartments are assumed to be "well-mixed") so that the partial differential equations 8.1–8.5 reduce to ordinary differential equations. The compartments exchange cells or cytokines with each other according to specified transport equations containing rate parameters. Such multi-compartment models can describe distributed systems with a spatial resolution that depends on the number of distinct compartments that are specified. These systems are usually readily solved numerically, but the number of equations involved is proportional to the number of compartments and can eventually become unwieldy. Furthermore, the system of equations must be specified on the basis of hypothesis and/or experimental data, along with both initial and boundary conditions, and then solved numerically. This is often a non-trivial task both from the perspective of determining how to parameterize the model and of obtaining a stable numerical solution. Fortunately, the latter issue is becoming less of a challenge through the availability of well-vetted computational tools such as MatLab™ that can serve a wide variety of applications.

Modeling the fibrotic lung can also become very computationally expensive depending on how much of the organ is being represented, and on the spatial resolution with which it is described. In the case of the whole organ, the boundary conditions are tortuous, comprising the external boundaries imposed by the pleural surface and the internal boundaries represented by the epithelial surfaces of the airspaces and endothelial surfaces of the pulmonary vasculature. There are also 7–8 orders of magnitude of scale separating the largest (whole lung) and smallest (single molecule) structures that are arguably relevant to PF, so including all of them in a model would amount to an enormous computational burden. One way to make the task more manageable is to consider only a portion of the lung, but this requires adjusting the boundary conditions accordingly, and it may leave out essential structures. Another approach is to coarse grain the entire system, but this limits spatial resolution, so some combination of the two approaches may be a reasonable compromise. These challenges highlight the importance of choosing a modeling approach that is best suited to the specific task at hand; there is no universally best model that can be applied to all questions related to PF, or any other disease.

An example of a multifactorial model of IPF pathophysiology based on many known cellular and cytokine interactions is shown in Fig. 8.2 [23]. This interaction network was formalized mathematically using a system of differential equations similar to those given above. In this scenario, following an initial injury to the alveolar epithelium, epithelial cells secrete factors such as monocyte chemotactic

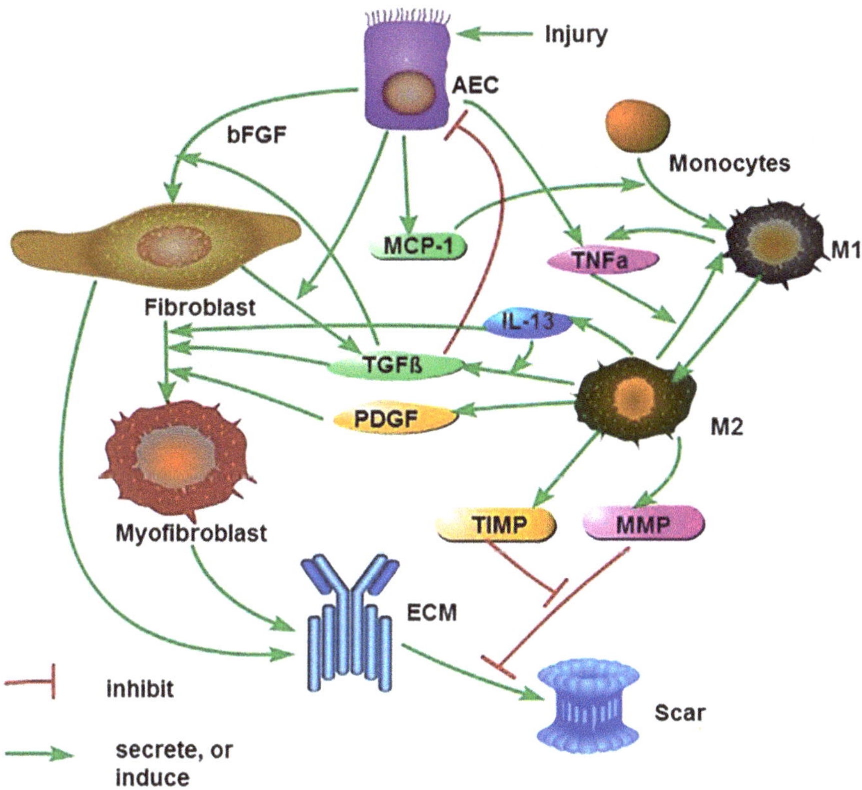

Fig. 8.2 Cell-protein interaction network in IPF. *AEC* alveolar epithelial cell, *M1, M2* macrophages, *bFGF* AEC-derived basic fibroblast growth factor, *MCP-1* monocyte chemotactic protein-1, *TNFα* tumor necrotic factor alpha, *TGFβ* transforming growth factor beta, *IL-13* interleukin 13, *PDGF* platelet-derived growth factor, *MMP* matrix metalloproteinase, *TIMP* tissue inhibitor of metalloproteinase, *ECM* extracellular matrix. (Reproduced with permission from Ref. [23])

protein-1 (MCP-1), which recruits monocytes from the circulation. The monocytes differentiate into M1 macrophages, which further transition into M2 macrophages. The epithelial cells together with the proteins secreted by the M2 macrophages activate fibroblasts, which differentiate into contractile myofibroblasts, which are the cells ultimately responsible for collagen deposition. However, this process is further complicated by the fact that many other cytokines, enzymes and their inhibitors are also involved. This signaling network was implemented on a simple cubic lattice representing the parenchyma in 3D. The model contains a total of 66 parameters whose values were obtained either directly from the literature or estimated from published data. Despite the very large number of adjustable parameters in this model, however, it is still just a crude approximation of reality.

The system of equations describing the above model was solved using a finite difference scheme in the spatial coordinates and a Runge-Kutta scheme in time, producing outputs that are in reasonable agreement with experimental data such as MMP and TIMP levels. The model can also be used as a *in silico* laboratory by studying what happens if some of the signaling pathways (represented by arrows in Fig. 8.2) are up- or down-regulated, simulating the effects of various drugs. Virtual inhibition of TNF-α, PDGF, or IL-13 does not produce any benefits. On the other hand, inhibition of TGF-β (simulating the effect of the FDA-approved drug pirfenidone) shows potential to stop the progression of IPF, and perhaps even reverse it [23], as illustrated in Fig. 8.3.

Although the simulated drug treatment results from this model are encouraging, it has several important limitations. For example, it has long been known that the physical properties of the ECM, such as its stiffness, influence numerous cellular

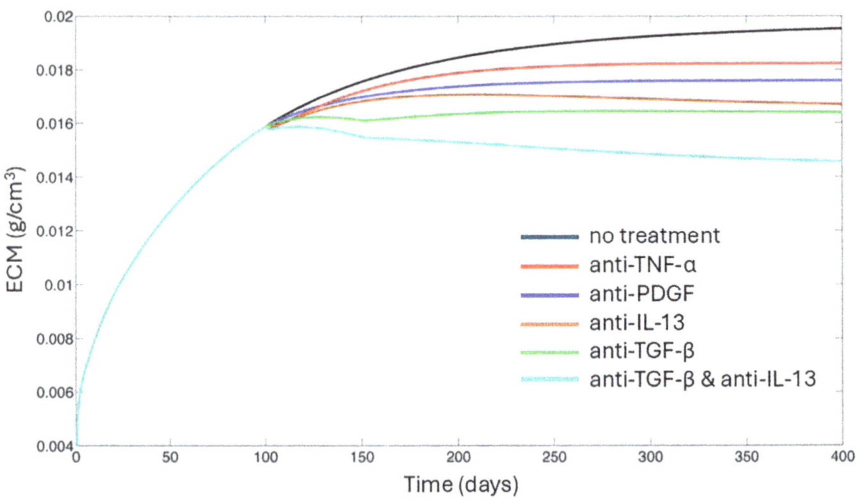

Fig. 8.3 Output of the differential equation model of the signaling network shown in Fig. 8.2. ECM density increases exponentially until day 100 when a treatment is simulated using one of the inhibitors shown in the inset. (Adapted with permission from Ref. [23])

processes [54]. This means that parenchymal stiffening is not only a consequence of IPF but also a driver of disease progression [19, 55]. The lung is a dynamic organ that is continually being stretched by the respiratory muscles during breathing, leading to the idea that mechanical forces may contribute to IPF progression via mechanotransduction that liberates TGF-β from the ECM [56, 57]. This notion is supported by the finding that mice treated with a drug that disrupts a mechanosensitive pathway was found to trigger myofibroblast apoptosis and protect against experimental PF [58]. Cytokine expression and cell density do not necessarily correlate with tissue structure and function. Since the above model [23] is unable to account for the highly heterogeneous mechanical microenvironment in the fibrotic lung, it cannot predict functional indexes such as FEV1 or FVC that ultimately derive from aberrant cell signaling and the subsequent remodeling of tissue architecture.

8.4 The Percolation Idea

Not all modeling has to invoke biologic mechanisms traceable to the activities of cells in order to have relevance for a pathological process such as PF. Emergent properties in complex systems often arise as a result of mechanisms operating at the level of system structure, regardless of the components that comprise that structure. A prime example is percolation [59], something that has far wider implications than simply making coffee.

Imagine a dedicated scientist working countless hours, desperately trying to create a computational model that sheds light on the progression of PF. Such an undertaking can only be sustained through the regular consumption of coffee, so our scientist switches on an Italian espresso maker stocked with water and coffee grounds. For a while, nothing happens, despite continuous heating, because when hot water vapor first makes contact with a column of coffee grounds, the pressure driving it is too low to force the water through the column. As the pressure builds, however, there suddenly comes a point where the water is forced to pick its way between the grounds all the way through the column, and coffee starts pouring into the pot. At this point, water has "percolated" through the column of grounds, and the espresso machine has undergone a major phase transition from serving no useful purpose to producing coffee, hopefully allowing the scientist to finally create a useful computational model.

As a more mathematically analyzable example, consider a square lattice in which neighboring nodes (those separated by one side of a square cell in the network) are connected with probability p by an electrical conductor. There is thus a probability $1 - p$ that two neighboring nodes are not connected. Suppose that the nodes at the top of the lattice are all attached to a copper plate, while those at the bottom are connected to a second plate. Now consider whether it is possible to pass an electric current from the top plate to the bottom plate. Conduction of current is possible only if there is at least one contiguous chain of inter-node conductors (bonds) that spans between the two plates (this conductive chain defines an unbroken pathway along which current can flow). If this condition is met, the entire lattice is a conductor. If not, it is an electrical insulator. Obviously, all the nodes are isolated when $p = 0$, and the lattice is an insulator. Similarly, all nodes are connected when $p = 1$, and the lattice is a conductor. Therefore, as p increases from 0 to 1 there comes a point known as the *percolation threshold* (when $p = p_c$), at which a contiguous pathway of conductors suddenly appears that percolates from the top to the bottom of the lattice. At this point, the lattice suddenly undergoes a phase transition from insulator to conductor. Because the placement of inter-node conductors in this model is probabilistic, the precise value of p_c will show some variability from one realization of the model to another, but p_c on a square lattice has an exact value of ½ whereas on a hexagonal lattice, the value is ~0.6527.

The percolating pathway does not, however, consist of an isolated string of conductors that progress across the lattice one after the other. Rather, it is part of a cluster of connected nodes that are generally much more numerous than the number of nodes participating directly in the percolating pathway. The number of inter-node conductors in this percolating cluster defines its size. There are other smaller clusters in the lattice at the percolation threshold that are isolated from the percolating cluster and which themselves do not percolate. In fact, multiple clusters appear over much of the range of $0 < p < 1$ and exhibit interesting properties as a function of p [59]. For example, when $p = p_c$, the percolating cluster is a self-similar fractal filled with holes of widely varying sizes. When p is near p_c, the distribution (Π) of cluster size S follows a power law such that $\Pi(S) \sim S^{-\tau}$ where τ is the Fisher exponent. Also, while the probability that a single bond belongs to the percolating cluster (called the strength of the cluster) is obviously zero for $p < p_c$, it is a power law for p just above p_c. These cluster attributes exhibit a remarkable feature known as universality, which means that while the actual value of p_c depends on the lattice structure (e.g., square versus hexagonal), the exponents in the power laws near p_c (e.g., τ) arise for any lattice structure within a space of a given dimension.

While the example of percolation given above might appear just a simple geometric game, it gives rise to behaviors characteristic of complex systems in general, including the ability to generate fractal structures and to undergo phase transitions. Indeed, percolation has been applied to variety of complex problems in the lung including gelation of bronchial mucus [60], the dynamics of airway opening [61], symptom progression in emphysema and pulmonary fibrosis [24], collagen stress distributions in the alveolar wall [26], and mechanical failure of the ECM [62, 63]. Furthermore, at a higher spatial scale, clusters of both low and high attenuation areas in CT images of IPF lungs also exhibit power law distributions [29], suggesting that a percolation process may be involved in the pathophysiology of parenchymal disease. In fact, spring-network models of parenchymal mechanics are well suited to the investigation of percolating tissue stiffness in much the same way that the percolation of conductors illuminates the appearance of global conductivity as described above.

To illustrate how percolation can link structure to function, consider the initially uniform prestressed hexagonal 2D spring network in Fig. 8.4 [24]. When all springs are soft and identical, the network is uniform (Fig. 8.4a, left-hand panel). When

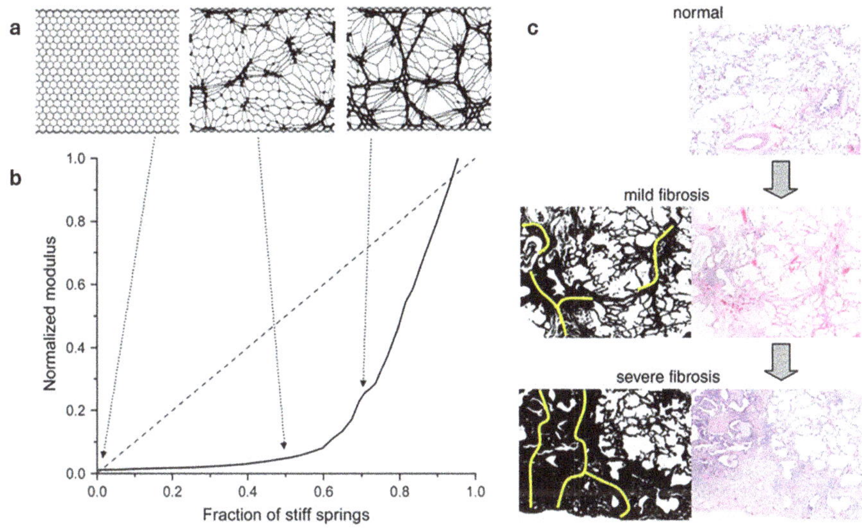

Fig. 8.4 (**a**) Hexagonal network configurations obtained from simulation of the progression of pulmonary fibrosis when (from left to right) 0, 50, and 67% of the springs have been stiffened. Thick lines show stiffened springs. (**b**) Normalized bulk modulus of the elastic network versus the fraction of springs randomly stiffened (equivalent to percolation probability) by a factor of 100. If all the spring constants were uniformly stiffened in a gradual manner from the baseline value of 1 to 100, the modulus would follow the dashed diagonal line. (**c**) Lung pathology illustrates the concepts of diffuse or focal disease and percolation at a microscopic level. Normal lung tissue (top right) shows numerous delicate alveolar walls and relatively uniform airspace diameters. Mild usual interstitial pneumonitis (UIP) (middle right) shows focal areas of cellular accumulation and fibrosis with adjacent areas of relatively normal lung, whereas severe UIP (bottom right) shows extensive dense fibrosis that spans the microscopic field. Next to the fibrosis specimens, the same images are shown in high-contrast black and white. Yellow lines indicate isolated islands of septal thickening in mild fibrosis and thickening that has completely percolated across the tissue in severe fibrosis. (Adapted with permission from Ref. [24])

about 50% of the springs have been randomly stiffened (Fig. 8.4a, middle panel), they exist in small, isolated islands that distort the surrounding normal springs. By the time about 66% of the springs have been stiffened, p just exceeds p_c and something new emerges: the islands suddenly coalesce to form pathways that span the entire network, causing even greater distortion in the remaining normal springs (Fig. 8.4a, right-hand panel). Figure 8.4b thus reveals the surprising functional consequence of structural percolation, namely the highly nonlinear way in which the bulk modulus of the network varies with the fraction of stiffened springs. Even though this fraction increases linearly along the horizontal axis, the modulus increases slowly at first and then, about halfway along, accelerates dramatically until every spring has been stiffened. This sudden change in trajectory corresponds to the percolation threshold where pathways of stiff spring first span across the network.

This simple example of percolation has potentially significant implications for PF if one takes the elastic network in Fig. 8.4a as a model of the lung parenchyma. The steady progression of underlying fibrotic disease at the microscale is represented by the horizontal axis in Fig. 8.4b, while the degradation of lung function, whole lung stiffness, characteristic of PF is represented by the vertical axis. The nonlinear relationship in this plot implies that underlying disease progression and clinical symptoms may become highly disconnected (as opposed to what would be the case if the relationship followed the dashed line in Fig. 8.4b). In other words, derangements in lung function, and hence clinical symptoms, may be minimal early in the course of PF, while symptoms will manifest suddenly and dramatically later in the disease course. Figure 8.4c supports this notion by showing histologic sections of lung tissue from a normal lung and from lungs with two different severities of PF in which thick-walled fibrotic tissue has replaced the normal alveolar architecture. Connectivity of the fibrotic clusters is indicated in adjacent black and white images, suggesting that the disease percolates at the microscale. The concept of percolation may thus offer an explanation for why PF can appear to remain dormant in its early stages before there is a sudden clinical decline leading to death [64].

The above percolation model is, of course, extremely simple and neglects a number of important aspects of actual PF pathophysiology, such as the fact that PF does not develop randomly at uncorrelated sites in the lung tissue. Characteristic scarring patterns in PF [65] indicate that, in fact, collagen deposition manifests as a relatively small number of larger fibrotic patches. This may reflect the existence of fixed sites of fibrogenesis that grow radially outward. Alternatively, it may be that once collagen is deposited at a particular location, the local increase in stiffness it engenders gives rise to a pro-fibrotic mechanical environment for the surrounding cells which themselves start to produce collagen in a positive feedback loop. Either case can be modeled by imposing an appropriate degree of spatial localization on the stiffening of the springs by, for example, randomly selecting a spring to stiffen and then proceeding to move in a random walk fashion to adjacent springs, stiffening each one along the way [25]. This process automatically generates a connected cluster, while the degree of spatial spreading in spring stiffening is determined by the number of steps the random walk is allowed to have before it is terminated and another walk begins from another randomly chosen spring. Depending on the nature of the random walk, this has a significant impact on the percolation threshold. Consider a square lattice of size 100 × 100 nodes that has 20,200 links. Random percolation in the square network is achieved when $p = p_c = 1/2$ [59] which means that at least 10,100 members need to be stiffened to reach p_c. However, a straight path, corresponding to a highly persistent walk, from one side of the network to the other, representing a fully deterministic process, requires only 100 stiff springs for percolation. Of course, an intermediate case with some moderate degree of local correlation might be more realistic than either the fully random or the fully deterministic percolation.

The combination of random walk and percolation can be implemented in either 2D [25] or 3D [27]. Interestingly, for regular percolation, p_c in 3D is significantly smaller than in 2D [66]. This means that a much smaller fraction of the lung tissue needs to become fibrotic in 3D to reach percolation via a correlated process, which is significant for PF because, of course, our lungs are 3D structures. Figure 8.5 compares fibrotic structures generated by the model in 3D for random walk lengths of 1 and 100, each with disease progression percentages of 1%, 5% and 20% corresponding to percolation probabilities of 0.01, 0.05 and 0.2. Figures 8.5a, c and e show the stiffened spring networks in 3D, while Figs. 8.5b, d, f show 2D cuts through the networks with the springs colored to indicate extent of local disease. These images demonstrate that with a random walk length of 1 (uncorrelated stiffening), fibrosis develops in isolated patches of 1 to 2 springs that lead to relatively homogeneous disease. In contrast, when the random walk includes 100 steps, fibrosis mostly develops along continuous paths leading to clumping of the stiffened springs (Figs. 8.5g–i). This has clinical implications since functional deterioration is influenced by the spatial pattern, not just the amount of tissue affected.

Percolation may thus help us understand the parenchymal patterns of fibrotic lesions characteristic of PF. These patterns are a classic example of emergent behavior because they arise as a result of a fibrogenic process operating on a spatial network. This demonstrates that simply understanding the activities of fibrogenic cells is not enough to fully understand clinical aspects of PF. On the other hand, the engine that ultimately drives the disease is cellular activity, to which we now turn.

8.5 Agent-Based and Structural Modeling of Honeycombing: The Role of Stiffness

As introduced in Chap. 6, agent-based modeling is naturally suited to the study of cellular dynamics because it explicitly represents cells as discrete entities (agents) that obey specific rules of behavior. These rules define how cells interact, move around in a tissue, divide, die, and so on, but they do not go so far as to link these behaviors specifically to intra-cellular mechanism such as second messenger signaling, gene expression, etc. The spring network models discussed in the previous section do not represent cell behavior, even though they can illuminate emergent consequences of this behavior such as percolation, because they cannot account for the bidirectional interactions that take place between the cells and their microenvironment. These interactions are critical because cells create and maintain the tissue environment, while the tissue exerts a major influence on cell behavior.

Cell-tissue interactions can be studied by combining spring network models of the parenchymal environment with agent-based models of cellular dynamics to create a hybrid Spring Network-Agent Based (SNAB) model. This modeling approach has been used to address the question of why the honeycomb structures seen in CT images of the fibrotic lung [67] usually first emerge next to the pleural membrane [29], while the central regions of the lung remain relatively spared [48]. Honeycombing is the term applied to mosaic patterns of enlarged airspaces

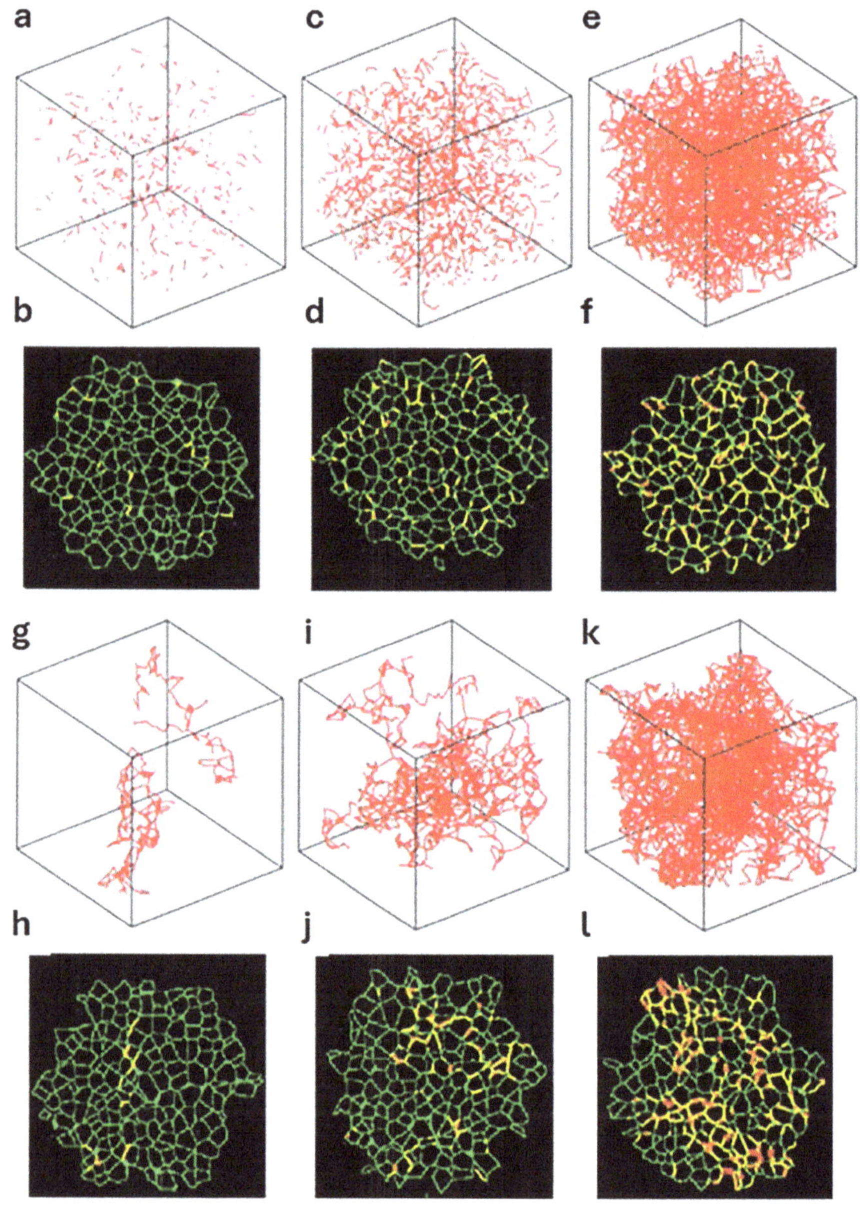

Fig. 8.5 Stiffened springs in the Amorphous Network (see Chap. 2). (**a**, **b**, **g**, and **h**) show 1% disease progression. (**c**, **d**, **i** and **j**) show 5% disease progression. (**e**, **f**, **k** and **l**) show 20% disease progression. (**a**—**f**) Agent step length of 1, i.e. random spring stiffening. (**g**—**l**) Agent step length of 100, showing clear clustering of disease progression. In the images (**b**, **d**, **f**, **h**, **j** and **l**), healthy walls are shown in green, 1/3 progression in yellow, 2/3 progression in orange, and complete disease in red. (Reproduced with permission from Ref. [27])

bordered by fibrotic tissue, and its genesis in IPF has been attributed to mechanical factors such as breathing-induced mechanical stresses [56, 68] and changes in the stiffness of the parenchymal ECM [18, 19]. This being the case, it is reasonable to suppose that the visceral pleura may be involved in the early pathogenesis of honeycombing given its high elastic modulus of about 50 kPa [69] compared to the 0.5–5 kPa typical of the alveolar parenchyma [70]. Furthermore, the cell most sensitive to and directly responsible for creating fibrotic tissue is the fibroblast [71], which is influenced by the local stiffness of the ECM [18, 19] and for which mechanotransduction is a major signaling pathway [17]. This leads to the possibility that when fibroblasts deposit collagen and other connective tissue macromolecules, the resulting alterations in tissue stiffness may in turn affect fibroblast activation, potentially creating a positive feedback loop that results in irreversible fibrosis. Slight initial variations in the stiffness of normal parenchyma could be differentially amplified by this fibrotic progression to produce the emergent spatial patterns we recognized as peripheral honeycombing.

To test this idea, a SNAB model was created in which some of the nodes of the spring network are occupied by agents conforming to a set of rules that have been determined experimentally to be typical of fibroblasts. Fibroblast migration is simulated by having each agent move, at each time step, across a spring to an adjacent node (with no possibility of immediate subsequent reversal). The choice of which spring to cross is made stochastically with probability weighted by the stiffness of the possible springs, since migration of human lung fibroblasts is enhanced on stiffer ECM [72] and is directed toward stiffer regions of tissue [73]. Since each node in the hexagonal spring network lies at the junction of three springs, one of which is disqualified by being the path at the previous time step, there are two springs available for an agent to cross at the current time step. If the spring constants of the two springs are k_1 and k_2, the two crossing probabilities are $k_1/(k_1 + k_2)$ and $k_2/(k_1 + k_2)$. Fibroblasts located in fibrotic tissue exhibit increased fibroblast contractility [72], proliferation [19], and gene expression [20], so each agent in the SNAB model is imbued with an activation level, a, that increases each time it crosses a spring having a pathologically increased stiffness defined as $k > k_{th}$ ($k_{th} = 2$). Conversely, when an agent crosses a spring having stiffness in the normal range ($k \leq k_{th}$), a is multiplied by 0.95, causing agents to become gradually deactivated when migrating on normal tissue [71]. When $k > k_{th}$, a is proportional to stiffness such that, after time step i, $a_{i+1} = Q_a(k - k_{th})$, with the proviso that $a_{i+1} < a_{max}$, where $Q_a = a_{max}/(k_{max} - k_{th})$ and k_{max} is the maximum stiffness allowed. This mimics the collagen deposition [14] and cross-linking [74] seen in PF. Thus, if k_i is the stiffness of a spring at time step i, and this spring is crossed by an agent with activation level a_i, then $k_{i+1} = k_i + Q_k a_i$, where Q_k is a constant such that $k_{i+1} \leq k_{max}$. Finally, agents randomly undergo apoptosis with a baseline probability $p_{apo} = 0.01$ that decreases linearly with increasing activity according to $p_{apo} = 0.01\left(1 - \frac{a}{a_{max}}\right)$, since fibroblasts exhibit increased resistance against apoptosis in IPF [75]. Whenever an agent undergoes apoptosis, it is immediately removed and replaced with a new naïve agent at a random location to represent replenishment of fibroblasts by cell division and recruitment from the vasculature.

The above SNAB model was implemented by randomly initializing its network springs with values of k distributed uniformly between 0.7 and 1.3 to reflect the natural heterogeneity of healthy parenchymal stiffness [70, 76]. The value of k for the springs in the pleura varied uniformly between 1 and 3. Agents at their initial levels of activation ($a_0 = 0$ or a given non-zero value from a uniform distribution) were placed randomly at 1% of the network nodes. Because agents become activated by high local stiffness, the initially activated agents were predominately located next to the relatively stiff pleura. The model was iterated according to the above rules and solved for mechanical equilibrium after every 16% increase in the bulk modulus of the network by minimizing the total strain energy of the system using simulated annealing [77]. Throughout this process, k was set to 0 for springs that became stretched beyond a set threshold to simulate tissue rupture. The sequential determination of mechanical equilibrium was repeated 6–8 times to simulate the development of mild fibrosis and continued for a total of 22–25 times to simulate severe fibrosis. The value of $k_{max} = 10$ was used to conform to atomic force microscopic measurements showing that regional lung stiffness in PF can reach values of 20 kPa [78].

Results from SNAB model simulations are shown in Fig. 8.6. Some of the springs in the network were removed to create holes of varying sizes representing airways to account for the distribution of low-attenuation areas (LAA) seen in CT images of normal lungs [79], and the model was initialized with $a_0 = 0$. In the early phase of the simulation exemplified by the left-hand panel in Fig. 8.6, the inactive agents (blue) are located predominately in regions of normal tissue having low stiffness (blue), while those few agents that have become active (yellow/red) reside in the stiffened regions (green/yellow). Later in the simulation exemplified by the right-hand panel in Fig. 8.6, stiff regions progressively engulf the tissue, beginning at the

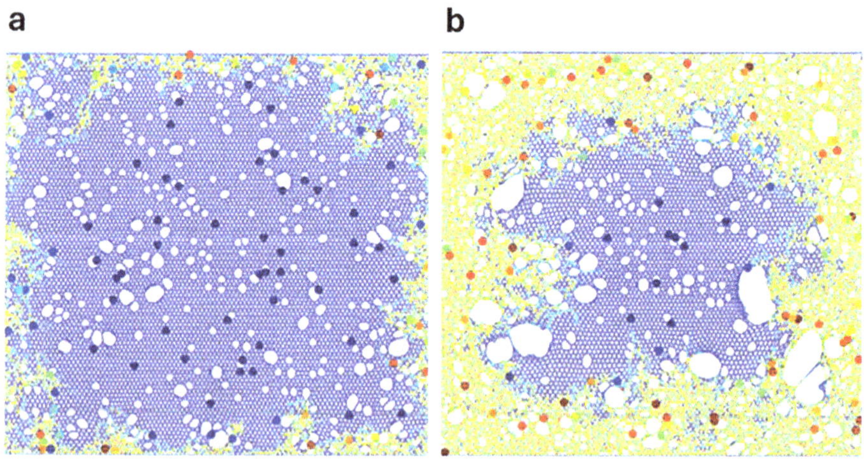

Fig. 8.6 Example images of the model in the early (**a**) and late (**b**) phases. The simulations began with initially unactivated agents (Model U: a_0=0). Network spring colors are related to local stiffness (blue: low, yellow/orange: high). Agents are shown as circles with colors indicating agent activity (dark blue: low, red: high). Note that activated agents are localized in the stiff regions of the network. (Reproduced with permission from Ref. [29])

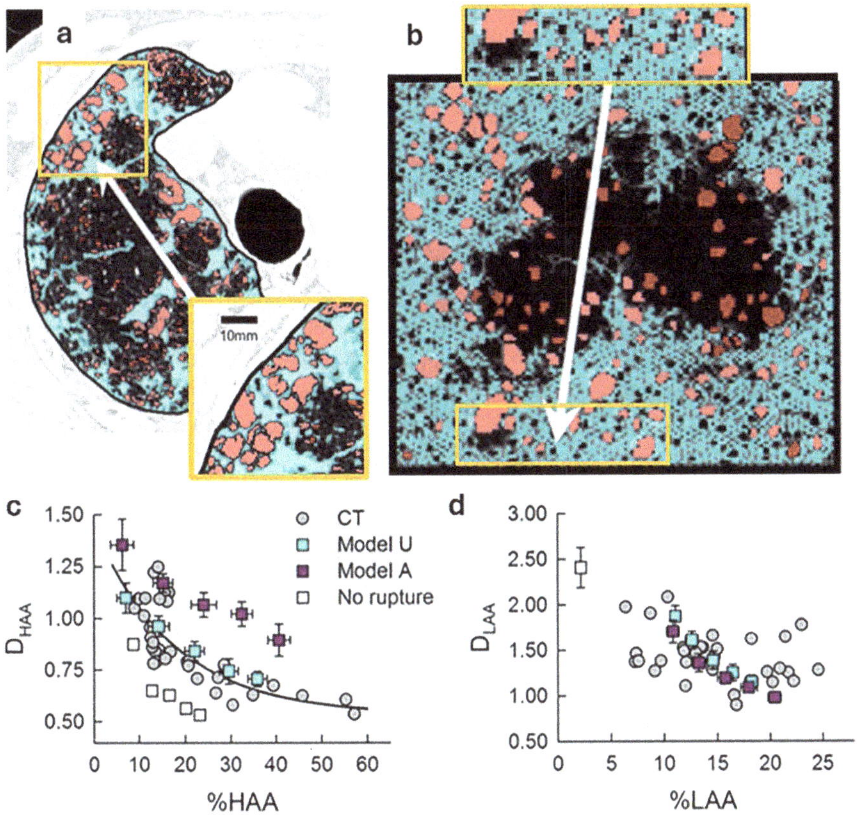

Fig. 8.7 Comparison of model results with clinical CT images. (**a**) Representative CT lung slice from a patient with IPF showing high attenuation areas (HAA: cyan), low attenuation areas (LAA) within 2 mm of an HAA (light red), and LAAs further than 2 mm away from an HAA (dark red). The inset shows an example of subpleural honeycombing. (**b**) Virtual CT image generated from the network model at the last stage of fibrosis development beginning with agents that were initially unactivated. The colors represent HAAs and LAAs as in panel **a**. (**c**) Exponents D_{HAA} of the HAA distributions as functions of overall degree of fibrosis (%HAA), and (**d**) exponents D_{LAA} of the distributions of LAAs as a function of total %LAA. The gray circles correspond to 36 CT images from human subjects. The colored squares correspond to simulated CT images when the model was run to produce progressive fibrosis. Simulations were performed with agents being initially unactivated (Model U in cyan), and initially activated (Model A). The single open square represents a simulation with Model U without allowing for rupture. To better visualize the general trend, a second order polynomial fit of the CT data (black line) is included. (Reproduced and adapted with permission from Ref. [29])

periphery, and the holes in the network become enlarged. The bulk modulus of the network also increases progressively despite larger holes forming in the network due to rupture.

Figure 8.7a shows a density color-coded CT image of the lung from an IPF patient. Figure 8.7b shows a synthetic CT image created by coarse-graining a late-stage simulation of a PF network from the SNAB model. Both images exhibit peripheral honeycombing and have similar distributions of high and low attenuation areas (Figs. 8.7c, d). Interestingly, both distributions follow a power law, although with varying exponents, which allows the distributions to be characterized by only a single quantity—the power law exponent. This also suggests that the model is able to capture essential aspects of the complex system dynamics involved in transitioning the lung from health to fibrotic pathology. Interestingly, better results are obtained when the model is initialized with inactive agents rather than agents assigned a finite level of activation. In addition, realistic values for the low attenuation area distributions are only obtained when rupture of springs under particularly high stress is allowed. This implies that the development of honeycombing requires tissue rupture, meaning that aberrant collagen deposition alone is not enough to account for the characteristic morphology of PF. This is a novel insight because tissue rupture is typically associated with emphysema [80] and not fibrosis.

The SNAB model thus suggests that stiffness-related collagen deposition and tissue failure under high stress conspire to give rise to the distinctive morphologic patterns seen both histologically and radiologically in PF. These processes develop as a result of tissue mechanics and cell activation that feed on each other in a positive feedback loop, resulting in irreversible and eventually fatal disease. Why this vicious cycle should suddenly develop in the lungs after decades of perfect health, as happens in the case of IPF, is still very unclear.

8.6 How Abnormal Stretch Turns Homeostasis into Fibrosis

The SNAB model discussed above [29] considers only a single cell type and a single cell signaling mechanism. It also neglects the possibility that PF may be initiated by some tissue-specific or blood-borne mediators that first makes their presence felt near the pleura. Such mediators could potentially induce the same kind of positive feedback process, with resultant peripheral honeycombing, that was attributed above to regional tissue stiffening. Persistent cyclic stretching of the lung parenchyma and its resident cells during breathing, which may be an important source of mechanotransduction, is also neglected in this model. Indeed, tissue strains such as occur during breathing can liberate activated TGF-β1 from extracellular matrix stores in fibrotic but not normal lungs because of the greater stiffness of fibrotic tissue [56]. Finally, and perhaps most important, the model has no mechanism by which the tissue can return toward homeostasis after starting down the fibrotic pathway.

Some of these shortcomings can be addressed with the inclusion of additional agent types and rules [28]. For example, suppose the spring network in the SNAB model is the 2D amorphous network introduced in Chap. 2, where each spring in the network has its own spring-constant k. The Young's modulus, Y, of a spring is then $Y = kL_0/A$, where A the cross-sectional area and L_0 is the resting length of the spring. Also suppose that whenever an agent traverses a spring, it evaluates its stiffness and then makes a contribution to the agent's activation level, a_k, according to

$$a_k = \frac{(k)^\beta}{(k)^\beta + 1} = \frac{\left(\frac{YA/Lo}{YA_t/Lo}\right)^\beta}{\left(\frac{YA/Lo}{YA_t/Lo}\right)^\beta + 1} \tag{8.7}$$

where A_t is the value of A at which the activity is 0.5 . If Y and L_0 are constants then Eq. 8.7 simplifies to

$$a_k = \frac{\left(\frac{A}{A_t}\right)^\beta}{\left(\frac{A}{A_t}\right)^\beta + 1} \tag{8.8}$$

where the Hill exponent β is set arbitrarily to 3. Equation 8.8 shows that the variable A can be used in lieu of k as an index of the stiffness of an individual spring.

The strain, ε, of the spring that an agent traverses also makes a contribution to agent activity by representing cell activation via the release of ECM-bound latent TGF-beta [18, 56], stretch-sensitive ion channels [81], and/or signaling pathways such as YAP and MRTF-A [82]. In analogy with Eq. 8.8, the strain-activation relationship is

$$a_\varepsilon = \frac{\left(\frac{\varepsilon}{\varepsilon_s}\right)^\gamma}{\left(\frac{\varepsilon}{\varepsilon_s}\right)^\gamma + 1} - c \tag{8.9}$$

where ε_s is the target (homeostatic) strain at steady-state and c is a constant that causes $a = 0$ when $\varepsilon = \varepsilon_s$, corresponding to homeostasis, and γ is arbitrarily set equal to 1.

The total activation of an agent when it traverses a spring of cross-sectional area A and strain ε is then a weighted sum of Eqs. 8.8 and 8.9 thus:

$$a = w_1 a_\varepsilon + w_2 a_k \tag{8.10}$$

where the weighting factors w_1 and w_2 determine the relative strengths of the strain-related and stiffness-related signaling pathways, respectively, and can be considered as reflecting genetic and/or environmental predispositions towards different forms of fibrosis.

For convenience, one can set $w_1 = 1$ and allow w_2 to vary. At homeostasis, $\varepsilon = \varepsilon_s$, $A = A_0$, and $a = 0$, which collectively determine the value of c. Under these conditions, the tissue is well below the fibrotic threshold, A_0 A_t, which means that $a_k \lll a_\varepsilon$. Cell activity during normal homeostatic maintenance is then driven mainly by cyclic stretch rather than ECM stiffness, which allows the usual substantial degrees of regional variability in parenchymal stiffness to exist without tissue normally becoming fibrotic [70]. This is equivalent to saying that tissue homeostasis is maintained so long as a perturbation in k, which can be achieved by changing A, is followed by a return of k to its equilibrium value.

At each time step, i, the agents update A for the spring they traverse according to

$$A_{i+1} = A_i + Q_A \rho_i a_i \tag{8.11}$$

where Q_A is a constant, and ρ_i is agent density defined as the number of agents in the network normalized by the number of springs. ρ_i can thus be thought of as the probability that an agent crosses a given spring in the network in a given time step. Since agent activity can be either positive or negative, $\Delta A_i = Q_A \rho_i a_i$ may also be positive representing collagen deposition, or negative representing tissue digestion. As the model simulation proceeds, if Eq. 8.11 produces the condition $A_i \leq 0$, the corresponding spring is considered to have ruptured. Such springs are then removed from the network, thereby eliminating them as migration pathways for agents. Agents with a close to zero are considered to represent fibroblasts, while those with a approaching the maximum possible value are taken to be myofibroblasts. The sources of myofibroblasts are still debated [83], but both fibroblasts and myofibroblasts have been observed to increase in number in PF, and it is possible that myofibroblasts arise via differentiation of fibroblasts.

As this model runs, the continually changing values of A throughout the network cause a change in its equilibrium configuration, which is determined at each time step by simulated annealing. Reductions in lung volume due to tissue stiffening, such as occurs in PF [84], are permitted by imposing fixed boundary forces rather than positions. Initial tissue injury is simulated by modifying the value of A for a given percent of springs in the network. After a given number of time steps, a synthetic histologic image can be created by drawing the network with edge thicknesses proportional to A.

Interestingly, this model has a stable fixed point in the sense that when a small perturbation (changing A modestly in a subset of springs) is imposed, it causes the bulk modulus of the network to deviate from its equilibrium value, yet the modulus returns to the fixed point as the model continues to run. This mimics the healing and return to homeostasis characteristic of living systems (see Chap. 6). However, when the perturbation is sufficiently large, the network becomes progressively fibrotic with no possibility of recovery. The model thus exhibits a bifurcation point below which it heals and beyond which it becomes increasingly fibrotic. Because of the stochastic nature of the model, however, this bifurcation is not precisely defined. For example, when 20% of the springs are initially stiffened five-fold, some injury configurations heal while others do not. The eventual outcome appears to be determined by whether some of the springs that are initially very close to the fibrotic threshold are pushed over the threshold by the evolving behavior of nearby springs. The behavior of the model is also affected by the value of the parameter w_2, which has a profound effect on the fibrotic phenotype as characterized by sparseness, honeycombing, and growth of dense fibrotic patches. Figure 8.8 demonstrates that the structural features created by this SNAB model are highly reminiscent of those seen in actual histologic images of the lung parenchyma at varying stages of IPF progression.

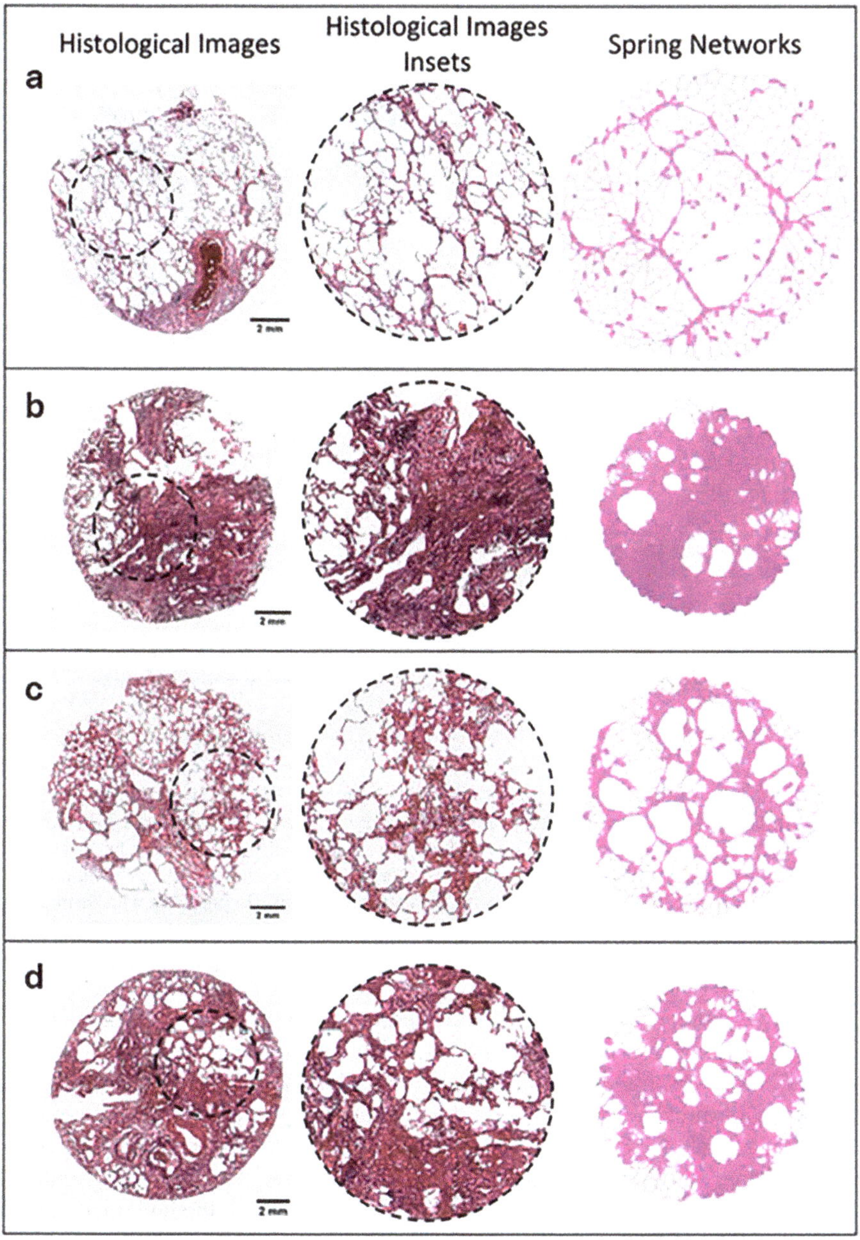

Fig. 8.8 The first column shows histological slices of IPF tissue with differing structures, while the second column shows enlarged sections from these slices. (Adapted with permission from Ref. [85].) The third column shows spring networks of varying sensitivity at different stages of progression that are similar in structure to the images on the same row. (Reproduced with permission from Ref. [28])

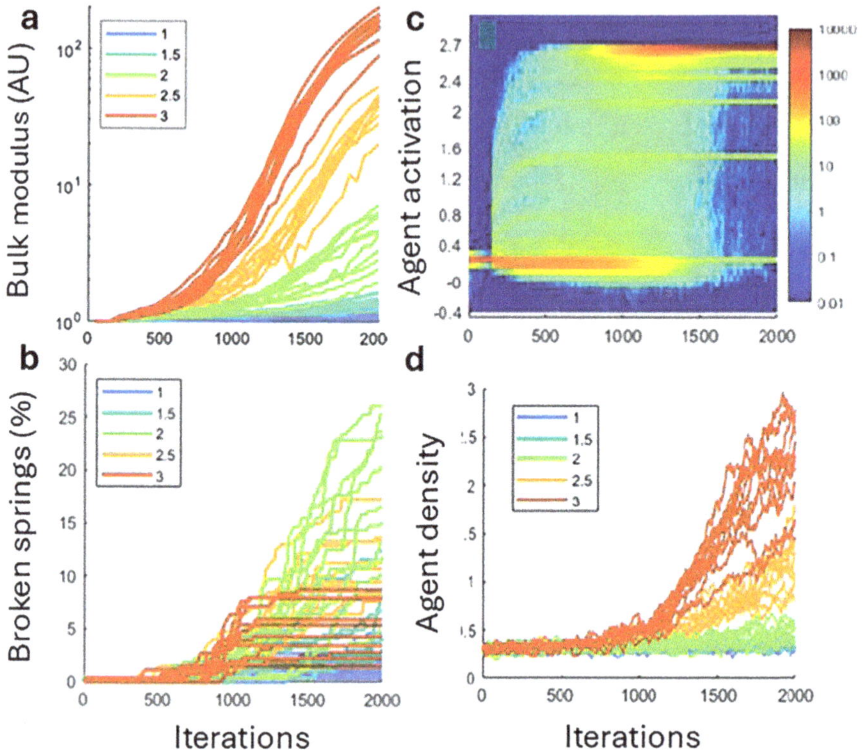

Fig. 8.9 Analysis of Mechanics in fibrotic network progression. Each network had 10% of the springs stiffened by 10× the initial cross-sectional area. Each color represents a different stiffness weight (w_2), and each sensitivity was tested with at least nine networks. (**a**) Network bulk modulus as progression occurs. (**b**) Percent ruptured springs as progression occurs. (**c**) Agent activation with $w_2 = 3$. (**d**) Agent density versus iteration for all network simulations. (Reproduced with permission from Ref. [28])

The progression of fibrosis in the model also causes a progressive elevation of its bulk modulus, which occurs more quickly in networks that are weighted toward a greater fibrotic response driven by greater values of w_2 (Fig. 8.9a). Interestingly, the percentage of broken springs is not monotonic but instead reaches a peak at intermediates values of w_2 (Fig. 8.9b). The model also displays a clear transition from a predominance of fibroblasts to myofibroblasts at $w_2 = 3$ (Fig. 8.9c), with an agent density that depends strongly on w_2 and correlates with the bulk modulus of the network (Fig. 8.9d). This illustrates the complex behaviors that can arise from what is still a relatively simple SNAB model.

Particularly insightful are the model predictions of a threshold level of fibrosis that separates recovery from irreversible disease. This threshold emerges from the competition between the stiffness and the strain responses. If the microscopic stiffness overpowers the local strain response, fibrosis is inevitable. Conversely, the macroscopic bulk modulus may indicate a highly fibrotic lung, yet the network may be able to heal if none of the spring stiffnesses locally exceed the fibrotic threshold. Clinically, this can occur following recovery from ARDS [86] and COVID-19 infection [87]. The position of the fibrotic threshold is determined by the point at which the stiffness response overpowers the strain response, which is presumably determined initially by genetic predisposition, but no doubt can be modulated by environmental factors. Thresholds such as this have been observed to influence differentiation of several precursor cell types into fibroblasts [18]. Importantly, what ultimately determines the fibrotic fate of this model is not necessarily how many springs have stiffness near the fibrotic threshold as much as whether there are a few springs that are close enough to the threshold to risk being pushed beyond it. This risk depends on the presence of nearby fibrotic regions that can maintain the springs in states of high stretch. This invites a misguided repair response aimed at reducing the strain by further stiffening the springs just beyond the threshold, from which they cannot recover.

8.7 Conclusions

The pathophysiology of PF involves not only the aberrant behavior of fibrogenic cells but also how these cells interact in the topographically complex landscape within which they live. The percolation idea is general and independent of the particular agent-based or differential equation-based approach being used to describe PF. Accordingly, modelers studying lung function deterioration, such as a sudden increase in bulk modulus, should always suspect that some sort of a percolation-like structural phase transition lurks behind any observed functional changes. Computational modeling based on SNAB models also reveal that emergent network behavior is a critical determinant of the PF phenotype that can only be understood by taking the spatial distribution of fibrotic lesions into account. This makes it clear that PF itself is an emergent phenomenon. The story does not end here, however, since many questions remain unanswered. This is particularly highlighted by the fact that there is still no cure for IPF. Nevertheless, current models remain extremely simplified representations of reality, despite the complex behaviors they exhibit. It is thus safe to assume that computational modeling will continue to play a significant role in advancing our understanding of fibrotic diseases of the lung as new measurement methodologies are developed and computational power increases.

8.8 Problems

1. Equations 8.1 and 8.2 describe the spatiotemporal dynamics of cell densities and cytokines driven by diffusion, proliferation, apoptosis, transitions to different cell types and actions of cytokines. In the following, assume that there are two cell types.

   (a) What is the necessary condition for a homogeneous equilibrium solution within a homogeneous medium in the absence of cytokines?

(b) How does the equilibrium condition change when cytokines are added? Assume that there are two cytokines each acting on one cell type and their concentration is constant.

2. The theory of percolation can be useful to describe many phenomena including PF. To model progressive fibrosis in a small lung parenchymal tissue strip, consider a rectangular lattice with M rows and N columns. Each square represents a single alveolus. The sides of the squares correspond to the septal walls modeled by linear springs each with spring constant k. The line elements can undergo fibrotic modifications via deposition of collagen, increasing k to $k_f \gg k$. The process of fibrosis will be assumed to attack each septal wall randomly with a given probability p.

   (a) Write a code that computes the stiffness of a 2D rectangular array with springs having arbitrary spring constants. Assume the network is uniaxially stretched in the horizontal x direction and, for simplicity, assume that the springs in the y direction do not contribute to the total stiffness K. Recall the formula to compute the total spring constant of springs in either series and parallel. Set up a matrix of springs with $M = 2$ and $N = 3$, fill it up with random values, calculate manually the resulting K, then compute this with your code and verify the results.
   (b) Create a larger square network with $M = 20$, $N = 200$. Set each spring to have an initial value of $k = 1$ and write a code that creates a loop that increases p from 0 to 1 in increments of 0.05. Stiffen each spring in the array by a factor $q = 100$ with probability $0 < p < 1$. This can be done by drawing a uniformly distributed random number r between 0 and 1 and if $r < p$, then the spring constant is multiplied by q. Next, for each value of p, compute the resulting total spring constant K of the network. Make sure that the spring constant matrix is initialized to 1 for each p. Visualize the network for each p by for example setting the colors to blue and red when $k = 1$ and $k = 100$, respectively. Plot K as a function of p. If the graph is noisy, repeat the simulations 10 to 100 times and average K for each p. In a square lattice, bond percolation occurs at $p_c = 0.5$. Do you see percolation-like behavior? If not, try to find the explanation. (Hint: since the stiffness is computed in uniaxial stretch, and the vertical springs don't count, percolation should be applied only to the horizontal springs.)
   (c) Here is a trick. Because of the assumption that vertical walls do not contribute to horizontal stiffness, it is not necessary to store all septal walls. Instead, simply set up a working array representing the horizontal walls in series, and compute its total spring constant K_1. Store this value, create a new realization in the same array, compute the resultant K_2 and add it to the previous one, i.e., $K = K_1 + K_2$. This method allows you to create much bigger systems. Verify this method by comparing the result to that obtained in part (b). Next, increase M to 1000 and N to 1,000,000. Repeat the calculations and plot K as a function of p. What are the general features of the graph? How does this curve compare to the case of the smaller system in part (b)?

(d) To study how size affects the $p - K$ curve, create a logarithmically spaced vector of the number of septal walls in series: $N = [5, 10, 20, \ldots, 10240]$ while keeping $M = 1000$. Also create a logarithmic scale for p going from 0.01 to 1 in 100 steps. Repeat the calculations of $p - K$ for each N. Normalize the spring constants so that $K = 1$ at $p = 1$, and plot all curves on a single graph. How does the normalized K approach unity as the system size increases? Find the value of p at which the normalized K exceeds 0.05 or 5% of the maximum total spring constant for each $p - K$ curve and plot this p as a function of N on a log-log graph. What can you observe?

(e) If you have reached this far, then you have created a model of percolation. As progressively more springs stiffen up by increasing p, the total K increases. This represents a model of developing fibrosis. Additionally, notice that having $M = 1000$ rows simply averages the $p - K$ curves. The average curve represents the approach toward a specific value, the percolation threshold. What is the percolation threshold in your model? If the model represents the fibrotic transition in the lung, what can you say about the course of the disease? How can you incorporate time into the model? If the increase in p is proportional to time, what can you say about the progression of the disease?

(f) Lastly, explain the limitations of the approach. In particular, is it reasonable to neglect the effects of vertical springs? If not, how can they be taken into account? Advanced readers may want to implement the simulated annealing algorithm from Chap. 4 (Problem 4) and compute the stiffness of the network numerically both in uniaxial and biaxial stretch. How does the theoretical stiffness (neglecting the vertical springs) compare to the numerical ones?

3. Implement an agent-based model on the square network. As the agents randomly migrate, they can deposit collagen, which is modeled by stiffening the network member they visit along their random walks.

   (a) Examine how the number of agents influences the pattern of fibrotic cluster formation. Compare the clustering in the agent-based model with that of the random percolation in Problem 2.

   (b) Add various features to the agent rules such as apoptosis, and cellular activity that stimulates collagen deposition as described in the text.

   (c) As a more involved exercise, implement soft boundaries. The network can be surrounded by a set of springs whose outer ends are fixed. Furthermore, a pressure boundary can then be obtained by calculating the forces per unit spring length along the circumference of the network. Setting up fixed external forces can then be applied to each of the boundary nodes to mimic the pleural pressure around the lung.

References

1. Nasser M, Larrieu S, Si-Mohamed S, Ahmad K, Boussel L, Brevet M, Chalabreysse L, Fabre C, Marque S, Revel D, et al. Progressive fibrosing interstitial lung disease: a clinical cohort (the PROGRESS study). Eur Respir J. 2021;57:2002718.
2. Jeganathan N, Sathananthan M. The prevalence and burden of interstitial lung diseases in the USA. ERJ Open Res. 2022;8:00630-2021.

3. Cottin V, Hirani NA, Hotchkin DL, Nambiar AM, Ogura T, Otaola M, Skowasch D, Park JS, Poonyagariyagorn HK, Wuyts W, Wells AU. Presentation, diagnosis and clinical course of the spectrum of progressive-fibrosing interstitial lung diseases. Eur Respir Rev. 2018;27:180076.
4. Nathan SD, Shlobin OA, Weir N, Ahmad S, Kaldjob JM, Battle E, Sheridan MJ, du Bois RM. Long-term course and prognosis of idiopathic pulmonary fibrosis in the new millennium. Chest. 2011;140:221–9.
5. Zheng Q, Cox IA, Campbell JA, Xia Q, Otahal P, de Graaff B, Corte TJ, Teoh AKY, Walters EH, Palmer AJ. Mortality and survival in idiopathic pulmonary fibrosis: a systematic review and meta-analysis. ERJ Open Res. 2022;8:00591-2021.
6. Moore BB, Moore TA. Viruses in idiopathic pulmonary fibrosis. Etiology and exacerbation. Ann Am Thorac Soc. 2015;12(Suppl 2):S186–92.
7. Lee JS. The role of gastroesophageal reflux and microaspiration in idiopathic pulmonary fibrosis. Clin Pulm Med. 2014;21:81–5.
8. Kitamura H, Ichinose S, Hosoya T, Ando T, Ikushima S, Oritsu M, Takemura T. Inhalation of inorganic particles as a risk factor for idiopathic pulmonary fibrosis--elemental microanalysis of pulmonary lymph nodes obtained at autopsy cases. Pathol Res Pract. 2007;203:575–85.
9. Mariscal-Aguilar P, Gomez-Carrera L, Bonilla G, Diaz-Almiron M, Gaya F, Carpio C, Zamarron E, Fernandez-Velilla M, Torres I, Esteban I, et al. Air pollution exposure and its effects on idiopathic pulmonary fibrosis: clinical worsening, lung function decline, and radiological deterioration. Front Public Health. 2023;11:1331134.
10. Steele MP, Schwartz DA. Molecular mechanisms in progressive idiopathic pulmonary fibrosis. Annu Rev Med. 2013;64:265–76.
11. Betensley A, Sharif R, Karamichos D. A systematic review of the role of dysfunctional wound healing in the pathogenesis and treatment of idiopathic pulmonary fibrosis. J Clin Med. 2016;6:2.
12. Wynn TA. Integrating mechanisms of pulmonary fibrosis. J Exp Med. 2011;208:1339–50.
13. Rodero MP, Khosrotehrani K. Skin wound healing modulation by macrophages. Int J Clin Exp Pathol. 2010;3:643–53.
14. Kirk JM, Da Costa PE, Turner-Warwick M, Littleton RJ, Laurent GJ. Biochemical evidence for an increased and progressive deposition of collagen in lungs of patients with pulmonary fibrosis. Clin Sci (Lond). 1986;70:39–45.
15. Jones MG, Andriotis OG, Roberts JJ, Lunn K, Tear VJ, Cao L, Ask K, Smart DE, Bonfanti A, Johnson P, et al. Nanoscale dysregulation of collagen structure-function disrupts mechano-homeostasis and mediates pulmonary fibrosis. elife. 2018;7:20180703.
16. Suki B, Stamenovic D, Hubmayr RD. Lung parenchymal mechanics. In: Fredberg JJ, Sieck GC, Gerthoffer WT, editors. Comprehensive physiology, the respiratory system, respiration mechanics: organ, cell, molecule, vol. 1. Wiley-Blackwell; 2011. p. 1317–51.
17. Duscher D, Maan ZN, Wong VW, Rennert RC, Januszyk M, Rodrigues M, Hu M, Whitmore AJ, Whittam AJ, Longaker MT, Gurtner GC. Mechanotransduction and fibrosis. J Biomech. 2014;47:1997–2005.
18. Hinz B. Tissue stiffness, latent TGF-beta1 activation, and mechanical signal transduction: implications for the pathogenesis and treatment of fibrosis. Curr Rheumatol Rep. 2009;11:120–6.
19. Liu F, Mih JD, Shea BS, Kho AT, Sharif AS, Tager AM, Tschumperlin DJ. Feedback amplification of fibrosis through matrix stiffening and COX-2 suppression. J Cell Biol. 2010;190:693–706.
20. Parker MW, Rossi D, Peterson M, Smith K, Sikstrom K, White ES, Connett JE, Henke CA, Larsson O, Bitterman PB. Fibrotic extracellular matrix activates a profibrotic positive feedback loop. J Clin Invest. 2014;124:1622–35.
21. Polak AG, Lutchen KR. Computational model for forced expiration from asymmetric normal lungs. Ann Biomed Eng. 2003;31:891–907.

22. Zappala CJ, Latsi PI, Nicholson AG, Colby TV, Cramer D, Renzoni EA, Hansell DM, du Bois RM, Wells AU. Marginal decline in forced vital capacity is associated with a poor outcome in idiopathic pulmonary fibrosis. Eur Respir J. 2010;35:830–6.
23. Hao W, Marsh C, Friedman A. A mathematical model of idiopathic pulmonary fibrosis. PLoS One. 2015;10:e0135097.
24. Bates JH, Davis GS, Majumdar A, Butnor KJ, Suki B. Linking parenchymal disease progression to changes in lung mechanical function by percolation. Am J Respir Crit Care Med. 2007;176:617–23.
25. Oliveira CL, Bates JH, Suki B. A network model of correlated growth of tissue stiffening in pulmonary fibrosis. New J Phys. 2014;16:065022.
26. Casey DT, Bou Jawde S, Herrmann J, Mori V, Mahoney JM, Suki B, Bates JHT. Percolation of collagen stress in a random network model of the alveolar wall. Sci Rep. 2021;11:16654.
27. Hall JK, Bates JHT, Casey DT, Bartolak-Suki E, Lutchen KR, Suki B. Predicting alveolar ventilation heterogeneity in pulmonary fibrosis using a non-uniform polyhedral spring network model. Front Netw Physiol. 2023;3:1124223.
28. Hall JK, Bates JHT, Krishnan R, Kim JH, Deng Y, Lutchen KR, Suki B. Elucidating the interaction between stretch and stiffness using an agent-based spring network model of progressive pulmonary fibrosis. Front Netw Physiol. 2024;4:1396383.
29. Wellman TJ, Mondonedo JR, Davis GS, Bates JHT, Suki B. Topographic distribution of idiopathic pulmonary fibrosis: a hybrid physics- and agent-based model. Physiol Meas. 2018;39:064007.
30. Suki B, Herrmann J, Bates JHT. An analytic model of tissue self-healing and its network implementation: application to fibrosis and aging. Front Physiol. 2020;11:583024.
31. Fry DL, Ebert RV, Stead WW, Brown CC. The mechanics of pulmonary ventilation in normal subjects and in patients with emphysema. Am J Med. 1954;16:80–97.
32. Pedersen OF, Butler JP. Expiratory flow limitation. Compr Physiol. 2011;1:1861–82.
33. Hyatt RE, Schilder DP, Fry DL. Relationship between maximum expiratory flow and degree of lung inflation. J Appl Physiol. 1958;13:331–6.
34. Murphy DM, Hall DR, Petersen MR, Lapp NL. The effect of diffuse pulmonary fibrosis on lung mechanics. Bull Eur Physiopathol Respir. 1981;17:27–41.
35. Paterniti MO, Bi Y, Rekic D, Wang Y, Karimi-Shah BA, Chowdhury BA. Acute exacerbation and decline in forced vital capacity are associated with increased mortality in idiopathic pulmonary fibrosis. Ann Am Thorac Soc. 2017;14:1395–402.
36. Reichmann WM, Yu YF, Macaulay D, Wu EQ, Nathan SD. Change in forced vital capacity and associated subsequent outcomes in patients with newly diagnosed idiopathic pulmonary fibrosis. BMC Pulm Med. 2015;15:167.
37. Machado FVC, Bloem AEM, Schneeberger T, Jarosch I, Gloeckl R, Winterkamp S, Franssen FME, Koczulla AR, Pitta F, Spruit MA, Kenn K. Relationship between body composition, exercise capacity and health-related quality of life in idiopathic pulmonary fibrosis. BMJ Open Respir Res. 2021;8:e001039.
38. Fry DL. Theoretical considerations of the bronchial pressure-flow-volume relationships with particular reference to the maximum expiratory flow volume curve. Phys Med Biol. 1958;3:174–94.
39. Lambert RK. Sensitivity and specificity of the computational model for maximal expiratory flow. J Appl Physiol Respir Environ Exerc Physiol. 1984;57:958–70.
40. Lambert RK. A new computational model for expiratory flow from nonhomogeneous human lungs. J Biomech Eng. 1989;111:200–5.
41. Lambert RK, Wilson TA, Hyatt RE, Rodarte JR. A computational model for expiratory flow. J Appl Physiol Respir Environ Exerc Physiol. 1982;52:44–56.
42. Bates JHT. Lung mechanics. An inverse modeling approach. Cambridge: Cambridge University Press; 2009.
43. Dawson SV, Elliott EA. Wave-speed limitation on expiratory flow-a unifying concept. J Appl Physiol Respir Environ Exerc Physiol. 1977;43:498–515.

44. Horsfield K, Dart G, Olson DE, Filley GF, Cumming G. Models of the human bronchial tree. J Appl Physiol. 1971;31:207–17.
45. Hedges KL, Tawhai MH. Simulation of forced expiration in a biophysical model, with homogeneous and clustered bronchoconstriction. J Biomech Eng. 2016;138:061008.
46. Tawhai MH, Hunter P, Tschirren J, Reinhardt J, McLennan G, Hoffman EA. CT-based geometry analysis and finite element models of the human and ovine bronchial tree. J Appl Physiol (1985). 2004;97:2310–21.
47. Barnea O, Abboud S, Guber A, Bruderman I. New model-based indices for maximum expiratory flow-volume curve in patients with chronic obstructive pulmonary disease. Comput Biol Med. 1996;26:123–31.
48. Nakagawa H, Nagatani Y, Takahashi M, Ogawa E, Tho NV, Ryujin Y, Nagao T, Nakano Y. Quantitative CT analysis of honeycombing area in idiopathic pulmonary fibrosis: correlations with pulmonary function tests. Eur J Radiol. 2016;85:125–30.
49. Sanjabi S, Zenewicz LA, Kamanaka M, Flavell RA. Anti-inflammatory and pro-inflammatory roles of TGF-beta, IL-10, and IL-22 in immunity and autoimmunity. Curr Opin Pharmacol. 2009;9:447–53.
50. Frangogiannis N. Transforming growth factor-beta in tissue fibrosis. J Exp Med. 2020;217: e20190103.
51. Sheppard D. Transforming growth factor beta: a central modulator of pulmonary and airway inflammation and fibrosis. Proc Am Thorac Soc. 2006;3:413–7.
52. Evans RA, Tian YC, Steadman R, Phillips AO. TGF-beta1-mediated fibroblast-myofibroblast terminal differentiation-the role of Smad proteins. Exp Cell Res. 2003;282:90–100.
53. Scheper T, Klinkenberg D, Pennartz C, van Pelt J. A mathematical model for the intracellular circadian rhythm generator. J Neurosci. 1999;19:40–7.
54. Discher DE, Janmey P, Wang YL. Tissue cells feel and respond to the stiffness of their substrate. Science. 2005;310:1139–43.
55. Freeberg MAT, Perelas A, Rebman JK, Phipps RP, Thatcher TH, Sime PJ. Mechanical feed-forward loops contribute to idiopathic pulmonary fibrosis. Am J Pathol. 2021;191:18–25.
56. Froese AR, Shimbori C, Bellaye PS, Inman M, Obex S, Fatima S, Jenkins G, Gauldie J, Ask K, Kolb M. Stretch-induced activation of transforming growth factor-beta1 in pulmonary fibrosis. Am J Respir Crit Care Med. 2016;194:84–96.
57. Hinz B. Mechanical aspects of lung fibrosis: a spotlight on the myofibroblast. Proc Am Thorac Soc. 2012;9:137–47.
58. Zhou Y, Huang X, Hecker L, Kurundkar D, Kurundkar A, Liu H, Jin TH, Desai L, Bernard K, Thannickal VJ. Inhibition of mechanosensitive signaling in myofibroblasts ameliorates experimental pulmonary fibrosis. J Clin Invest. 2013;123:1096–108.
59. Stauffer D, Aharony A. Introduction to percolation theory. 2nd ed. London/Washington, DC: Taylor & Francis; 1992.
60. McCullagh CM, Jamieson AM, Blackwell J, Gupta R. Viscoelastic properties of human tracheobronchial mucin in aqueous solution. Biopolymers. 1995;35:149–59.
61. Barabasi AL, Buldyrev SV, Stanley HE, Suki B. Avalanches in the lung: a statistical mechanical model. Phys Rev Lett. 1996;76:2192–5.
62. Black LD, Allen PG, Morris SM, Stone PJ, Suki B. Mechanical and failure properties of extracellular matrix sheets as a function of structural protein composition. Biophys J. 2008;94:1916–29.
63. Ritter MC, Jesudason R, Majumdar A, Stamenovic D, Buczek-Thomas JA, Stone PJ, Nugent MA, Suki B. A zipper network model of the failure mechanics of extracellular matrices. Proc Natl Acad Sci USA. 2009;106:1081–6.
64. Martinez FJ, Safrin S, Weycker D, Starko KM, Bradford WZ, King TE Jr, Flaherty KR, Schwartz DA, Noble PW, Raghu G, et al. The clinical course of patients with idiopathic pulmonary fibrosis. Ann Intern Med. 2005;142:963–7.

65. Hoyle GW, Li J, Finkelstein JB, Eisenberg T, Liu JY, Lasky JA, Athas G, Morris GF, Brody AR. Emphysematous lesions, inflammation, and fibrosis in the lungs of transgenic mice overexpressing platelet-derived growth factor. Am J Pathol. 1999;154:1763–75.
66. Zhukov DO, Andrianova EG, Lesko SA. The influence of a network's spatial symmetry, topological dimension, and density on its percolation threshold. Symmetry. 2019;11:920.
67. Akira M, Sakatani M, Ueda E. Idiopathic pulmonary fibrosis: progression of honeycombing at thin-section CT. Radiology. 1993;189:687–91.
68. Carloni A, Poletti V, Fermo L, Bellomo N, Chilosi M. Heterogeneous distribution of mechanical stress in human lung: a mathematical approach to evaluate abnormal remodeling in IPF. J Theor Biol. 2013;332:136–40.
69. Melo E, Cardenes N, Garreta E, Luque T, Rojas M, Navajas D, Farre R. Inhomogeneity of local stiffness in the extracellular matrix scaffold of fibrotic mouse lungs. J Mech Behav Biomed Mater. 2014;37:186–95.
70. Yi E, Sato S, Takahashi A, Parameswaran H, Blute TA, Bartolak-Suki E, Suki B. Mechanical forces accelerate collagen digestion by bacterial collagenase in lung tissue strips. Front Physiol. 2016;7:287.
71. Marinkovic A, Liu F, Tschumperlin DJ. Matrices of physiologic stiffness potently inactivate idiopathic pulmonary fibrosis fibroblasts. Am J Respir Cell Mol Biol. 2013;48:422–30.
72. Asano S, Ito S, Takahashi K, Furuya K, Kondo M, Sokabe M, Hasegawa Y. Matrix stiffness regulates migration of human lung fibroblasts. Physiol Rep. 2017;5:e13281.
73. Lo CM, Wang HB, Dembo M, Wang YL. Cell movement is guided by the rigidity of the substrate. Biophys J. 2000;79:144–52.
74. Reiser KM, Tryka AF, Lindenschmidt RC, Last JA, Witschi HR. Changes in collagen cross-linking in bleomycin-induced pulmonary fibrosis. J Biochem Toxicol. 1986;1:83–91.
75. Moodley YP, Misso NL, Scaffidi AK, Fogel-Petrovic M, McAnulty RJ, Laurent GJ, Thompson PJ, Knight DA. Inverse effects of interleukin-6 on apoptosis of fibroblasts from pulmonary fibrosis and normal lungs. Am J Respir Cell Mol Biol. 2003;29:490–8.
76. Sicard D, Haak AJ, Choi KM, Craig AR, Fredenburgh LE, Tschumperlin DJ. Aging and anatomical variations in lung tissue stiffness. Am J Physiol Lung Cell Mol Physiol. 2018;314:L946–55.
77. Cavalcante FS, Ito S, Brewer KK, Sakai H, Alencar AM, Almeida MP, Andrade JS Jr, Majumdar A, Ingenito EP, Suki B. Mechanical interactions between collagen and proteoglycans: implications for the stability of lung tissue. J Appl Physiol. 2005;98:672–9.
78. Liu F, Tschumperlin DJ. Micro-mechanical characterization of lung tissue using atomic force microscopy. J Vis Exp. 2011;54:2911.
79. Mishima M, Hirai T, Itoh H, Nakano Y, Sakai H, Muro S, Nishimura K, Oku Y, Chin K, Ohi M, et al. Complexity of terminal airspace geometry assessed by lung computed tomography in normal subjects and patients with chronic obstructive pulmonary disease. Proc Natl Acad Sci USA. 1999;96:8829–34.
80. Kononov S, Brewer K, Sakai H, Cavalcante FS, Sabayanagam CR, Ingenito EP, Suki B. Roles of mechanical forces and collagen failure in the development of elastase-induced emphysema. Am J Respir Crit Care Med. 2001;164:1920–6.
81. Murata N, Ito S, Furuya K, Takahara N, Naruse K, Aso H, Kondo M, Sokabe M, Hasegawa Y. Ca2+ influx and ATP release mediated by mechanical stretch in human lung fibroblasts. Biochem Biophys Res Commun. 2014;453:101–5.
82. Cui Y, Hameed FM, Yang B, Lee K, Pan CQ, Park S, Sheetz M. Cyclic stretching of soft substrates induces spreading and growth. Nat Commun. 2015;6:6333.
83. Moore MW, Herzog EL. Regulation and relevance of Myofibroblast responses in idiopathic pulmonary fibrosis. Curr Pathobiol Rep. 2013;1:199–208.
84. Plantier L, Cazes A, Dinh-Xuan AT, Bancal C, Marchand-Adam S, Crestani B. Physiology of the lung in idiopathic pulmonary fibrosis. Eur Respir Rev. 2018;27:170062.
85. Tanabe N, McDonough JE, Vasilescu DM, Ikezoe K, Verleden SE, Xu F, Wuyts WA, Vanaudenaerde BM, Colby TV, Hogg JC. Pathology of idiopathic pulmonary fibrosis assessed

by a combination of microcomputed tomography, histology, and immunohistochemistry. Am J Pathol. 2020;190:2427–35.
86. Herridge MS, Tansey CM, Matte A, Tomlinson G, Diaz-Granados N, Cooper A, Guest CB, Mazer CD, Mehta S, Stewart TE, et al. Functional disability 5 years after acute respiratory distress syndrome. N Engl J Med. 2011;364:1293–304.
87. Lorx A, Baglyas S, Podmaniczky E, Valko L, Gal J, Suki B. Lung mechanics during recovery of a non-invasively ventilated patient with severe COVID-19 pneumonia. Respir Physiol Neurobiol. 2022;306:103960.

Chapter 9
Chronic Obstructive Pulmonary Disease

Abstract Chronic Obstructive Pulmonary Disease (COPD) is a chronic and progressively debilitating fatal lung disease. Although much is now known about COPD pathogenesis and progression, current treatments are not curative. Computational modeling has a significant role to play in elucidating the link between structure and function in COPD, and in suggesting new approaches to therapy. Following a brief introduction to the pathogenesis and progression of COPD, this chapter describes several modeling approaches that have been proposed to address the biomechanical and pathological aspects of the disease. Little modeling has been proposed to describe airway inflammation and mucus production. Solid mechanics of the alveolar tissue has demonstrated the presence of stress concentrations. Various computational network models of the alveoli with and without gravity incorporate the observation that mechanical stresses can rupture the enzymatically weakened fibers and septal walls, which redistributes these stresses and increases the risk of further rupture in nearby alveoli. This positive feedback process leads to the characteristic microscale airspace enlargement as well as giant low-attenuation super clusters in CT images of COPD patients. Agent-based modeling can be invoked to incorporate more specific pathobiology of cellular mechanotransduction that remodels the alveolar tissue, allowing rupture to take place.

Keywords Cigarette smoke · Flow limitation · Mechanical stress · Mechanical force · Stretch · Rupture · Failure · Network effects · Agent-based modeling

9.1 Introduction

Chronic Obstructive Pulmonary Disease (COPD) is a chronic and progressively debilitating lung disease that is ultimately fatal. In many ways, COPD is worst for long-term survivors because of the way it destroys quality of life. The risk factors for COPD are both environmental and genetic. Cigarette smoke is the most common cause, so heavy smokers are at particular risk, although second-hand smoke exposure also plays a role [1], even in utero [2]. Inhalation of particulate matter from

B. Suki, J. H. T. Bates, *Mathematical Modeling of the Healthy and Diseased Lung*,
https://doi.org/10.1007/978-3-031-95322-4_9

outdoor pollution [3] and indoor cooking [4] are additional significant risk factors. Finally, COPD has a strong familial link [5], and is genetically determined in individuals deficient in the enzyme alpha-1-antitrypsin [6].

COPD is ranked the third leading cause of death worldwide [7], so its morbidity and mortality [7–11] place an enormous burden on public health [7]. Although much is now known about COPD pathogenesis and progression, current treatments are not curative. Drug therapies are restricted to management with bronchodilators and corticosteroids, so the development of new drugs [12] and strategies to eliminate COPD [7] has become a global health priority. Smoking cessation is the most obvious medical strategy, but even if that were to be successful everywhere tomorrow, the detrimental health effects of years of smoking in an enormous number of people would take a very long time to wash out from the population. In the meantime, computational modeling has a significant role to play in elucidating the link between structure and function in COPD, and in suggesting new approaches to therapy.

Following a brief introduction to the pathogenesis and progression of COPD, this chapters describes several modeling approaches that have been proposed to address the biomechanical and pathological aspects of the disease. These include airway mechanics, solid mechanics of the tissue, network models of the alveoli with gravity as well as agent-based modeling.

9.2 Brief Introduction to COPD Pathogenesis and Progression

COPD comprises two key components, chronic bronchitis and emphysema. Chronic bronchitis is defined as "*productive cough of more than 3 months occurring within a span of 2 years*" [13]. This condition results from long-term exposure of the pulmonary airways to irritants that include allergens, pollutants and microbes. These irritants perturb the airway epithelial layer and, over time, destroy cell-cell adhesions and cause cell death [14]. As a consequence, pathogens enter the airway wall tissue where they cause incessant stimulation of the immune system. This leads to chronic oxidative stress, inflammation, ciliary dysfunction and mucus hyperproduction [15–17]. Eventually, a vicious cycle of injury and abnormal epithelial repair develops, which keeps the airway wall in a chronic inflammatory state [14]. Excess mucus secreted by an increasing number of goblet cells becomes viscous to the point that it cannot be effectively cleared by cough and the beating of weakened cilia [18]. Accumulating mucous debris then occludes the small airways, as evidenced by a reduction in FEV1 (the volume of air exiting the lung during the first second of a maximal forced expiration) [19]. Cigarette smoke exacerbates the situation by causing the airway smooth muscle to contract and further narrow the airways [20]. Ciliary length is also shortened in smokers compared to non-smokers, which may further compromise mucociliary clearance [21].

Emphysema, the other main component of COPD, is a chronic lung condition that according to the American Lung Association damages "*the lung tissue and alveoli or tiny air sacs. Over time, this damage causes the tiny air sacs to break and create one big air pocket instead of small ones. When the air sacs break, the big air pockets trap stale air in the damaged tissue, and this stops oxygen from easily moving into your blood stream.*" Thus, emphysema slowly but progressively restructures the distal alveolar regions of the lung so that the airspaces become enlarged and the reduced number of septal walls become thickened with increased elastin and collagen [22]. However, this increased protein deposition does not cause the septa to become stiffer. Instead, septal stiffness is reduced [23] due to the disorganized fibrous microenvironment [24] and altered proteoglycan content [25] of the remodeled tissue. This tissue remodeling is thought to be driven by the persistent presence of various inflammatory cells and their secreted cytokines and enzymes [26]. The enzymes weaken the collagen and elastic fibers, which predispose the alveolar septal walls to mechanical failure when faced with the stresses of breathing and coughing [27], resulting in a complex and highly heterogeneous airspace structure [28]. Tissue remodeling and rupture also reduce the inward elastic recoil of the lung itself, which in turn allows the chest wall to recoil outward and thus increase residual volume and functional residual capacity [29]. Such massive changes in the structure and function of the lung significantly compromise exercise capacity, even when emphysema is relatively mild [30].

The earliest anatomic evidence of COPD is usually the disappearance of small airways [31]. The state of the airways is not independent of the state of the parenchyma, however, because the airways are embedded within, and thus attached to, the alveolar tissue. Excessive outward tethering forces by the alveoli on the walls of the small airways could potentially damage either of these delicate structures, but this is unlikely to pertain in COPD where wall stiffness is reduced [23]. It seems more likely that reduced outward tethering due to inflammation-associated degradation of alveolar and airways walls [27] leads to airway collapse. In addition, cigarette smoke causes airway smooth muscle contraction [20], which generates additional stresses on septal walls that are then likely to rupture, especially if already weakened by enzymatic digestion [32]. Some of the inflammatory cell types responsible for this action, such as M1-like macrophages and neutrophils, may have migrated to the alveolar parenchyma from the airway tissue. Thus, the tissues in the walls of the small airways and the alveoli are not independent of each other, causing chronic bronchitis and emphysema to be intimately linked, as opposed to being entities that contribute separately to COPD.

Whether the airways, the parenchyma or both can be considered the primary driver of COPD progression remains unclear. However, the temporal nature of COPD is punctuated by episodes of acute exacerbation characterized by a sudden and marked worsening of symptoms (increased cough and sputum production, shortness of breath, and wheezing). Such exacerbations frequently require a change in medication [33], and can be triggered by bacterial [34] or viral [35] infection, prior hospitalization, and cold temperatures [36]. When exacerbations are severe enough, they lead to structural changes indicative of irreversible airspace enlargement that can be seen in follow-up CT images [37]. These changes are associated with a decline in lung function [38] indicative of advancing emphysema. When COPD patients with an acute exacerbation are hospitalized, which is frequently the case, they have an increased mortality risk; 23% die in the hospital, and 45% within the following 12 months [39, 40]. This suggests that exacerbations are key and often accelerating drivers of disease progression. The pathophysiology of COPD is thus complex and insidious, and greatly in need of further elucidation.

9.3 Chronic Bronchitis: Where Are the Airway Models?

Narrowing of the lumen of an airway as a result of mucus secretion and/or airway smooth muscle contraction, even by a relatively small amount, can cause a major increase in its resistance to airflow (see Sect. 3.2.2). Such narrowing never occurs homogenously throughout the airway tree, so some branches may be relatively spared while others are completely occluded. The overall resistance of the tree is always elevated, however, and those branches that close off completely cause gas to be trapped distally, which results in hyperinflation of the lung. These distributed airway effects and their consequences seem obvious targets for computational modeling. A recent computational fluid dynamics study simulated airflow in 3D short sections of the small airways obtained from micro-CT imaging of human lungs and reported that, compared to normal lungs, COPD airways have increased resistances and experience greater pressures and wall shear stresses [41]. Nevertheless, computational efforts to characterize airflow have so far largely focused on asthma (Chap. 7). Many of the models that have been constructed in the context of asthma, some highly detailed and personalized [42], seem to be directly applicable to the description of altered lung function in chronic bronchitis. It is therefore curious that few activities in this direction have been pursued to date.

The pathophysiology of chronic bronchitis is, of course, not identical to that of asthma, so some modification of existing asthma-related models is required to make them applicable to chronic bronchitis. Some of these modifications are relatively minor. For example, while excess mucus is a feature of allergic asthma and can be modeled by thickening the airway lining [43], the nature and volume of the mucus are not the same as in chronic bronchitis, so existing asthma models would have to be adjusted accordingly. The biggest challenge in translating asthma-related models of the lung to chronic bronchitis, however, is presented by the reduced parenchymal tethering experienced by bronchitic airways as a result of alveolar tissue destruction. This renders the airways excessively collapsible and thus more likely to be sites of expiratory flow limitation (see Sect. 8.2). Indeed, the presence of flow limitation is a better predictor of patient dyspnea than other pulmonary function indices such as FEV1 [44]. Flow limitation is an inherently nonlinear phenomenon, and its consequences must be calculated in the time domain. When the lung can be reasonably viewed as a linear dynamic system, as is usually the case for determinations of mechanical input impedance in the asthmatic lung, lung function can be modeled much more simply in the frequency domain (See Sect. 3.5). Incorporating the effects of expiratory flow limitation into an anatomically consistent and personalized tree model is thus nontrivial, although not impossible [45] (also see Fig. 8.1).

Measuring input impedance using the forced oscillation technique, also called oscillometry [46], can nevertheless be very useful in chronic bronchitis. Here, small-amplitude oscillations in flow at frequencies above those of spontaneous breathing are applied at the airway opening while a subject breathes normally. Input impedance is determined from the relationships between the applied flow oscillations and the resulting oscillations in airway opening pressure as a function of frequency [46]. When the airways are patent, the oscillating flow travels deep into the distal reaches of the lung. The mechanical properties of the lung periphery are then reflected in the measured impedance. In contrast, in the presence of flow limitation, the oscillating flow cannot travel past sites where limitation occurs. Since flow limitation usually manifests during expiration, this gives rise to characteristic differences in the reactive component of impedance during inspiration versus expiration that are diagnostic of the condition [47]. In fact, the effects of flow limitation on impedance are reminiscent of those caused by extensive airway closure in asthma [48].

The effects of flow limitation on impedance have so far been modeled using a simple lumped parameter electrical analog [47]. A fully detailed treatment of the subject would require a model incorporating computational fluid dynamics in an anatomically and functionally accurate model of the airway tree, which would represent an enormous undertaking. In addition, there is currently a dearth of published literature on the mathematical/computational description of airway-resident cellular responses to particulate exposure, as well as cell proliferation, airway smooth muscle contraction and excess mucus hypersecretion that ensues. Computational multi-scale modeling of chronic bronchitis is thus an area rife with possibilities for elucidating pathophysiology and developing clinical investigative tools.

9.4 Solid Mechanical Stresses

Mechanical factors undoubtedly contribute to the progression of emphysema [27, 49–51]. Our understanding of how this happens is facilitated by computationally modeling the mechanical stresses in the lung parenchyma. Chapter 4 introduced mechanical models that account for the macroscopic elastic behavior of the lung parenchyma predominately as a network phenomenon. Now it is time to move to a lower level of length scale and consider the distribution of mechanical stresses within the alveolar septal walls, since this is the primary site of the rupture events that lead to the characteristic morphology of emphysema.

The microstructure of the septal wall and the effects of emphysema on its stress distribution have been investigated using a 2D continuum-type model [52]. The geometry of the model is based on an electron microscopic image of the lung

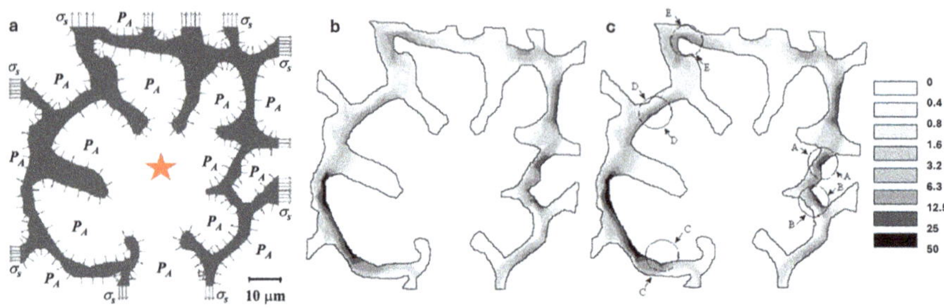

Fig. 9.1 (**a**) Two-dimensional digitized image serving as a continuum model of a typical alveolar region. The location marked by a red asterisk appears to be an alveolar duct seen in cross section. The pressure in the airspaces is alveolar pressure (P_A). The tissue is maintained at a given state of inflation by uniform normal stresses (σ_s) that are transmitted by the connecting septal walls. (**b**) Predicted principal stresses (in kPa) assuming normal tissue properties. (**c**) Predicted principal stresses assuming emphysematous tissue properties. (Reproduced with permission from [52])

parenchyma, similar to that shown in Fig. 2.3. The image selected for analysis (Fig. 9.1) contains several alveolar sacs and was digitized at a resolution of approximately 1 μm. The resulting structure is exposed to uniform normal stress, as indicated in Fig. 9.1a, to simulate biaxial stretching of the tissue. The material properties of the alveolar tissue are assumed to be isotropic and non-linearly elastic based on uniaxial length-tension measurements in lung strips [53] that were fitted with the constitutive equation [54].

$$\sigma = \sigma_0\left(\lambda^{\alpha_1} - \lambda^{-\alpha_2}\right) \tag{9.1}$$

where σ is the Cauchy stress, λ is the stretch ratio (see Chap. 4), and the exponents $\alpha_1 = 8$ and $\alpha_2 = -6.4$ were obtained from curve fitting. Note that λ is unitless so that σ and σ_0 both have units of pressure. It is assumed that $\sigma_0 = 1$ cmH$_2$O, so it does not have to appear explicitly in the following modulus equations [52]. The incremental Young's modulus is the derivative of Eq. 9.1 with respect to strain thus:

$$Y(\lambda) = \frac{d\sigma}{d\epsilon} = \frac{d\sigma}{d\lambda} = \alpha_1\lambda^{\alpha_1 - 1} + \alpha_2\lambda^{-\alpha_2 - 1}. \tag{9.2}$$

The shear modulus (G) is computed using linear elasticity theory (see text following Eq. 4.21) to be

$$G(\lambda) = \frac{Y(\lambda)}{2(1+\nu)} = \frac{\alpha_1\lambda^{\alpha_1 - 1} + \alpha_2\lambda^{-\alpha_2 - 1}}{2(1+\nu)} \tag{9.3}$$

where $\nu = 0.4$ is the Poisson's ratio that is assumed to be a constant [55].

To compute the mechanical stresses in the model, a mesh of 513 quadrilateral and triangular elements was generated from the image in Fig. 9.1, where the mechanical properties of the elements are given by Eqs. 9.1, 9.2 and 9.3. The boundary conditions on the model are the alveolar pressure (P_A), the traction stresses arising from the connecting septal walls (σ_s), and constraints to avoid rigid body translation. At baseline, corresponding to a lung volume of functional residual capacity (FRC), λ is set equal to 1. To simulate lung inflation, P_A is gradually increased while the corresponding planar stresses are calculated numerically using a finite element scheme. This was done using constitutive parameters appropriate both for a normal lung and for a lung with emphysema.

The results of these simulations (Figs. 9.1b, c) show that stress is highly heterogeneous even within a single alveolus, with some areas of concentrated stress being 20-fold above the average for the tissue. Stress is essentially negligible along the free septal walls that separate individual alveoli, which is reasonable given that these walls are apparently not involved in stress transmission. In contrast, stress is concentrated along the curved internal alveolar walls that back onto adjoining alveoli. These regions of high stress concentration (circled regions in Fig. 9.1c) may thus correspond to preferential sites of tissue failure, which could be where emphysema begins. This model is limited, however, by being two dimensional, which may, for example, explain the lack of stress in the free septal walls (Figs. 9.1b, c). This model also assumes that tissue stiffness increases in emphysema, but recent evidence suggests that the opposite may be the case [23, 56]. In addition, the mechanical properties of the tissue are heterogeneous between different alveoli and heterogeneously nonlinear within a single wall [57], which no doubt further amplifies the stress differences during inflation.

A somewhat similar finite element approach combined the constitutive laws of Eqs. 9.1, 9.2 and 9.3 with a 3D rhombic dodecahedron model of the parenchymal architecture [58] within which a small airway was explicitly included. The main conclusions drawn from this model are similar to those of the model illustrated in Fig. 9.1, namely that stress heterogeneity is amplified in alveolar walls affected by emphysema. Neither model, however, can explain the airspace enlargement that is so characteristic of emphysema. This turns out to be a network effect driven by force-based tissue rupture.

9.5 Enter the Mechanical Force

In 1971, John B. West published a hypothesis paper [51] in which he used the best computational tools available at the time to calculate the spatial distribution of mechanical stress in a lung-shaped elastic solid supported around its surface by pleural pressure and loaded by its own weight. The stresses predicted by this model are eerily similar to the spatial distribution of lesions seen in centrilobular emphysema. In particular, the stresses are the greatest at the lung apex, which coincides with the greatest prevalence of emphysematous regions. West concluded that mechanical stresses must somehow contribute to disease pathogenesis and even speculated that while airway inflammation is expected to be distributed throughout the lung, "*mechanical failure with consequent dilatation happens first in the regions of highest expanding stresses*" [51]. This study appears to have been the first serious computational modeling effort to interpret disease localization, and in many ways it launched the field of computational lung biomechanics.

With the advancement of CT imaging, it has become possible to characterize the state of a COPD lung non-invasively in fine detail. Traditionally, the clinical assessment of CT images is performed via subjective scoring by experts based on visual impression, but objective measures of the condition of the parenchymal tissue are now being increasingly derived through digital image analysis, particularly in the research setting. For example, CT images of the emphysematous lung are characterized by enhanced low attenuation areas (LAA) within which the Hounsfield unit of each image pixel is less than −960 (Hounsfield units are −1000 for air, 0 for water, and 1000 for bone). The size, S, of an LAA cluster is equal to the number of pixels it contains. It turns out [28] that the cumulative probability distribution of S follows a power law of the form

$$P(S \geq X) = KX^{-D} \tag{9.4}$$

with exponent D and a proportionality constant K. As emphysema progresses, the value of D decreases from 1.68 in healthy subjects to about 0.5 in severe COPD patients [28]. Importantly, the total %LAA (sum of S normalized by the size of the lung field as a percentage) was found not to be different in early emphysema compared to health, yet D was significantly smaller than in a normal lung. This shows that, as with pulmonary fibrosis (Chap. 8), it is often the physical arrangement of structural components rather than their relative amounts that most determine tissue function.

These findings can be interpreted using an elastic spring network model of the parenchyma in which individual springs are allowed to rupture if they carry a force larger than a specified threshold. When the network is allowed to incur various degrees of emphysematous damage and then coarse-grained to create virtual CT images, the LAAs in the images are found to conform to Eq. 9.4. The model also shows that particularly large forces are carried by springs in the network that bridge between adjacent damaged regions of size S_1 and S_2 that correspond to LAA clusters (Fig. 9.2a). These bridge springs (e.g. Fig. 9.3b) are more likely to rupture than less stressed springs, and when they do, the two neighboring LAA regions coalesce into a single region of size $S_3 = S_1 + S_2$ (Fig. 9.2b). This process retains the power law shape of the LAA distribution. However, because the only addition to the LAA is the narrow bridge of tissue, the total LAA% barely increases while 2 small clusters are eliminated and one larger cluster is created, effectively decreasing D. This network model thus explains how the total %LAA can remain unchanged while D changes in early emphysema. A similar study using a finite element analysis of a network of identical two-node linear beam elements came to a similar conclusion [59]. Personalized models have also been created by mapping LAA clusters from CT images onto spring networks so as to preserve LAA cluster size and shape in the network [60].

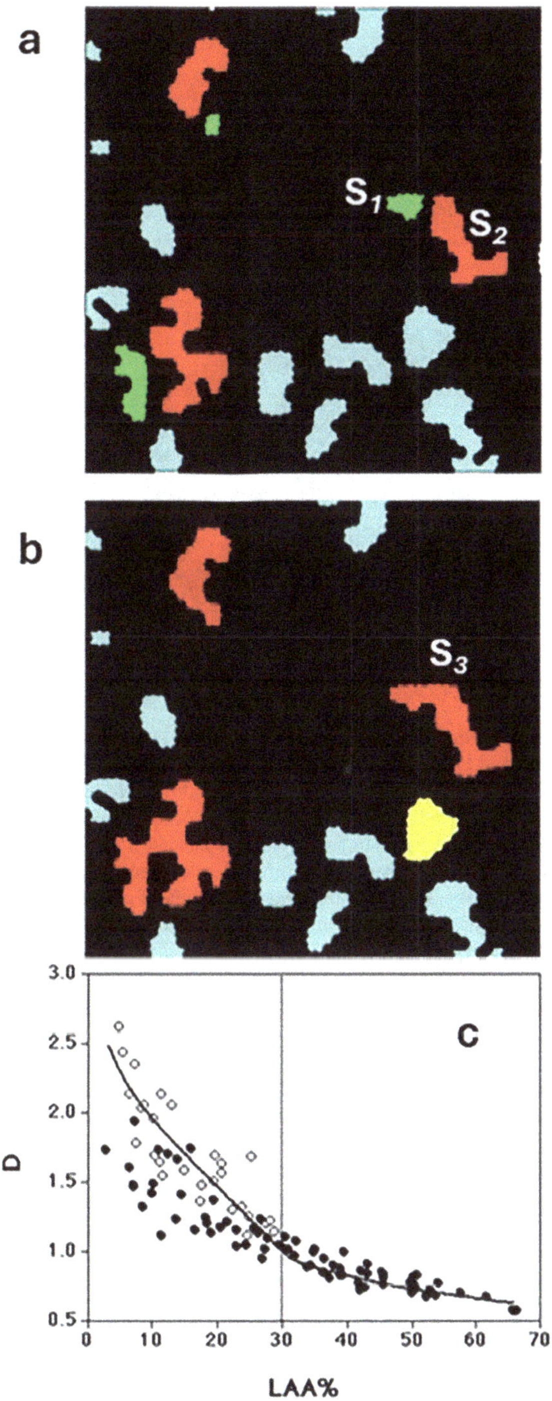

Fig. 9.2 Progression of early emphysema represented in a square network of 500 × 500 nodes connected by linearly elastic springs. Coarse graining was used to create virtual CT images. The different colors represent different low attenuation area (LAA) clusters in which springs and nodes were removed from the network due to mechanical failure. Black corresponds to healthy parenchyma. (**a**) Zoom into a 50 × 50 area of the network. The three red clusters and the smaller neighboring green clusters are separated by normal tissue. Notice the small green and red clusters with their sizes $S_1$and S_2. (**b**) The same network after the elimination of springs having tension greater than 80% of the maximum tension. For example, the tension in the alveolar walls separating the red and green clusters in A was high, so these walls were eliminated, which caused the green cluster S_1 to be subsumed into its larger neighbor S_2 to create the composite cluster S_3. This leads to a change in the LAA distribution such that the exponent of the power law decreases. (**c**) Exponent (Eq. 9.4) of the cumulative distribution of cluster size as a function of the percent total LAA (LAA %) in normal (open symbols) and emphysematous (filled symbols) lungs. Solid line is model simulation. (Adapted with permission from [28], Copyright (1999) National Academy of Sciences, U.S.A.)

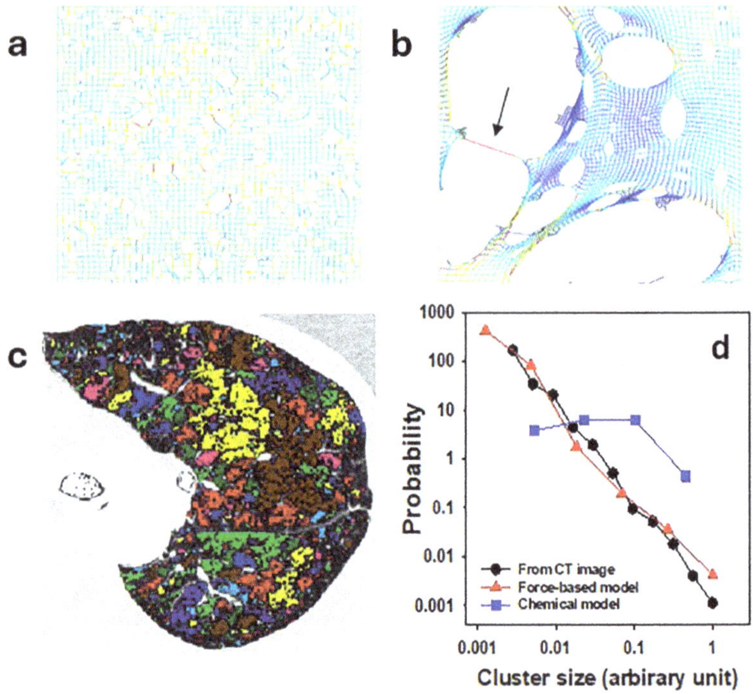

Fig. 9.3 Simulations of the progression of emphysema based on the spring network model. (**a**) Chemical breakdown of tissue simulated by randomly elimination of springs from the model. (**b**) Force-based elimination of springs. Note the heterogeneous distribution of defect clusters and the high forces (red) around the perimeter of the clusters especially in regions where a "thin wall" separates two clusters (arrow). (Color represents force in **a** and **b**.) (**c**) Color-coded CT image from a patient with emphysema. The different colors represent contiguous low attenuation areas. Note the heterogeneity of cluster sizes and their spatial distribution. (**d**) Size distributions of defect clusters from chemical (blue) and force-based (red) models as well as the size distribution of low attenuation area clusters from C (black). (Reproduced with permission from [49])

The genesis of LAA cluster distributions has been investigated from the perspective of how the CT image evolves as emphysema develops [37, 61]. Working from an initial image taken in a given patient when the disease is first diagnosed, pixels can be modified according to various algorithms to simulate personalized emphysema progression. Comparison of these prediction images with subsequent CT images from the same patient allows the best algorithm to be identified. This approach supports the assumption that force-based tissue rupture causes small LAA clusters to coalesce into larger clusters. It also suggests that small LAA clusters often merge into nearby larger ones, presumably due to the high mechanical stresses surrounding the larger clusters [61].

The possibility that tissue rupture can occur at locations of high stress within parenchymal tissue by the mechanical forces produced during normal breathing has been confirmed experimentally [27]. It has also been shown that cutting springs at random locations in a 2D rectangular network, similar to complete elimination of tissue by chemical digestion, does not produce a power law distribution of LAA clusters [49], whereas force-based cutting produces cluster size distributions consistent with patient data even when the identical number of springs are cut in both cases (Fig. 9.3). For example, in both Fig. 9.3a, b, the total number of springs eliminated from the networks is 13%, yet there is a spectacular difference in the structures due to the methods of spring removal. Furthermore, when a critical spring such as that in Fig. 9.3b breaks, the entire network can be affected dramatically. This is reminiscent of the large and lasting negative effects that an acute COPD exacerbation can have on lung structure and function [37]. Indeed, if a chain of expanded airspaces percolates across the entire parenchymal tissue, lung compliance can potentially become infinite [62]. This corresponds to a percolation threshold that, similar to pulmonary fibrosis (Sect. 8.4), explains why there can be a disconnect between the progression of underlying tissue disease and the progression of clinical symptoms.

The most clinically important functional indicator of emphysema is lung compliance (or its inverse, elastance). Lung compliance, the change in volume divided by the corresponding change in transpulmonary pressure, is what determines how inflated the lung becomes when subjected to the boundary conditions imposed by pleural pressure. The effects of parenchymal tissue destruction on lung compliance can be investigated by imposing a uniform negative pressure around the borders of a 2D spring network and observing how its compliance changes following the application of various spring rupture algorithms [63]. Computationally, the network compliance C can be obtained by increasing the border pressure by a small amount, determining the amount by which this causes the area of the network to increase, and then taking the ratio of the change in area to the change in boundary pressure. Several different mechanisms can lead to airspace enlargement, although with different statistical characteristics. Progressive equal weakening of all springs mimics uniform digestion of tissue fibers and results in uniform expansion, randomly cutting springs simulates chemical or enzymatic digestion and leads to near uniform expansion, and cutting springs based on how much stress they carry simulates force-based tissue destruction resulting in the large heterogeneities described above [27].

A convenient parameter reflecting network structure is the mean area-weighted airspace diameter (D_2), which encapsulates both airspace enlargement and heterogeneity [64]. Figure 9.4 shows the relationships between C and D_2 predicted by the three mechanisms of airspace enlargement described above. Also shown are experimental data from an elastase-treated mouse model of emphysema [63], which clearly agree most closely with the model predictions of force-based spring cutting. This demonstrates that neither uniform softening nor random rupture of the tissue produce realistic degrees of structural heterogeneity, suggesting that alveolar tissue heterogeneity is strongly linked to lung function decline.

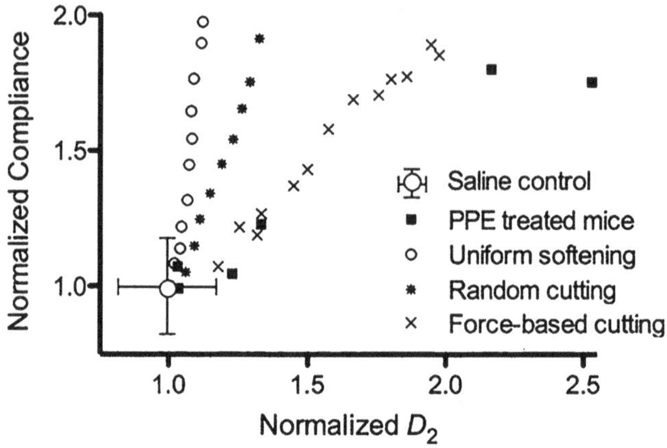

Fig. 9.4 Comparison of structure-function relationships predicted by the network model together with experimental data. The data have been normalized by the mean value of respiratory compliance and the mean area-weighted airspace diameter (D_2), an index of airspace enlargement and heterogeneity, from control mice. Each black square corresponds to data from a single mouse. The symbols corresponding to the various model predictions have been normalized by the prediction of the baseline model. (Reproduced with permission from [63])

9.6 Using Elastic Networks to Model Emphysema Progression and Therapy

Now that computational modeling has established the critical role of mechanical stress in emphysema progression, it can be used to dig more deeply into the mechanisms by which stress mediates tissue destruction, and potentially how such destruction might be modulated or perhaps even reversed. This is a speculative area but one of undeniable importance because emphysema remains an incurable and progressive disease.

Recall from Chap. 4 that proteoglycans stabilize collagen and elastic fibers in the parenchymal tissue during lung inflation by hindering their alignment with the direction of stress (see Fig. 4.14). This suggests that remodeling of proteoglycans in emphysema might affect disease progression [65]. This idea has been tested experimentally in mice after 30 days of treatment with porcine pancreatic elastase by observing how parenchymal stiffness and alveolar stability are influenced by varying tonicity conditions that affect the stiffness of proteoglycans [66] but not elastic [67] and collagen [68] fibers. The relative increases in stiffness and alveolar shape distortion following a reduction in tonicity have been found to be elevated in emphysema, as is the content of glycosaminoglycans (side chains attached to the proteoglycan core protein) in alveolar walls [65].

These findings have been interpreted with the aid of a 2D hexagonal spring network model having fixed boundaries, where each spring is prestressed and represents an alveolar wall. The internal nodes are free to move, but the network is further stabilized by having adjacent walls resist a change in the angle between them through the

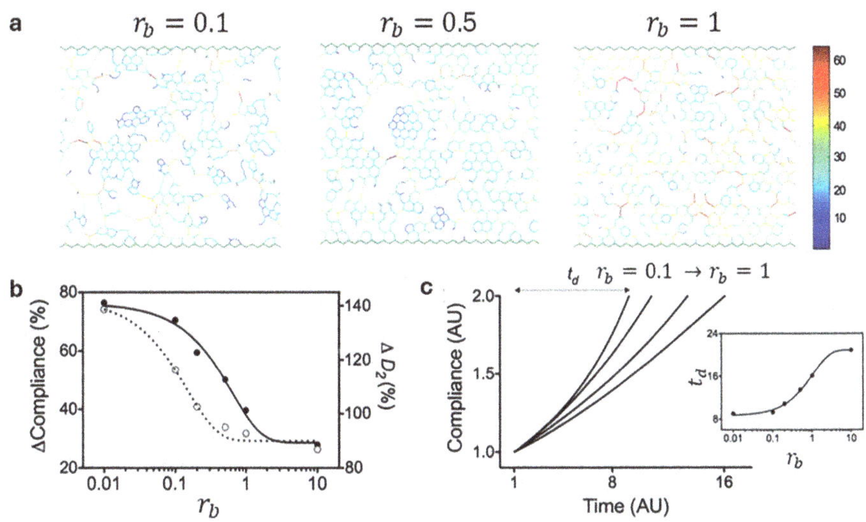

Fig. 9.5 Network model simulations showing the effects of elastic opposition to changes of angle between neighboring springs, via the bond-bending parameter r_b, on tissue network compliance and the area weighted mean diameter (D_2) during mechanical force-induced breakdown of the network. (**a**) Representative images of network configurations are shown after eight iterations for different values of r_b. The color of a spring is proportional to the tensile force it carries (blue: low force; red: high force). (**b**) Sensitivity plot showing the % increase in compliance (ΔCompliance, filled symbols) and D_2 (ΔD_2, open symbols) between the first and eighth iteration as a function of r_b. The solid and dotted lines are curve fits. (**c**) Network compliance as a function time (iteration number) for several values of r_b. Simulations were carried out until the compliance of the network doubled. The inset shows the doubling time (t_d) as a function of r_b. (Reproduced with permission from [65])

addition of a bending elasticity term to the total network energy as in Eq. 4.51. The parameter that governs this additional behavior is the elastic bond-bending constant r_b, which reflects the extent to which the proteoglycans resist compression and shear [69]. Figure 9.5 shows the predicted effects that proteoglycans have on tissue structure and function during forced-based tissue destruction. As springs are broken, (Fig. 9.5a), both C and the structural index D_2 increase at rates that depend inversely on r_b (Fig. 9.5b). However, the heterogeneity of the emphysematous network also depends strongly on the value of r_b (Fig. 9.5a), as does the time, t_d, required for C to double (Fig. 9.5c, inset). Taken together, this suggests that an elastic impediment to angle change between alveolar walls protects the network from deterioration.

These modeling results have potential clinical implications by showing how it might be possible to reduce the rate of emphysema progression. Indeed, experimental studies have suggested a protective role for aerosolized hyaluronan, a proteoglycan-associated constituent of the ECM. In the emphysematous lung, hyaluronan is presumed to shield elastic fibers from enzymatic degradation [70, 71]. Moreover, a short-term clinical trial has shown that hyaluronan reduces elastic fiber breakdown without significant side effects [72]. The power of computational modeling is that it offers a network level explanation for these experimental findings. By stabilizing alveoli, an increase in the amount and stiffness of proteoglycans reduces the spatial heterogeneity of mechanical forces within the alveolar network, which in turn reduces the forces on the tissue fibers and slows their rates of degradation.

Another intervention for treating emphysema that has been informed by computational modeling is lung volume reduction [73]. The rationale for this procedure is that because the emphysematous lung is significantly larger than normal, which causes relatively spared portions of the parenchyma to collapse, removing the most damaged portions of the tissue will allow the remainder of the lung to re-expand and function more normally [74]. Lung volume reduction can be achieved either through surgical removal of targeted lung regions [75] or less invasively by bronchoscopically instilling lung sealants that block small airways and eliminate collateral ventilation pathways, ultimately collapsing the treated region through absorption atelectasis [76, 77]. By restoring lung elastic recoil in emphysema, lung volume reduction reduces dyspnea, improves quality of life, and may even increase survival [78, 79]. Although beneficial in many cases [80, 81], the outcome of lung volume reduction in any given patient is difficult to predict, so it is recommended that patients be carefully phenotyped before the procedure, and an optimal treatment strategy be designed by a multidisciplinary team [82].

Understanding structure-function relationships before and after lung volume reduction has been facilitated by a computational model of a prestressed 2D hexagonal network of linear springs in a gravitational field [83]. As above, the progression of emphysema in this model is simulated through the probabilistic force-based cutting of springs, followed every five iterations by the determination of the network configuration having the minimal (elastic plus gravitational) energy. To simulate lung volume reduction surgery, the upper 30% of the network area was removed and the remainder stretched to cover the original area. To simulate bronchoscopic lung volume reduction, network nodes around the perimeter of a larger airspace were moved toward the geometric center of the circumscribed region, causing the area of the region to be reduced by specified percentages.

During simulated emphysema progression, tissue destruction was prevalent in the upper part of the network due to the added stress from the weight of the lung (Fig. 9.6a). Figure 9.6b shows visually similar structural consequences of surgical (LVRS) versus bronchoscopic (bLVR) lung volume reduction when initiated from the same emphysematous network configuration. Figure 9.6c shows that network compliance increases with the number of cut springs, consistent with the progressive enlargement of airspaces seen in Fig. 9.6a. Both LVRS and bLVR yield comparable short-term and long-term improvements in C as characterized by similar immediate drops following the procedures and the subsequent gradual recoveries, respectively (Fig. 9.6c). Nevertheless, LVRS and bLVR are not always equally effective [83], demonstrating that a patient's lung structure at time of treatment has an impact on which treatment strategy should be employed.

Further examination of network structure in a set of different realizations representing intersubject variability offers a clue as to what may be behind the differences in LVRS versus bLVR outcomes. When the coefficient of variation of

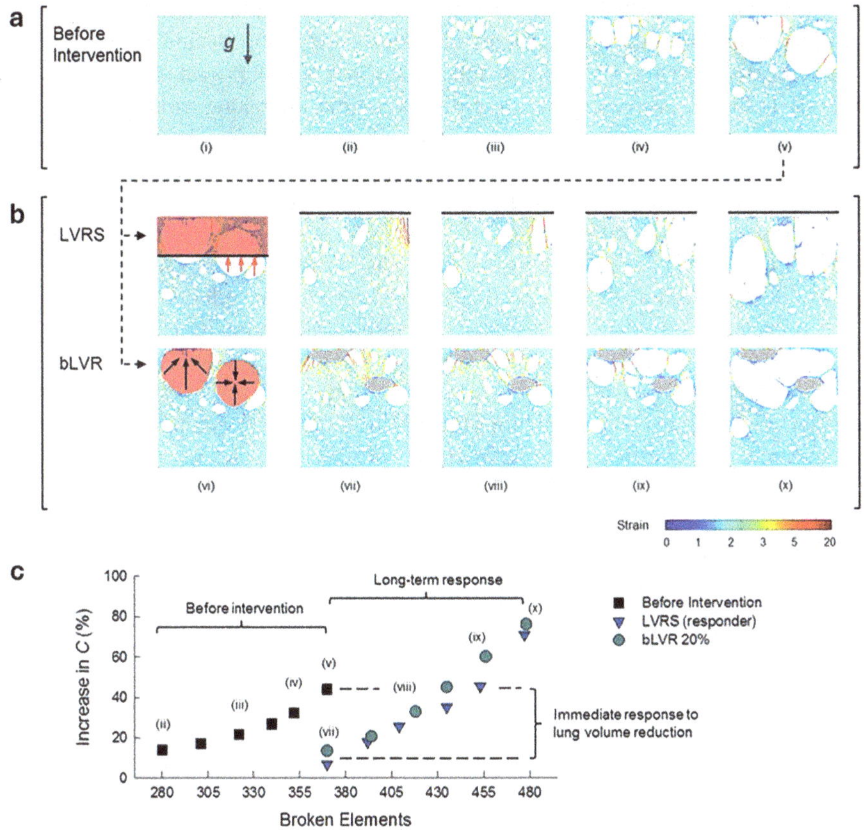

Fig. 9.6 Network modeling of lung volume reduction. (**a**) Representative simulation of emphysema progression before intervention, and (**b**) comparison of lung volume reduction techniques in a responder network. (*i*) Initial network configuration representing healthy tissue (direction of gravity shown by arrow). (*ii*) Disease progression initiated by randomly breaking ~1% of spring elements. (*iii–v*) Gradual network destruction representing tissue failure during emphysema progression. (*vi*) Illustrations of lung volume reduction techniques initiated from the same disease condition. For LVRS, the upper 30% of the network (red shaded region) was removed and intersected regions were stretched (red arrows) to form a continuous upper border (solid line). For bLVR, affected regions (red shaded areas) were reduced (black arrows) to 20% of their original sizes. (*vii–x*) Disease progression following either LVRS or bLVR (reduced regions shown in grey). Color bar represents strain distribution among individual spring elements. (**c**) Changes in compliance, *C* for the representative network shown. (Reproduced with permission from [83])

airspace size in the region just below the segment of surgically removed lung is small, there is more complete removal of emphysematous tissue, so the network responds well to LVRS. These networks may correspond to the "responder" type of subject. Conversely, when the spatial heterogeneity of emphysematous legions is large, removal of the upper portion of the lung is less effective at removing abnormal tissue, so the network responds better to bLVR that can target concentrations of emphysema more equitably throughout the lung. This suggests that a corresponding analysis of pre-procedure CT images in patients may help decide which patient should receive which procedure. The other important insight provided by this model analysis is that *C* increases more rapidly over time after lung volume reduction compared to before, consistent with clinical observation [84]. The reason for this may be that when the most emphysematous regions of a lung are removed, the increased stresses in the remaining tissue following re-inflation cause an increased rate of force-based tissue rupture.

The spring network model also offers a means of predicting how much quality of life can be improved and survival extended by lung volume reduction. If quality of life is reflected in how far *C* increases above normal, and mortality is marked by the point at which *C* exceeds a lethal threshold, then the timing of each is provided by predictions such as those shown in Fig. 9.6c. Figure 9.7 shows a diagrammatic representation of the model predictions that illustrates these potential benefits of lung volume reduction. Despite the accelerated progression of emphysema in the re-expanded lung tissue following LVRS, or bLVR when the latter reduces targeted tissue volume by 20%, both interventions restore *C* to near baseline values, which lengthens survival by about 50% and improves quality of life. However, these

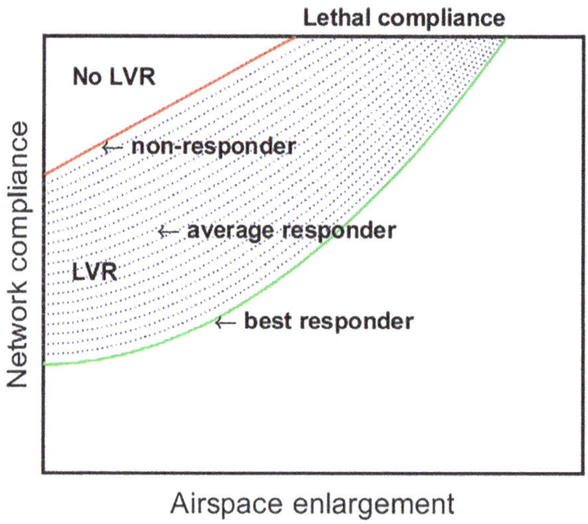

Fig. 9.7 Estimation of survival and quality of life. Each line corresponds to a single network representing the trajectory of a subject. Green and red curves represent changes in compliance, *C*, in the best responder and in a non-responder, respectively, to lung volume reduction (LVR). Predicted survival was estimated as the number of broken network elements, expressed as airspace enlargement, needed to reach a 60% increase in the lethal value of *C* from baseline. The combined relative benefit index, which represents a clinical measure of quality of life, was defined as the area between the red and any of the blue dotted or green curves and the lethal value of *C* such that larger values correspond to greater improvements following lung volume reduction

benefits are much reduced when bLVR produces only a 40% reduction in tissue volume. Results such as these demonstrate that network modeling approaches have the potential to assist in medical intervention planning including patient selection as well as prediction of survival and quality of life.

9.7 Modeling Alveolar Structure in 3D

Although the 2D parenchymal models discussed above have yielded important insights into the nature and progression of emphysema, they are limited in their capacities to answer certain questions related to the 3D nature of mechanical equilibrium, especially when tissue breakdown caused by emphysema creates complex geometries [85, 86]. Micro-CT imaging now makes it possible to visualize alveolar structure in 3D [87–89] and has shown, for example, that an early sign of emphysema in elastase-treated mice is an increase in the standard deviation of alveolar airspace size [87]. Micro-CT images can thus inform the construction of 3D models of the alveolar parenchyma, such as those based on space-filling polyhedra [90, 91] (see Chap. 2). For example, models containing a relatively small number of alveoli, together with alveolar ducts implemented using octahedra [92] or dodecahedra [93], have been used to predict the pressure-volume curve and elastic moduli of the lung. Accounting for large-scale network effects, however, such as the coalescence of defective airspaces into large clusters requires a model comprising a correspondingly large number of alveoli.

A large-scale 3D model of the parenchyma has been constructed as a pre-strained space-filling tessellation of polyhedral unit cells including both cubic and truncated octahedral polyhedra that represent individual alveoli [94] (Fig. 9.8a). Enzymatic tissue destruction in this model is mimicked by randomly removing faces that separate two adjoining polyhedra (Fig. 9.8b). Spatially correlated destruction is achieved by removing faces that carry the greatest mechanical forces (Fig 9.8c). A mixture of random and force-based destruction can also be imposed (Fig. 9.8d). As with the 2D models above (Fig. 9.3), these different scenarios produce different outcomes. For example, the 3D model shows that spatially correlated force-based destruction results in a much faster rate of decline in bulk modulus (change in pressure divided by the relative change in volume) with enhanced heterogeneity compared to the purely random elimination of faces (Fig. 9.8e). The model thus allows tissue destruction to be studied as a function of various airspace geometries, levels of pre-strain, and relative degrees of random and force-based spring cutting. The springs in the model can also be assigned linear or nonlinear elastic behaviors. In all cases, the change in the bulk modulus of the 3D alveolar network can be reliably estimated from the first four moments of the airspace volume distribution [94]. These findings support the notion that functional decline in emphysema, as measured by an increase in lung compliance or a decrease in bulk modulus, is not related so much to the amount of tissue destroyed but more to the microscale spatial pattern of the destruction process.

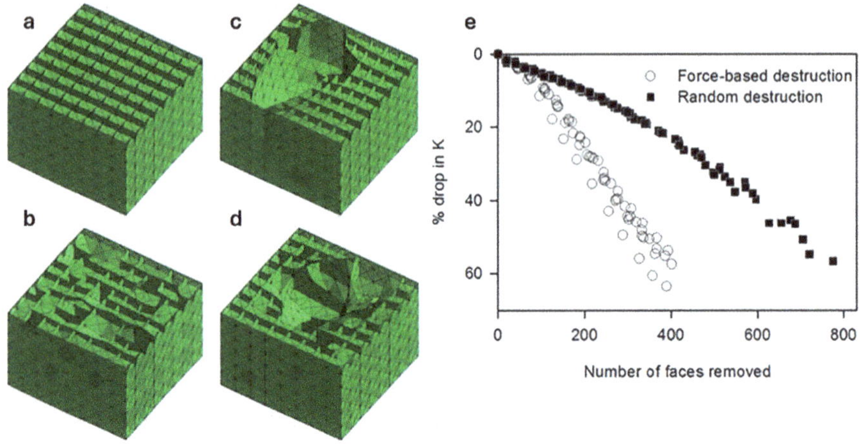

Fig. 9.8 Networks of cubic cells with the top removed to reveal their internal structure. (**a**) Initial geometry. (**b**) Result of the random destruction pattern. (**c**) Result of the force-based destruction pattern. (**d**) Result of the mixed destruction pattern. Networks **b**–**d** have the same drop in bulk modulus, but their internal structures are very different. (**e**) Percent drop in bulk modulus (K) of the network as faces are removed randomly or based on the force. Note that for the same number of faces removed (mimicking the extent of tissue destruction), the decline in elasticity can be very different depending on the pattern of destruction. (Reproduced with permission from [94])

The 2D and 3D spring network models described above collectively support the notion that mechanical forces imposed on the lung parenchyma by pleural pressure, breathing and gravity are the primary drivers behind the progression of airspace enlargement. The models also reveal how this progression occurs. Since parenchymal forces are not uniform throughout the tissue, some alveolar septa are under more stress, and thus more likely to rupture, than others. When a high-stress wall ruptures, it transfers the stress it was bearing onto neighboring tissue elements, which elevates their risk of rupture, and so on. Localized clusters of parenchymal holes at site of rupture then begin to coalesce to form larger holes that continue to grow with time. Eventually, the largest such cluster percolates across the spatial domain of the lung to create a “Super Cluster”. This has been confirmed in CT images [95] showing power-law distributions of low-attenuation volumes together with one or two outlier Super Clusters (Fig. 9.9a). The development of 3D low-attenuation is thus a sequential process in which spatially distributed rupture events lead to the development of Super Clusters in a rich-get-richer fashion because larger holes have greater surrounding tissue stresses, making it more likely the hole will expand. Interestingly, this is a process reminiscent of the mechanism that has been proposed as an explanation for quasilinear viscoelasticity (Sect. 5.4), which is also characterized by power-law behavior (see Fig. 5.8).

The probability of a Super Cluster forming in the parenchyma increases with the percentage of low-attenuation voxels (Fig. 9.9b), and correlates strongly with disease severity and clinical indices of impaired lung function. These include reductions in FEV1 and the carbon monoxide diffusing capacity of the lung. The explanation

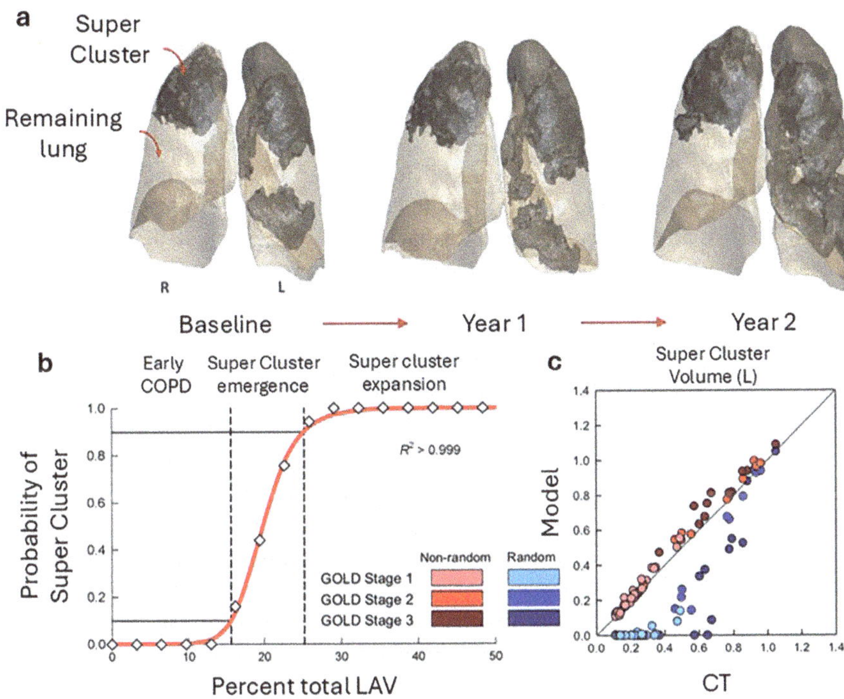

Fig. 9.9 (**a**) Super Cluster progression in the same patient at baseline and at annual follow-up. Boundaries are shown by shaded regions. Notice that the Super Cluster at the bottom of the left lung (L) at baseline becomes part of the larger Super Cluster in Year 2. (**b**) The probability of finding a Super Cluster increased with the total percent of low-attenuation voxels (LAV). Dashed lines represent the tenth and 90th percentiles. The solid red line is a three-parameter sigmoidal fit. (**c**) Personalized computational simulations using the random (blue shades) and non-random (red shades) models comparing the predicted Super Cluster volumes to those measured in CT images. The color shadings correspond to different Global Initiative for Chronic Obstructive Lung Disease (GOLD) stages. (Reproduced with permission from [95])

for the emergence of Super Clusters is again informed by a personalized computational modeling approach, beginning with a 3D clinical CT scan from an emphysema patient. The image voxels in the lung are all assigned healthy Hounsfield numbers with the exception of 2–7% that are chosen randomly and assigned reduced (emphysematous) Hounsfield numbers to serve as seeds for further progression. The progression itself is simulated either by making a random selection of the remaining healthy voxels emphysematous, or by convolving the image with a 3D Gaussian filter that causes the initial emphysematous seed areas to expand so that emphysematous regions become spatially correlated. Both progression strategies continued until the percentage of low-attenuation voxels in the simulated image match that in the original patient CT image. Only the correlated progression strategy produces Super Clusters and distributions of low-attenuation volumes matching those seen in the patient CT image [95] (Fig. 9.9c). Furthermore, in patients with mild emphysema, the random progression strategy does not result in any Super Clusters forming at all. This demonstrates that that Super Cluster formation is not due solely to the spatial constraints of the finite volume of the lung, but instead relies critically on the outward growth of low-attenuation regions driven by local mechanical forces. This is consistent with the "lung at-risk" view of emphysema in which disease progression occurs within a penumbra of healthy tissue surrounding existing diseased regions [96]. This personalized spatial modeling of super cluster emergence thus exemplifies how computational modeling can play a key role in the discovery of clinically relevant mechanisms related to disease progression.

9.8 Modeling Septal Wall and Single Fiber Mechanics

The computational models described so far in this chapter have considered COPD over a range of different length scales from that of the whole lung, as seen in a CT image of the chest, down to the level of the individual alveolar septal wall. The mechanical properties of a single wall, however, are themselves emergent properties arising from the ensemble behavior of the wall components, which represents a further decrease in length scale. Figure 9.1 suggests that there can be significant regional variation in stress within a single septal wall even when the wall tissue is homogeneous. In reality, the wall tissue is not homogeneous at the level of the structural fiber since these include both collagen and elastin, which have very different mechanical properties. Accordingly, one might expect heterogeneity to be even greater than that indicated in Fig. 9.1.

The accepted explanation for the strain-stiffening behavior of lung tissue is that the relative soft elastin fibers bear most of the stress at low strain, while as strain increases, progressively more of the much stiffer collagen fibers are recruited into the load-bearing role (see Chap. 4). Accordingly, the nonlinear mechanical behavior of individual alveolar walls [53] has been modeled computationally by representing the collagen and elastin fibers as two colocalized 2D networks. Again, nonlinear stress–strain behavior emerges from the interactions between the networks [97]. When the septal wall is stretched, the wavy collagen fibers become recruited, and the main stress transmission pathway consists of the stretched elastic fibers and the percolating collagen fibers (Fig. 9.10). In COPD, both fiber types can be weakened by enzymes in a stress-dependent way (see below). Additionally, the mechanical behavior of the septal wall can be modulated by factors normally present in tissue such as the proteogylcans that resist changes in angle between adjacent fibers to extents that vary with the number of water molecules they sequester [69].

The stress–strain behavior of the alveolar wall can also be affected by pathologic processes such as digestion of fibers by proteases released by inflammatory cells. This causes fiber stiffness to decrease, and has been shown to occur more rapidly when fibers are statically stretched versus when they are unstressed [98]. The reason for this stretch effect appears to be a change in the kinetics of how enzymes interact with binding sites on the fiber surface [50], possibly because stretch affects the

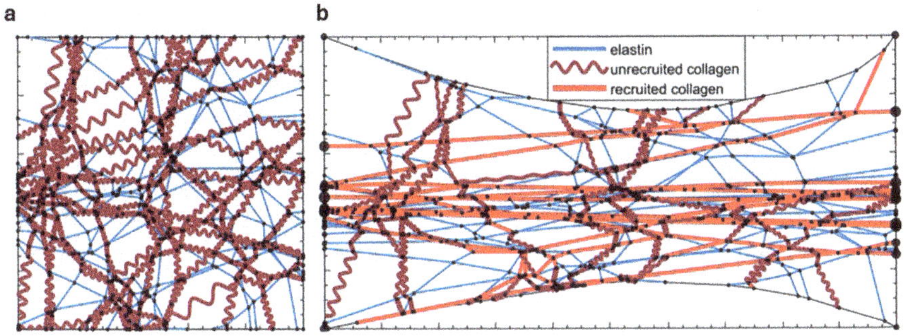

Fig. 9.10 Alveolar network model of the normal septal wall before (**a**) and after (**b**) stretching. Elastic fibers are shown in blue and collagen fibers in red. The dark red represents unrecruited collagen fibers. The contour length of an unrecruited fiber is equal to its resting length when straightened. Bright red is recruited collagen. Black circles indicate nodes where two fibers are cross-linked, with the diameter of the circle representing the net force on the node. (Adapted with permission from [97])

accessibility of the enzymes to their binding sites. This has been modeled by representing a fiber as a set of springs arranged in series with each spring corresponding to a parallel combination of elastin molecules [99]. The fiber is exposed to enzyme particles by allowing them to perform a random walk along the fiber surface where they hop probabilistically from one binding site to the next. Each time an enzyme particle leaves a binding site, it decreases fiber stiffness by a small factor $\gamma = 0.995$ corresponding to cleaving a single molecule. If the model fiber is under a constant force, F, a reduction in stiffness results in an increase in fiber length. This, in turn, reveals previously hidden binding sites on the fiber surface, making the fiber more vulnerable to further digestion.

This model predicts that that fiber stiffness decays exponentially from an initial stiffness value of Y_0 according to $Y(t) = Y_0 e^{-t/\tau}$ with a time constant τ that decreases with increasing F on the fiber according to

$$\tau = \frac{\gamma}{1-\gamma}\left[\frac{1}{P_{on}} + \frac{1}{P_{off}}\right]\frac{A}{c} \tag{9.5}$$

where A is a constant proportional to the size of the fiber, c is enzyme concentration, and P_{on} and P_{off} are the probabilities of an enzyme particle binding and unbinding, respectively. P_{on} also depends inversely on F so that τ decreases quickly as F increases. The factor $\gamma/(1-\gamma)$ in Eq. 9.5 depends on the fiber structure, the factor in the square bracket characterizes the chemistry between the fiber and the enzyme, and c/A is the density of enzyme particles relative to the average binding site density which is controlled by cellular processes. Equation 9.5 thus implies that enzymatic degradation of elastic fibers in the alveolar septal wall is increased in regions of high stress as well as during inflammatory flareups that enhance enzyme release from immune cells.

Stress-strain curves have also been measured in collagen fibers from bovine lungs during digestion by bacterial collagenase while the fibers were exposed to various mechanical loads [100], again showing that stress increases the rate of digestion as reflected in a decrease in fiber stiffness. This has been modeled by representing a collagen fiber as a bundle of wavy (flaccid) but non-interacting fibrils, each with an identical linear stress-strain relationship. An individual fibril in the model ruptures with a probability P_r that depends on the force F_f carried by the fibril and its normalized modulus $\frac{Y_{f,n}}{Y_{f,0}}$ according to

$$P_r = 1 - \frac{Y_{f,n}}{Y_{f,0}} + e^{\frac{F_f}{qF_{f,r}}} - 1 \tag{9.6}$$

where $q = 4$ based on experimental data. The exponential term in Eq. 9.6 is derived from avalanche-like failures in fracture processes [101]. $F_{f,r}$ is a reference force defined as the average fibril force at 10% fiber strain. After a fibril fails, its force is set to zero. Motivated by Bell's model [102], the digestion probability applied to each fibril is given by

$$P_d = \frac{B}{w_f} e^{sF_f} \tag{9.7}$$

where w_f is the fibril's waviness, defined as the ratio of the contour length and the end-to-end distance, and was determined from scanning electron microscopy data. Experimental observations led to $B = 0.02$ and $s = 10^{-9}$. Enzymatic digestion in the model reduces fibril modulus at time step n according to $Y_{f,\,n} = \gamma\, Y_{f,\,n-1}$ ($\gamma = 0.998$) with probability P_d ($n = 0, \ldots, 60$), where $Y_{f,\,0} = 100$ MPa. The fiber stress is obtained by summing the forces on the fibrils that have not ruptured and normalizing this sum by the cross-sectional area of the fiber. This model is able to reproduce the results of several different digestion protocols including the recruitment of wavy fibrils during stretching of a fiber (Fig. 9.11a). In particular, the application of force

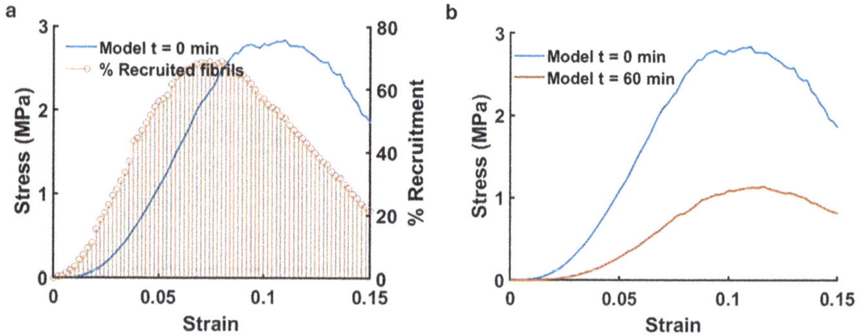

Fig. 9.11 (**a**) Simulated stress-strain curve with the corresponding percentage of recruited fibrils superimposed. (**b**) Stress-strain curves from the computational model of a fiber before ($t = 0$ min) and after ($t = 60$ min) digestion. (Reproduced with permission from [100])

during digestion significantly reduces the failure stress of the fiber (Fig. 9.11b). This is also the case if the force on the model fiber is intermittent, perhaps mimicking the stresses in the lung during coughing. Since the matrix metalloprotease MMP-9 is a key collagen-digesting enzyme that is over-expressed in the lung during COPD exacerbations [103], it is tempting to speculate that the mechanical forces of coughing can rupture the digested fibrils, fibers and septal walls in a percolating process that will eventually show up as an irreversible increase in LAA on CT images [37].

9.9 Adding Cells to the Models

The computational modeling described above reveals some of the intricate interactions between mechanical forces at various length scales from that of the whole lung down to fibrils in fibers and digestive enzymes. These interactions are further complicated by the complex network effects of stress transmission. It is equally important, however, to consider the dynamics of cellular behavior and their effects on the inflammatory microenvironment for the development of a comprehensive model of emphysema pathogenesis. Differential equations as described in Chap. 8 represent one approach to modeling such cellular behavior. Here, the various cell types involved appear as variables in the differential equations, while numerous parameters define exactly how the equations behave. Finding the values of the parameters is usually the principal challenge with this approach to modeling because there are often a large number of them, with little data to guide their evaluation. Consequently, it is frequently necessary to resort to guesswork based on what seems reasonable in terms of model behavior, informed by *a priori* biological knowledge.

One such model of emphysema pathogenesis [104] contains 15 coupled nonlinear differential equations that describe a network of interactions between several cell types (macrophages, fibroblasts, neutrophils), the chemicals they release, and the amounts of elastin and collagen in the tissue. The released chemicals comprise several distinct classes. The proinflammatory cytokines are tumor necrosis factor alpha (TNFα), the interleukins IL-1 and IL-8, and transforming growth factor beta (TGFβ). Also released are the anti-inflammatory cytokines IL-10, MMP-9, and elastase. The interaction network described by this model is shown in Fig. 9.12. Disease is initiated when inhaled cigarette smoke particles deposit on the septal walls causing the alveolar epithelial cells to rapidly release cytokines that recruit naïve macrophages and activate fibroblasts. The naïve macrophages then transition into interchangeable M1 and M2 macrophage subtypes that eventually activate neutrophils and secrete MMP-9. Neutrophil elastase and MMP-9 damage the elastic and collagen fibers, respectively, reducing parenchymal stiffness.

This system of differential equations has been implemented on a spatially extended 2D grid of tissue. The model is initialized with a small number of randomly placed macrophages that proceed to randomly move around the grid. When smoke particles impinge on the epithelium, the equations determine the local inflammatory

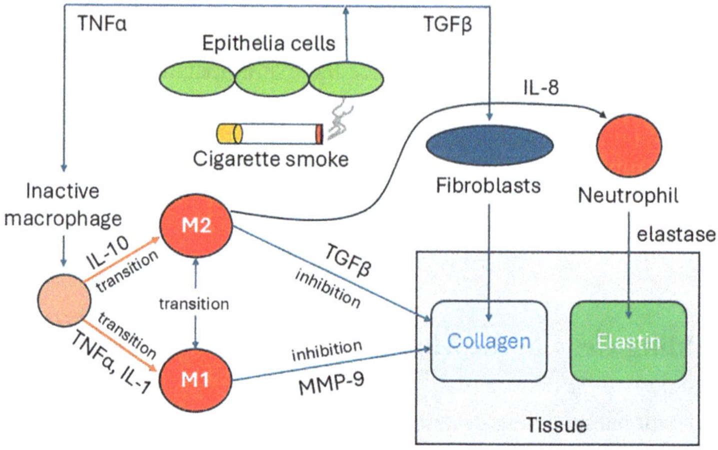

Fig. 9.12 A signaling network model of emphysema. Cigarette smoke triggers epithelial release of cytokines such as TNFα and TGFβ, which then attract inactive undifferentiated macrophages and activate fibroblasts, respectively. The inactive macrophages differentiate into M1 and M2 subtypes via both inflammatory and anti-inflammatory signals such as IL-1, TNFα and IL-10. The M1 and M2 macrophages can transition into each other while activating MMPs and recruiting neutrophils and fibroblasts. The MMPs digest collagen whereas neutrophil elastase cleaves elastin, both of which affect the regional mechanical properties of the tissue. (Reproduced with permission from [104])

response and the local elastin and collagen content of the tissue. Local stiffness is then computed as the weighted sum of the elastin and collagen contents. Periodically, a finite element analysis is used to compute the material properties of the tissue and its regional strain map. Cell death at each step in this process is linked to local tissue death, which mimics emphysema progression and the decrease this causes in mean tissue density. Since the tissue is represented as a continuum, however, this model cannot easily simulate mechanical stress-related tissue degradation to the point of creating emphysematous holes and hence airspace enlargement. Also, by ignoring the effects of local tissue stiffness and strain on cell behavior, it neglects the fact that virtually all cell types in the lung are mechanosensitive.

Many of the shortcomings of the above model could be remedied by using the same signaling network equations with a spring network model of the parenchymal tissue. This type of approach has been represented by a model that combines agent-based modeling of cell dynamics with a physics-based spring network model of the tissue (a so-called SNAB model, as described in Sect. 8.5) [105]. In this model, a simplified version of the scheme shown in Fig. 9.12 is imposed on a 2D spring network representation of the parenchyma. The network also includes small airways because the disappearance of small airways is a major feature of early COPD progression [31]. The springs corresponding to the airway walls are tenfold stiffer than those in the parenchyma. The model also includes four types of agents: (1) a fibroblast-like agent (FLA) that deposits or digests collagen in response to both stretch and stiffness, (2) an inflammatory cell-like agent (ILA) that represents both macrophages and neutrophils, (3) a contractile agent representing airway smooth

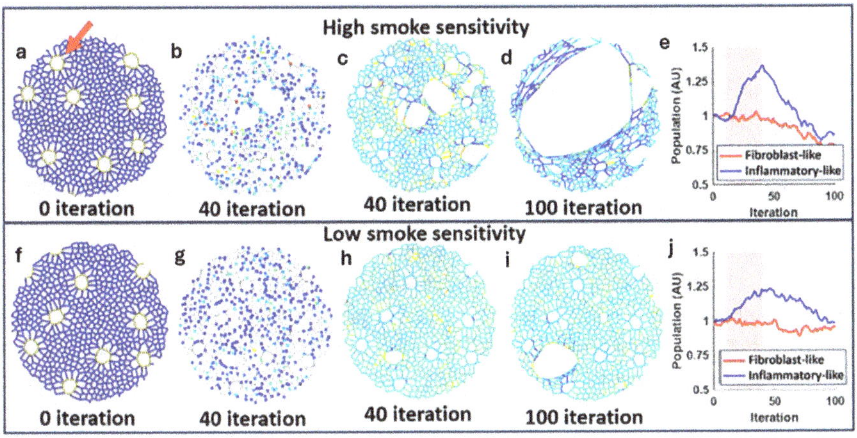

Fig. 9.13 Emphysema progression with high (top) and low (bottom) sensitivity to smoke exposure. (**a**) and (**f**) show initial networks (blue) with embedded airways (red arrow). (**b**) and (**g**) show ILAs with activity increasing from blue to red after 40 iterations. (**c**) and (**h**) show networks corresponding to (**b**) and (**g**); colors from blue to red show increasing stretch on springs. (**d**) and (**i**) show networks after 100 iterations. (**e**) and (**j**) show time variations of FLA and ILA populations. Periods of smoke exposure are shown by the pink regions. (Kindly provided to us by Joseph K. Hall)

muscle that causes bronchoconstriction in response to cigarette smoke, and (4) an agent representing epithelial cells (ECs) that cover the springs and which are activated by cigarette smoke. The FLAs and ILAs migrate around the network and are activated by cigarette smoke causing them to release enzymes that digest the parenchymal tissue. They also have fading activation memories. FLAs respond to stretch caused by airway narrowing by secreting collagen to stiffen the airway walls in an attempt to maintain homeostasis. ILAs that migrate over activated ECs also become activated, which initiates enzyme release and digestion.

Model simulations begin with the random deposition of cigarette smoke particles on springs. The competition between the stiffening caused by the FLAs and the digestion of septal walls caused by the ILAs then leads to either a return to homeostasis or progressive septal weakening that ends in rupture. Figure 9.13a shows the pre-stressed network with a fixed boundary at the start of a simulation. Figure 9.13b shows the ILA distribution over the network after 40 iterations, while Fig. 9.13c shows the corresponding progression of emphysema. After 100 iterations, the tissue destruction becomes reminiscent of human emphysema (Fig. 9.13d), with most airways obliterated [31]. The combination of ILA activity and the highly heterogeneous mechanical force on the network thus drives progression [27, 28, 49]. A potential emphysema drug treatment was also simulated by reducing the sensitivity of the ILAs to smoke exposure, which lowered the amount of enzyme they release (Fig. 9.13, bottom row). This could conceivably be achieved by inhibiting MMP-1 and/or MMP-9 production, resulting in significantly reduced tissue destruction (Fig. 9.13h, i), and a return to near-healthy populations of ILAs and FLAs.

9.10 Conclusions

COPD is an insidious disease that involves myriad multiscale complexities. Anatomically and biologically informed computational models are eminently suited to linking structure to function through biological and physical mechanisms. Indeed, these models have been able to provide some important insights in COPD pathophysiology. Nevertheless, there is plenty of room for further improvement.

There have so far been relatively few attempts to model the development and physiological consequences of chronic bronchitis. Fortunately, the extensive literature on lung modeling related to asthma offers numerous opportunities for translation to the bronchitic lung, even though this is complicated by the need to account for the nonlinear phenomenon of expiratory flow limitation. There is now ample experimental and computational evidence that mechanical forces in the parenchyma are key factors driving emphysema progression, so emphysema modeling must be grounded in a consideration of the stress and strain fields that exist throughout the lung. This has been pursued at several levels of length scale from that of the whole lung in a gravitational field to down to the level of a single alveolar septal wall and the embedded fibers. Spring network models of the parenchymal tissue, for example, demonstrate that the characteristic radiologic features of emphysema do not develop as a result of spatially random tissue failure, but rather reflect sequences of force redistribution in the parenchyma that result from mechanical failure occurring preferentially in regions of high stress, a specific network effect that continuum models are not capable of reproducing.

Despite these modeling successes, however, most efforts to date have failed to account for the significant interplay between the altered airway function that occurs in bronchitis and the tissue destruction that occurs in emphysema. The two conditions are linked by airway-parenchymal interdependence, which means that altered airway mechanics can affect the stress and strain fields in the parenchyma. Conversely, reduced parenchymal tethering of the airway wall by its parenchymal attachments leads to chronic airway collapse. Models of the future will need to specifically incorporate airway-parenchymal interdependence in order to fully recapitulate the spatially distributed nature of COPD progression. Future models will also need to be personalized by having geometries and boundary conditions, and possibly even exacerbation histories, that can be adapted to a given patient. Such models could be useful not only to optimize interventions such as lung volume reduction, but they could also predict the subsequent rate of lung function decline and hence quality of life and survival. Finally, the molecular events involved in COPD pathogenesis have so far been modeled in only rudimentary ways. As new data emerge from various omics projects, incorporation of patient-specific signaling pathways will be possible. Perhaps with the assistance of machine learning, this can be expected to help elucidate why COPD outcomes have such a large intra-patient variability and predict drug responses. It is an ambitious goal to irradicate COPD from the world [7], but computational modeling provides a powerful and integrative tool that can help.

9.11 Problems

1. Simulate the enzymatic degradation of elastic fibers. First, assume that the fiber is a continuum, represented by a single spring, and its stiffness decreases exponentially from an initial value of $Y_0 = 1$ with a time constant τ given by Eq. 9.5. Assume a fixed ratio of the concentration of enzymes and binding sites so that $\frac{A}{c} = \text{const}$. The structural damage parameter $\gamma = 0.995$ is also constant. Baseline values can be set as $P_{on} = 0.3$, $P_{off} = 0.5$, $A = 0.1$, and $c = 0.1$. Simulate how $Y(t)$ decays between $t = 0$ and 1000 for several parameter combinations including P_{on}, P_{off}, A and γ. This can be done by simulating $Y(t)$ for various values of a parameter around the assumed baseline (e.g., 50%, 75%, 125% and 150%). For which parameter do you find $Y(t)$ to be most sensitive?
2. Create a chain of linear springs in series with 100 segments. Each spring has a constant of $k = 1$ and the system has a length to cross-sectional area ratio of 100 so that $Y_0 = 1$. For this system, the baseline decrease in $Y(t)$ can be obtained from Problem 1 above. Study the effects of a force-dependent P_{on} on the digestion process.

    (a) Set up a random digestion process as follows. For each time point t and each spring along the chain, update the spring constant as $k' = \gamma k$ with a probability of P_{on}. Repeat this process N_r times for each time point. Find the value for N_r such that, given a baseline P_{on}, the spring system shows a decaying Y nearly identical to that of the baseline case.

    (b) Implement an exponential force-dependent binding process given by $P_{on} = 0.3 + \left(1 - \exp\left\{-\frac{\lambda F}{k}\right\}\right)$ where $\lambda = 1$ is a space constant and k is the local spring constant, expressing the idea that when the force on the fiber is high, binding sites unfold and become available, accelerating binding and cleaving. Compute the evolution of $Y(t)$ for $F = 0$, 0.01, 0.05, 0.1 and 0.2. If your model is correct, the digestion should be accelerated by increasing force. Explore whether the decay is exponential. If not, explain why. Determine how the mean and coefficient of variation of spring constants vary along the chain over time and with force. Explain the results.

    (c) As an advanced exercise, implement an agent-based model where the agents are the enzyme molecules carrying out a random walk in the neighborhood of the fiber. An enzyme can bind with probability P_{on}, unbind with P_{off} or just walk around the fiber. When it hops off the fiber, the local spring constant is reduced as in part (b) above. Simulate the process as a function of enzyme concentration and force. More details can be found in ref. [99].

3. Assume that a collagen fiber is composed of a set of fibrils modeled as linear springs arranged in parallel. Implement the rupture and digestion probabilities given by Eqs. 9.6 and 9.7, respectively. Code up the model and reproduce the graphs in Fig. 9.11. Explore the model by varying its parameters. In particular, the data in ref. [100] display very large sample-to-sample variability. How might you explain this variability in terms of the model parameters?

4. Using simulated annealing to find the equilibrium configuration of either the square or the hexagonal network. Write a code to mimic the progression of emphysema as follows.
   (a) Implement random elimination of springs as well as mechanical force-based elimination of springs. For the latter, first remove a small percentage of springs. Find the equilibrium configuration and then rank order the springs according to the force they carry. Remove the top 1% of springs carrying the largest forces. Solve the network for equilibrium. Find again the top 1% of springs carrying the largest forces. Continue this for a few steps and examine how the network structure evolves. When approximately 13% of springs are removed, the random and the force-based models should produce networks similar to those in Fig. 9.3a, b, respectively.
   (b) Determine the areas enclosed by the network polygons and calculate their area distributions. These distributions should be similar to those in Fig. 9.3d.
   (c) Implement a soft boundary condition and repeat (a) and (b).

References

1. Chen P, Li Y, Wu D, Liu F, Cao C. Secondhand smoke exposure and the risk of chronic obstructive pulmonary disease: a systematic review and meta-analysis. Int J Chron Obstruct Pulmon Dis. 2023;18:1067–76.
2. Beyer D, Mitfessel H, Gillissen A. Maternal smoking promotes chronic obstructive lung disease in the offspring as adults. Eur J Med Res. 2009;14(Suppl 4):27–31.
3. Ryu MH, Murphy S, Hinkley M, Carlsten C. COPD exposed to air pollution: a path to understand and protect a susceptible population. Chest. 2024;165:836–46.
4. Ortiz-Quintero B, Martinez-Espinosa I, Perez-Padilla R. Mechanisms of lung damage and development of COPD due to household biomass-smoke exposure: inflammation, oxidative stress, MicroRNAs, and gene polymorphisms. Cells. 2022;12:67.
5. Li LS, Paquet C, Johnston K, Williams MT. "What are my chances of developing COPD if one of my parents has the disease?" A systematic review and meta-analysis of prevalence of co-occurrence of COPD diagnosis in parents and offspring. Int J Chron Obstruct Pulmon Dis. 2017;12:403–15.
6. DeMeo DL, Silverman EK. Alpha1-antitrypsin deficiency. 2: genetic aspects of alpha(1)-antitrypsin deficiency: phenotypes and genetic modifiers of emphysema risk. Thorax. 2004;59: 259–64.
7. Stolz D, Mkorombindo T, Schumann DM, Agusti A, Ash SY, Bafadhel M, Bai C, Chalmers JD, Criner GJ, Dharmage SC, et al. Towards the elimination of chronic obstructive pulmonary disease: a lancet commission. Lancet. 2022;400:921–72.
8. Global strategy for the diagnosis, management and prevention of chronic obstructive pulmonary disease. In Global initiative for chronic obstructive lung disease. www.goldcopd.org/2013; 2013.
9. Anzueto A. Impact of exacerbations on COPD. Eur Respir Rev. 2010;19:113–8.
10. Groenewegen KH, Schols AM, Wouters EF. Mortality and mortality-related factors after hospitalization for acute exacerbation of COPD. Chest. 2003;124:459–67.
11. Hurd S. The impact of COPD on lung health worldwide: epidemiology and incidence. Chest. 2000;117:1S–4S.

12. Belz DC, Putcha N, Alupo P, Siddharthan T, Baugh A, Hopkinson N, Castaldi P, Papi A, Mannino D, Miravitlles M, et al. Call to action: how can we promote the development of new pharmacologic treatments in COPD? Am J Respir Crit Care Med. 2024;210:1300–7.
13. Kim V, Criner GJ. Chronic bronchitis and chronic obstructive pulmonary disease. Am J Respir Crit Care Med. 2013;187:228–37.
14. Raby KL, Michaeloudes C, Tonkin J, Chung KF, Bhavsar PK. Mechanisms of airway epithelial injury and abnormal repair in asthma and COPD. Front Immunol. 2023;14:1201658.
15. Bals R, Hiemstra PS. Innate immunity in the lung: how epithelial cells fight against respiratory pathogens. Eur Respir J. 2004;23:327–33.
16. Rogers DF. The airway goblet cell. Int J Biochem Cell Biol. 2003;35:1–6.
17. Shaykhiev R, Zuo WL, Chao I, Fukui T, Witover B, Brekman A, Crystal RG. EGF shifts human airway basal cell fate toward a smoking-associated airway epithelial phenotype. Proc Natl Acad Sci USA. 2013;110:12102–7.
18. Rose MC, Voynow JA. Respiratory tract mucin genes and mucin glycoproteins in health and disease. Physiol Rev. 2006;86:245–78.
19. Button B, Anderson WH, Boucher RC. Mucus hyperconcentration as a unifying aspect of the chronic bronchitic phenotype. Ann Am Thorac Soc. 2016;13(Suppl 2):S156–62.
20. Sathish V, Freeman MR, Long E, Thompson MA, Pabelick CM, Prakash YS. Cigarette smoke and estrogen signaling in human airway smooth muscle. Cell Physiol Biochem. 2015;36: 1101–15.
21. Leopold PL, O'Mahony MJ, Lian XJ, Tilley AE, Harvey BG, Crystal RG. Smoking is associated with shortened airway cilia. PLoS One. 2009;4:e8157.
22. Vlahovic G, Russell ML, Mercer RR, Crapo JD. Cellular and connective tissue changes in alveolar septal walls in emphysema. Am J Respir Crit Care Med. 1999;160:2086–92.
23. Kim JH, Schaible N, Hall JK, Bartolak-Suki E, Deng Y, Herrmann J, Sonnenberg A, Behrsing HP, Lutchen KR, Krishnan R, Suki B. Multiscale stiffness of human emphysematous precision cut lung slices. Sci Adv. 2023;9:eadf2535.
24. Finlay GA, O'Donnell MD, O'Connor CM, Hayes JP, FitzGerald MX. Elastin and collagen remodeling in emphysema. A scanning electron microscopy study. Am J Pathol. 1996;149: 1405–15.
25. van Straaten JF, Coers W, Noordhoek JA, Huitema S, Flipsen JT, Kauffman HF, Timens W, Postma DS. Proteoglycan changes in the extracellular matrix of lung tissue from patients with pulmonary emphysema. Mod Pathol. 1999;12:697–705.
26. Goldklang M, Stockley R. Pathophysiology of emphysema and implications. Chronic Obstr Pulm Dis. 2016;3:454–8.
27. Kononov S, Brewer K, Sakai H, Cavalcante FS, Sabayanagam CR, Ingenito EP, Suki B. Roles of mechanical forces and collagen failure in the development of elastase-induced emphysema. Am J Respir Crit Care Med. 2001;164:1920–6.
28. Mishima M, Hirai T, Itoh H, Nakano Y, Sakai H, Muro S, Nishimura K, Oku Y, Chin K, Ohi M, et al. Complexity of terminal airspace geometry assessed by lung computed tomography in normal subjects and patients with chronic obstructive pulmonary disease. Proc Natl Acad Sci USA. 1999;96:8829–34.
29. Verbeken EK, Cauberghs M, Mertens I, Clement J, Lauweryns JM, Van de Woestijne KP. The senile lung. Comparison with normal and emphysematous lungs. 2. Functional aspects. Chest. 1992;101:800–9.
30. Chin RC, Guenette JA, Cheng S, Raghavan N, Amornputtisathaporn N, Cortes-Telles A, Webb KA, O'Donnell DE. Does the respiratory system limit exercise in mild chronic obstructive pulmonary disease? Am J Respir Crit Care Med. 2013;187:1315–23.
31. McDonough JE, Yuan R, Suzuki M, Seyednejad N, Elliott WM, Sanchez PG, Wright AC, Gefter WB, Litzky L, Coxson HO, et al. Small-airway obstruction and emphysema in chronic obstructive pulmonary disease. N Engl J Med. 2011;365:1567–75.

32. Booth S, Hsieh A, Mostaco-Guidolin L, Koo HK, Wu K, Aminazadeh F, Yang CX, Quail D, Wei Y, Cooper JD, et al. A single-cell atlas of small airway disease in chronic obstructive pulmonary disease: a cross-sectional study. Am J Respir Crit Care Med. 2023;208:472–86.
33. Global strategy for the diagnosis, management and prevention of chronic obstructive pulmonary disease. In Global initiative for chronic obstructive lung disease. www.goldcopd.org; 2011.
34. Stockley RA, O'Brien C, Pye A, Hill SL. Relationship of sputum color to nature and outpatient management of acute exacerbations of COPD. Chest. 2000;117:1638–45.
35. Seemungal TA, Harper-Owen R, Bhowmik A, Jeffries DJ, Wedzicha JA. Detection of rhinovirus in induced sputum at exacerbation of chronic obstructive pulmonary disease. Eur Respir J. 2000;16:677–83.
36. Donaldson GC, Seemungal T, Jeffries DJ, Wedzicha JA. Effect of temperature on lung function and symptoms in chronic obstructive pulmonary disease. Eur Respir J. 1999;13:844–9.
37. Tanabe N, Muro S, Hirai T, Oguma T, Terada K, Marumo S, Kinose D, Ogawa E, Hoshino Y, Mishima M. Impact of exacerbations on emphysema progression in chronic obstructive pulmonary disease. Am J Respir Crit Care Med. 2011;183:1653–9.
38. Donaldson GC, Seemungal TA, Bhowmik A, Wedzicha JA. Relationship between exacerbation frequency and lung function decline in chronic obstructive pulmonary disease. Thorax. 2002;57:847–52.
39. Garcia-Aymerich J, Serra Pons I, Mannino DM, Maas AK, Miller DP, Davis KJ. Lung function impairment, COPD hospitalisations and subsequent mortality. Thorax. 2011;66:585–90.
40. Hajizadeh N, Goldfeld K, Crothers K. What happens to patients with COPD with long-term oxygen treatment who receive mechanical ventilation for COPD exacerbation? A 1-year retrospective follow-up study. Thorax. 2015;70:294–6.
41. Mohamady YK, Geudens V, De Fays C, Zapata M, Hagrass O, Aversa L, Vermant M, Jin X, Willems L, Gyselinck I, et al. Computational fluid dynamics of small airway disease in chronic obstructive pulmonary disease. EBioMedicine. 2025;114:105670.
42. Bordas R, Lefevre C, Veeckmans B, Pitt-Francis J, Fetita C, Brightling CE, Kay D, Siddiqui S, Burrowes KS. Development and analysis of patient-based complete conducting airways models. PLoS One. 2015;10:e0144105.
43. Wagers S, Lundblad LK, Ekman M, Irvin CG, Bates JH. The allergic mouse model of asthma: normal smooth muscle in an abnormal lung? J Appl Physiol (1985). 2004;96:2019–27.
44. Eltayara L, Becklake MR, Volta CA, Milic-Emili J. Relationship between chronic dyspnea and expiratory flow limitation in patients with chronic obstructive pulmonary disease. Am J Respir Crit Care Med. 1996;154:1726–34.
45. Hedges KL, Tawhai MH. Simulation of forced expiration in a biophysical model, with homogeneous and clustered bronchoconstriction. J Biomech Eng. 2016;138:061008.
46. Bates JH, Irvin CG, Farre R, Hantos Z. Oscillation mechanics of the respiratory system. Compr Physiol. 2011;1:1233–72.
47. Dellaca RL, Santus P, Aliverti A, Stevenson N, Centanni S, Macklem PT, Pedotti A, Calverley PM. Detection of expiratory flow limitation in COPD using the forced oscillation technique. Eur Respir J. 2004;23:232–40.
48. Tgavalekos NT, Venegas JG, Suki B, Lutchen KR. Relation between structure, function, and imaging in a three-dimensional model of the lung. Ann Biomed Eng. 2003;31:363–73.
49. Suki B, Lutchen KR, Ingenito EP. On the progressive nature of emphysema: roles of proteases, inflammation, and mechanical forces. Am J Respir Crit Care Med. 2003;168:516–21.
50. Suki B, Sato S, Parameswaran H, Szabari MV, Takahashi A, Bartolak-Suki E. Emphysema and mechanical stress-induced lung remodeling. Physiology (Bethesda). 2013;28:404–13.
51. West JB. Distribution of mechanical stress in the lung, a possible factor in localisation of pulmonary disease. Lancet. 1971;1:839–41.

52. Gefen A, Elad D, Shiner RJ. Analysis of stress distribution in the alveolar septa of normal and simulated emphysematic lungs. J Biomech. 1999;32:891–7.
53. Sugihara T, Martin CJ, Hildebrandt J. Length-tension properties of alveolar wall in man. J Appl Physiol. 1971;30:874–8.
54. Elad D, Kamm RD, Shapiro AH. Tube law for the intrapulmonary airway. J Appl Physiol (1985). 1988;65:7–13.
55. Lai-Fook SJ. Elasticity analysis of lung deformation problems. Ann Biomed Eng. 1981;9:451–62.
56. Suki B, Bates JHT, Bartolak-Suki E. Remodeling of the aged and emphysematous lungs: roles of microenvironmental cues. Compr Physiol. 2022;12:3559–74.
57. Beltran G, Navajas D, Garcia-Aznar JM. Mechanical modeling of lung alveoli: from macroscopic behaviour to cell mechano-sensing at microscopic level. J Mech Behav Biomed Mater. 2022;126:105043.
58. de Ryk J, Thiesse J, Namati E, McLennan G. Stress distribution in a three dimensional, geometric alveolar sac under normal and emphysematous conditions. Int J Chron Obstruct Pulmon Dis. 2007;2:81–91.
59. Diciotti S, Nobis A, Ciulli S, Landini N, Mascalchi M, Sverzellati N, Innocenti B. Development of digital phantoms based on a finite element model to simulate low-attenuation areas in CT imaging for pulmonary emphysema quantification. Int J Comput Assist Radiol Surg. 2017;12:1561–70.
60. Yuan Z, Herrmann J, Murthy S, Peters K, Gerard SE, Nia HT, Lutchen KR, Suki B. A personalized spring network representation of emphysematous lungs from CT images. Front Netw Physiol. 2022;2:828157.
61. Tanabe N, Muro S, Sato S, Tanaka S, Oguma T, Kiyokawa H, Takahashi T, Kinose D, Hoshino Y, Kubo T, et al. Longitudinal study of spatially heterogeneous emphysema progression in current smokers with chronic obstructive pulmonary disease. PLoS One. 2012;7: e44993.
62. Bates JH, Davis GS, Majumdar A, Butnor KJ, Suki B. Linking parenchymal disease progression to changes in lung mechanical function by percolation. Am J Respir Crit Care Med. 2007;176:617–23.
63. Hamakawa H, Bartolák-Suki E, Parameswaran H, Majumdar A, Lutchen KR, Suki B. Structure–function relations in an elastase-induced mouse model of emphysema. Am J Respir Cell Mol Biol. 2011;45:517–24.
64. Parameswaran H, Majumdar A, Ito S, Alencar AM, Suki B. Quantitative characterization of airspace enlargement in emphysema. J Appl Physiol. 2006;100:186–93.
65. Takahashi A, Majumdar A, Parameswaran H, Bartolak-Suki E, Suki B. Proteoglycans maintain lung stability in an elastase-treated mouse model of emphysema. Am J Respir Cell Mol Biol. 2014;51:26–33.
66. Buschmann MD, Grodzinsky AJ. A molecular model of proteoglycan-associated electrostatic forces in cartilage mechanics. J Biomech Eng. 1995;117:179–92.
67. Chalmers GW, Gosline JM, Lillie MA. The hydrophobicity of vertebrate elastins. J Exp Biol. 1999;202:301–14.
68. Fratzl P, Daxer A. Structural transformation of collagen fibrils in corneal stroma during drying. An x-ray scattering study. Biophys J. 1993;64:1210–4.
69. Cavalcante FS, Ito S, Brewer KK, Sakai H, Alencar AM, Almeida MP, Andrade JS Jr, Majumdar A, Ingenito EP, Suki B. Mechanical interactions between collagen and proteoglycans: implications for the stability of lung tissue. J Appl Physiol. 2005;98:672–9.
70. Cantor JO, Cerreta JM, Ochoa M, Ma S, Liu M, Turino GM. Therapeutic effects of hyaluronan on smoke-induced elastic fiber injury: does delayed treatment affect efficacy? Lung. 2011;189: 51–6.
71. Cantor JO, Turino GM. Can exogenously administered hyaluronan improve respiratory function in patients with pulmonary emphysema? Chest. 2004;125:288–92.

72. Cantor JO, Ma S, Liu X, Campos MA, Strange C, Stocks JM, Devine MS, El Bayadi SG, Lipchik RJ, Sandhaus RA, Turino GM. A 28-day clinical trial of aerosolized hyaluronan in alpha-1 antiprotease deficiency COPD using desmosine as a surrogate marker for drug efficacy. Respir Med. 2021;182:106402.
73. Shah PL, Herth FJ, van Geffen WH, Deslee G, Slebos DJ. Lung volume reduction for emphysema. Lancet Respir Med. 2017;5:147–56.
74. Franzen DP, Weder W. Lung volume reduction-a great chance for emphysema patients! J Thorac Dis. 2018;10:S2696–7.
75. Criner GJ, Cordova F, Sternberg AL, Martinez FJ. The National Emphysema Treatment Trial (NETT) part II: lessons learned about lung volume reduction surgery. Am J Respir Crit Care Med. 2011;184:881–93.
76. Kramer MR, Refaely Y, Maimon N, Rosengarten D, Fruchter O. Bilateral endoscopic sealant lung volume reduction therapy for advanced emphysema. Chest. 2012;142:1111–7.
77. Ingenito EP, Tsai LW, Suki E, Hoffman A. Fibroblast growth Factor-1 therapy for advanced emphysema—a new tissue engineering approach for achieving lung volume reduction. J Bronchol Interv Pulmonol. 2006;13:114–23.
78. Naunheim KS, Wood DE, Mohsenifar Z, Sternberg AL, Criner GJ, DeCamp MM, Deschamps CC, Martinez FJ, Sciurba FC, Tonascia J, et al. Long-term follow-up of patients receiving lung-volume-reduction surgery versus medical therapy for severe emphysema by the National Emphysema Treatment Trial Research Group. Ann Thorac Surg. 2006;82:431–43.
79. Deslee G, Mal H, Dutau H, Bourdin A, Vergnon JM, Pison C, Kessler R, Jounieaux V, Thiberville L, Leroy S, et al. Lung volume reduction coil treatment vs usual care in patients with severe emphysema: the REVOLENS Randomized Clinical Trial. JAMA. 2016;315:175–84.
80. El Gohary D, Eltomey MA, Mohamed AS, Farahat A. Abd El Zaher AH: Bronchoscopic lung volume reduction by instillation of fibrinogen and thrombin in COPD patients with homogenous emphysema. BMC Pulm Med. 2024;24:86.
81. Miller JD, Malthaner RA, Goldsmith CH, Goeree R, Higgins D, Cox PG, Tan L, Road JD, Canadian Lung Volume Reduction Surgery Study. A randomized clinical trial of lung volume reduction surgery versus best medical care for patients with advanced emphysema: a two-year study from Canada. Ann Thorac Surg. 2006;81:314–20; discussion 320–311.
82. Koster TD, Dijk MV, Slebos DJ. Bronchoscopic lung volume reduction for emphysema: review and update. Semin Respir Crit Care Med. 2022;43:541–51.
83. Mondonedo JR, Suki B. Predicting structure-function relations and survival following surgical and bronchoscopic lung volume reduction treatment of emphysema. PLoS Comput Biol. 2017;13:e1005282.
84. Gelb AF, McKenna RJ Jr, Brenner M, Epstein JD, Zamel N. Lung function 5 yr after lung volume reduction surgery for emphysema. Am J Respir Crit Care Med. 2001;163:1562–6.
85. Butler JP, Oldmixon EH, Hoppin FG Jr. Dihedral angles of septal "bend" structures in lung parenchyma. J Appl Physiol. 1996;81:1800–6.
86. Oldmixon EH, Butler JP, Hoppin FG Jr. Lengths and topology of alveolar septal borders. J Appl Physiol. 1989;67:1930–40.
87. Parameswaran H, Bartolak-Suki E, Hamakawa H, Majumdar A, Allen PG, Suki B. Three-dimensional measurement of alveolar airspace volumes in normal and emphysematous lungs using micro-CT. J Appl Physiol. 2009;107:583–92.
88. Vasilescu DM, Klinge C, Knudsen L, Yin L, Wang G, Weibel ER, Ochs M, Hoffman EA. Stereological assessment of mouse lung parenchyma via nondestructive, multiscale micro-CT imaging validated by light microscopic histology. J Appl Physiol. 2013;114:716–24.
89. Watz H, Breithecker A, Rau WS, Kriete A. Micro-CT of the human lung: imaging of alveoli and virtual endoscopy of an alveolar duct in a normal lung and in a lung with centrilobular emphysema—initial observations. Radiology. 2005;236:1053–8.
90. Fung YC. A model of the lung structure and its validation. J Appl Physiol. 1988;64:2132–41.

91. Hall JK, Bates JHT, Casey DT, Bartolak-Suki E, Lutchen KR, Suki B. Predicting alveolar ventilation heterogeneity in pulmonary fibrosis using a non-uniform polyhedral spring network model. Front Netw Physiol. 2023;3:1124223.
92. Denny E, Schroter RC. The mechanical behavior of a mammalian lung alveolar duct model. J Biomech Eng. 1995;117:254–61.
93. Kimmel E, Budiansky B. Surface tension and the dodecahedron model for lung elasticity. J Biomech Eng. 1990;112:160–7.
94. Parameswaran H, Majumdar A, Suki B. Linking microscopic spatial patterns of tissue destruction in emphysema to macroscopic decline in stiffness using a 3D computational model. PLoS Comput Biol. 2011;7:e1001125.
95. Mondonedo JR, Sato S, Oguma T, Muro S, Sonnenberg AH, Zeldich D, Parameswaran H, Hirai T, Suki B. CT imaging-based low-attenuation super clusters in three dimensions and the progression of emphysema. Chest. 2019;155:79–87.
96. Bhatt SP, Bodduluri S, Hoffman EA, Newell JD Jr, Sieren JC, Dransfield MT, Reinhardt JM, Investigators CO. Computed tomography measure of lung at risk and lung function decline in chronic obstructive pulmonary disease. Am J Respir Crit Care Med. 2017;196:569–76.
97. Casey DT, Bou Jawde S, Herrmann J, Mori V, Mahoney JM, Suki B, Bates JHT. Percolation of collagen stress in a random network model of the alveolar wall. Sci Rep. 2021;11:16654.
98. Jesudason R, Sato S, Parameswaran H, Araujo AD, Majumdar A, Allen PG, Bartolak-Suki E, Suki B. Mechanical forces regulate elastase activity and binding site availability in lung elastin. Biophys J. 2010;99:3076–83.
99. Araujo AD, Majumdar A, Parameswaran H, Yi E, Spencer JL, Nugent MA, Suki B. Dynamics of enzymatic digestion of elastic fibers and networks under tension. Proc Natl Acad Sci USA. 2011;108:9414–9.
100. Deng Y, Herrmann J, Wang Y, Nguyen M, Hall JK, Kim JH, Smith ML, Lutchen KR, Bartolák-Suki E, Suki B. Physiological mechanical forces accelerate the degradation of bovine lung collagen fibers by bacterial collagenase. Sci Rep. 2024;14:29008.
101. Zapperi S, Ray P, Stanley HE, Vespignani A. Avalanches in breakdown and fracture processes. Phys Rev E. 1999;59:5049–57.
102. Bell GI. Models for the specific adhesion of cells to cells. Science. 1978;200:618–27.
103. Mercer PF, Shute JK, Bhowmik A, Donaldson GC, Wedzicha JA, Warner JA. MMP-9, TIMP-1 and inflammatory cells in sputum from COPD patients during exacerbation. Respir Res. 2005;6:151.
104. Ceresa M, Olivares AL, Noailly J, Gonzalez Ballester MA. Coupled immunological and biomechanical model of emphysema progression. Front Physiol. 2018;9:388.
105. Hall JK, Bates JH, Krishnan R, Lutchen KR, Suki B. A microscale personalized spring network model predicts small airway disappearance in chronic obstructive pulmonary disease (abstract). Am J Respir Crit Care Med. 2024;209:A6654.

Chapter 10
Acute Lung Injury

Abstract Acute lung injury, and its severe form, acute respiratory distress syndrome (ARDS), affect thousands of people per year in the US. Management of ARDS frequently involves administration of supportive mechanical ventilation. Mechanical ventilation itself, however, can make matters worse by causing ventilator-induced lung injury (VILI). Modeling plays a vital role in quantitatively defining the link between mechanical ventilation and VILI production by predicting the lung tissue damage caused by over-distension (volutrauma) and repetitive recruitment of derecruited lung units (atelectrauma). VILI can be predicted in an overall sense using a single compartment model that embodies both alveolar tissue distension and recruitment of closed lung units. VILI development can be modeled more accurately by associating a virtual trajectory with each independent lung unit. Tissue overdistention and repetitive recruitment also act synergistically, which can be explained on the basis of a rich-get-richer mechanism. Failure of the blood-gas barrier by repetitive recruitment can be explained on the basis of a multi-hit mechanism in which multiple intercellular adhesions must fail before plasma derived material starts leaking into the alveoli. Models based on these mechanisms explain how VILI develops over time during injurious mechanical ventilation and the critical role played by pulmonary surfactant dysfunction.

Keywords Acute respiratory distress syndrome · Ventilator-induced lung injury · Volutrauma · Atelectrauma · Over-distension · Recruitment and derecruitment · Virtual trajectory · Rich-get-richer · Multi-hit · Blood-gas barrier

10.1 Introduction

The pulmonary blood-gas barrier is a very thin (~1 μm) membrane with a very large area (~10 m^2) across which oxygen and carbon dioxide diffuse to meet the gas exchange needs of the body (Sect. 2.3). Despite its extreme thinness, the physical integrity of the blood-gas barrier keeps alveolar air and capillary blood separated from each other. Any fluid that makes its way into the alveolar spaces is efficiently cleared by the lymphatic system [1], while pulmonary surfactant prevents injurious surface tension from developing at the alveolar air-liquid interface [2]. The

B. Suki, J. H. T. Bates, *Mathematical Modeling of the Healthy and Diseased Lung*,
https://doi.org/10.1007/978-3-031-95322-4_10

blood-gas barrier is nevertheless a delicate structure susceptible to damage from a number of sources including the physical stresses and strains of mechanical ventilation, caustic materials such as inhaled noxious gases and aspirated gastric contents, and the inflammatory consequences of an over-exuberant immune response to infection [2]. When damage occurs, the integrity of the blood-gas barrier can become compromised such that fluid and possibly proteins from the plasma can start to leak into the alveolar airspaces [2]. This can rapidly overwhelm the fluid clearing ability of the lymphatic system, leading to accumulation of plasma-derived material in the airspaces. Eventually this material will begin to interfere with the ability of the pulmonary surfactant to lower surface tension at the air-liquid interface [2]. When this happens, the ventilation-induced stresses on the parenchyma increase, potentially leading to a vicious cycle of worsening tissue injury.

When acute lung injury becomes severe enough, it is defined as *acute respiratory distress syndrome* (ARDS). This is a frequently fatal condition that affects tens of thousands of people per year in the US alone, many of whom since 2020 were victims of Covid. The diagnosis of ARDS is based primarily on two clinical signs that are readily linked to damage of the blood-gas barrier [3]. One of these signs is the appearance of bilateral opacities on chest x-ray or computed tomography, reflecting a diffuse accumulation of fluid in the airspaces (provided this cannot be attributed to heart failure). The other sign is the ratio of alveolar partial pressure of oxygen (PaO_2) to inspired oxygen fraction (FiO_2), a PaO_2/ FiO_2 ratio < 300 being pathognomonic for ARDS. There are currently no medical therapies for ARDS. Treatment is supportive, often in the intensive care unit, with positive-pressure mechanical ventilation playing a central role.

While mechanical ventilation is intended to be a life-saving intervention, it can actually make things worse by causing additional injury to the lung (even spontaneous breathing can sometimes cause damage when the lung is acutely injured) [4–6]. This *ventilator-induced lung injury* (VILI) is thought to occur through two distinct mechanisms known as *volutrauma* and *atelectrauma* [7]. The most intuitively obvious of these is volutrauma, which refers to the damage caused by over-distention of the parenchymal tissues to the point where they begin to fail mechanically. This is something that can occur in any material if subjected to sufficient strain. Of course, the parenchyma is designed by Nature to be distended by large amounts during breathing, so volutrauma is difficult to produce in a healthy organ, but an already injured lung is much more susceptible to being further injured by over-distension. Limiting tissue distension is the obvious antidote to this problem, and indeed the use of lower-than-normal tidal volumes has become the standard of care during mechanical ventilation of ARDS patients [8].

Atelectrauma is an injury mechanism specific to the lung that is associated with collapse of alveoli (a phenomenon known as atelectasis) and occlusion of small airways by secretions and exudates [7, 9]. Such *derecruitment* events do not normally occur to any significant extent in a healthy lung during either spontaneous breathing or mechanical ventilation (although this can begin to happen in elderly people). In an injured lung, however, derecruitment may occur cyclically with each expiration. This means that the derecruited alveoli and airways must be forcibly reopened with each inspiration, a process that generates large stresses at sites of reopening. Such stresses are highly damaging to the pulmonary epithelium and endothelium, and can rapidly damage the blood-gas barrier [10, 11]. The presence of chronic atelectasis also means that regions of established alveolar collapse will be juxtaposed with regions of normal open alveoli. When open regions expand during inspiration, the alveolar walls near the borders with collapsed regions experience excessive strain and thus become a nidus for injury development and propagation. Such regions have been referred to as *stress concentrators*, and essentially lead to volutrauma at the micro scale [7]. Preventing or reversing lung derecruitment during mechanical ventilation is conventionally achieved by raising lung volume through the application of *positive end-expiratory pressure* (PEEP) [4, 7]. This stents the alveoli and small airways open, but at the expense of greater overall distension of the parenchyma.

10.2 The Role of Modeling in VILI Reduction

Ventilating the injured lung so as to minimize the chances of causing VILI is a question of balance. On the one hand, reducing end-inspiratory lung volume protects against volutrauma, but it increases the risk of derecuitment and thus atelectrauma. On the other hand, increasing end-expiratory lung volume protects against derecruitment, but this increases the risk of volutrauma. The two strategies are thus at odds with each other. This implies there should exist some optimum combination of end-expiratory and end-inspiratory lung volumes (Fig. 10.1) that

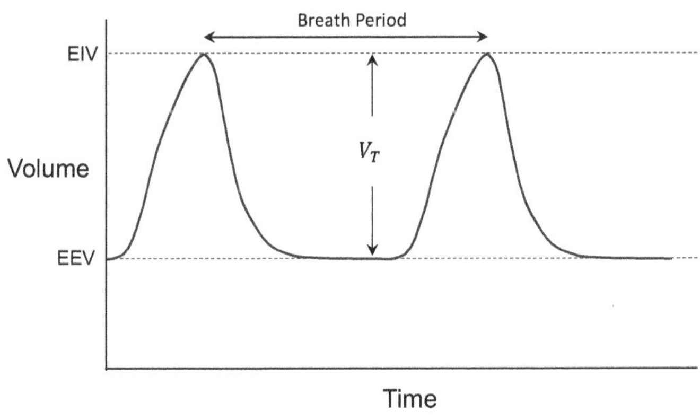

Fig. 10.1 Lung volume waveform during mechanical ventilation. Decreasing end-inspiratory lung volume (EIV) mitigates against volutrauma, while increasing end-expiratory volume (EEV) does the same for atelectrauma. Tidal volume (V_T) can be modulated along with breathing frequency (inverse of the Breath Period) to maintain a required alveolar minute ventilation. There are also limitless possible volume waveform shapes compatible with a given EEV, EIV and V_T

causes the least amount of VILI. The matter is complicated, however, by the fact that these two strategies must be jointly implemented to satisfy the constraint that the gas exchange needs of the patient are met. This constraint can be expressed as the need to have a given volume of fresh gas reach the alveolar regions of the lung each minute, known as the *alveolar minute ventilation*, expressed as

$$\dot{V}_A = f(V_T - V_D) \tag{10.1}$$

where V_T is tidal volume, V_D is the dead space of the lung (equal to the volume of the conducting airways and unperfused alveolar regions), and f is the number of breaths per minute. Equation 10.1 shows that V_T and f can vary over wide physiologically reasonable ranges while leaving $\dot{V}_A$ unaffected. The shapes of the inspiratory and expiratory volume waveforms can also vary significantly for a given V_T and f. In addition, end-expiratory lung volume can be varied over a wide range through the adjustment of PEEP. The minimally injurious ventilator waveform for a given patient thus exists within a very large space of possibilities.

The above considerations show that identifying and implementing the minimally injurious ventilation strategy for a given patient lends itself to being expressed as a problem in feedback control, as illustrated in Fig. 10.2. That is, given some function Ω that expresses the rate at which VILI develops in the lungs of the patient in question for a given ventilation strategy, it should be possible to use a computational model to determine the strategy that keeps Ω at or close to its minimum value. Ω changes with the injury status of the lungs, however, so some means of continually monitoring injury status is required so that the model can be updated and Ω adjusted accordingly.

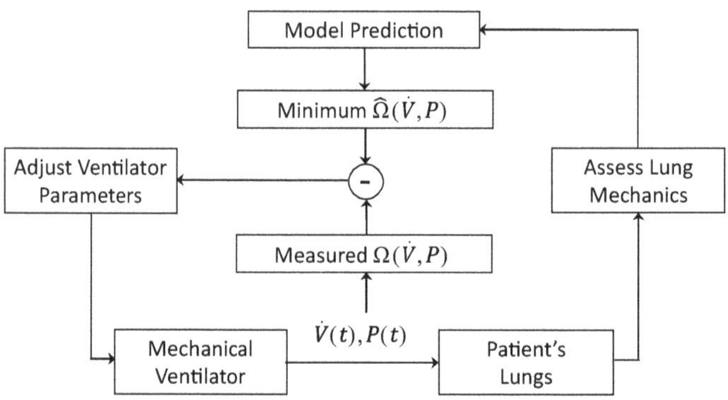

Fig. 10.2 A mechanical ventilator produces flow $\dot{V}(t)$ and pressure $P(t)$ at the patient's airway opening. These signals can be used to calculate an injury cost function $\Omega(\dot{V}, P)$ on the basis of a computational model embodying the mechanisms of VILI and the current mechanical state of the patient's lungs. The model is used to search for the minimum value of $\widehat{\Omega}(\dot{V}, P)$, which is then implemented by the mechanical ventilator. The ventilator settings are adjusted continually as the model adapts to the changing lung mechanics of the patient

In order to be useful clinically, Ω needs to be a function of quantities that can be tracked rapidly enough so that a response to a patient's changing condition can be implemented in real time. At the present time, these quantities are essentially limited to those reflective of lung mechanical function, namely airway pressure (P) and flow ($\dot{V}$). Arterial blood gases and x-ray imaging also provide valuable information relevant to ARDS, but they are not suited to the kind of continuous monitoring needed to respond rapidly when clinical conditions change, and so they are not as easily incorporated into an automatic servo-control algorithm. We can thus say that $\Omega = \Omega(\dot{V}, P)$. Computational modeling has a central role to play in minimizing $\Omega(\dot{V}, P)$ (Fig. 10.2).

The most challenging aspect of realizing the plan depicted in Fig. 10.2 is determining how to calculate $\Omega(\dot{V}, P)$. Evaluating this cost function from measurements of $P(t)$ and $\dot{V}(t)$ (t is time) made at the bedside can only be done to the extent that the biophysical processes involved in the generation of volutrauma and atelectrauma are reflected in $P(t)$ and $\dot{V}(t)$. It is far from obvious how these biophysical processes are linked individually and in combination to the generation of VILI. As always when simulating a physical system, the way forward is to start with a simple model that plausibly makes this link in mechanistic terms, and then add complexities as mandated by the data.

10.3 Compartment Models of VILI

Volutrauma arises when the lung tissues are stretched past a dangerous limit, while atelectrauma arises when lung regions that derecruit during each expiration are forcibly recruited again during the subsequent inspiration. Figure 10.3 depicts a very simple model of the lung that embodies these two processes [12]. The model has a single alveolar compartment served by a single airway. Unlike the conventional single-compartment model of lung mechanics (Sect. 3.3.1), however, the model in Fig. 10.3 has two mechanical degrees of freedom because its alveolar compartment can expand in two orthogonal directions.

Expansion of the alveolar compartment in the vertical direction stretches the spring denoted E_{tis} in Fig. 10.3, which generates elastic recoil in the parenchymal tissues. The expansion of this spring is determined by alveolar pressure, P_A, according to

$$P_A = E_{tis} y \tag{10.2}$$

where the explicit t dependence has been left out because this equation describes only static elastic behavior. The spring constant E_{tis} represents lung elastance, which increases with increasing y. Assume that the relationship between E_{tis} and y conforms to the commonly used exponential relationship [13]

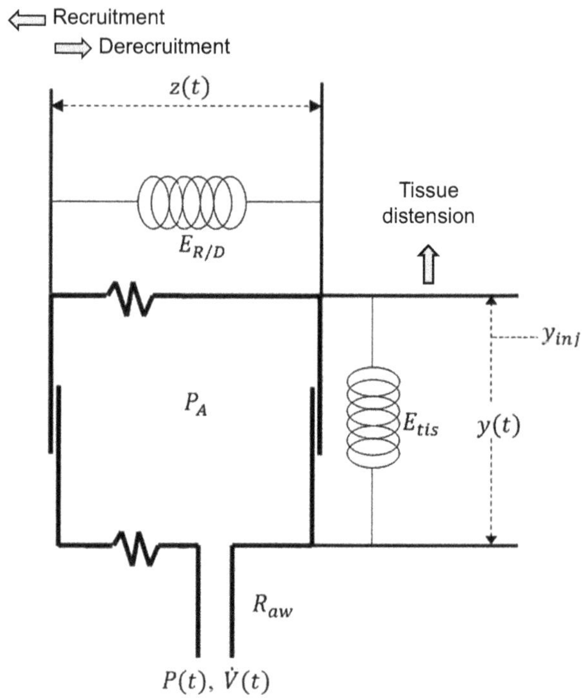

Fig. 10.3 A model of the lung consisting of a single alveolus served by a single airway. The alveolar compartment can expand in two orthogonal directions—vertically via the independent movement of the upper versus lower parts of the compartment, and horizontally via the bellows action of both parts. Tissue distension is represented by the variable $y(t)$ which generates elastic recoil in the spring E_{tis}. Volutrauma begins once $y(t)$ exceeds y_{inj}. Recruitment is represented by an increase in the variable $z(t)$, and is resisted by the spring $E_{R/D}$

$$y = A - Be^{-KP_A}. \tag{10.3}$$

Rearranging Eq. 10.3 shows that Eq. 10.2 becomes

$$P_A = \frac{1}{K}\ln\left(\frac{B}{A-y}\right). \tag{10.4}$$

As well as being able to expand vertically, the alveolar compartment in Fig. 10.3 is also able to expand horizontally, as quantified by the variable $z(t)$. Pure lung recruitment is represented by an increase in $z(t)$ when $y(t)$ remains constant. This results an increase in lung volume even though there is no change in the degree of distension of the units that are already open. (In fact, when recruitment of closed lung units occurs in avalanches, it is possible for the lung to have a negative elastance over small regions of its pressure-volume relationship [14].) Likewise, derecruitment is represented by a decrease in $z(t)$. This model assumes that the amount of recruited (open) lung is purely a function of pressure, and that this amount is controlled by a single spring with spring constant $E_{R/D}$.

Most recruitment and derecruitment takes place over an intermediate pressure range [15]. This can be accounted for by making the spring $E_{R/D}$ very stiff at both high and low pressures, and much less stiff over the intermediate pressure range. In other words, z has a sigmoidal dependence on P_A. There are numerous ways of expressing such a dependence, a convenient one being the logistic function [16] (e.g., Eq. 4.1)

$$z = \alpha + \frac{(\gamma - \alpha)}{1 + e^{\frac{P_{1/2} - P_A}{\tau}}} \tag{10.5}$$

where α is the fraction of permanently recruited (i.e., non-collapsible) lung, $(1 - \gamma)$ is the fraction of lung that is unrecruitable, $(\gamma - \alpha)$ is the fraction of potentially recruitable lung, and $0 < \alpha < \gamma < 1$. $P_{1/2}$ is the value of P_A at which half the recruitable lung is open.

Alveolar volume, V, is determined by both y and z thus:

$$V(t) = y(t)z(t) \tag{10.6}$$

so flow into the compartment is

$$\dot{V}(t) = y(t)\dot{z}(t) + \dot{y}(t)z(t) \tag{10.7}$$

where $y(t)$ and $z(t)$ are given by Eqs. 10.3 and 10.5, respectively, and their time derivatives are

$$\dot{y}(t) = BKe^{-KP_A}\dot{P}_A \tag{10.8}$$

and

$$\dot{z}(t) = \frac{\dot{P}_A(t)}{\tau}\left[\frac{(\gamma - \alpha)e^{\frac{P_{1/2} - P_A(t)}{\tau}}}{\left(1 + e^{\frac{P_{1/2} - P_A(t)}{\tau}}\right)^2}\right]. \tag{10.9}$$

P_A is related to airway pressure (P) by

$$P(t) - P_A(t) = R_{aw}\dot{V}(t) \tag{10.10}$$

where R_{aw} is the resistance of the single airway serving the alveolar compartment (Fig. 10.3). Equations 10.6, 10.7, 10.8, 10.9 and 10.10 comprise the set of coupled ordinary differential equations that define the behavior of this model. These equations can be integrated numerically to simulate model behavior when it is driven by a prescribed $\dot{V}(t)$ or $P(t)$ signal [16].

This model can also be used to simulate the development of VILI. Suppose, for example, that volutrauma occurs at a rate proportional to the amount by which the strain, y, in spring E_{tis} exceeds some injury threshold y_{inj}. Define a new variable $y'(t)$ as being equal to $y(t)$ whenever $y(t) > y_{inj}$ and equal to y_{inj} otherwise. The total amount of tissue damage due to volutrauma caused by a duration T of mechanical ventilation is then

$$\Omega_{voltrau} = K_{voltrau} \int_0^T \left[y'(t) - y_{inj} \right] dt \tag{10.11}$$

where $K_{voltrau}$ is a constant of proportionality. Further suppose that atelectrauma accumulates at a rate proportional to the rate at which the spring $E_{R/D}$ lengthens during recruitment (since injury is caused by recruitment and not derecruitment). Define a new variable $z'(t)$ as being equal to $z(t)$ whenever $dz(t)/dt > 0$ and zero otherwise. The total amount of tissue damage due to atelectrauma caused by a duration T of mechanical ventilation is then

$$\Omega_{atelectrau} = K_{atelectrau} \int_0^T \frac{dz'(t)}{dt} dt \tag{10.12}$$

where $K_{atelectrau}$ is another constant of proportionality.

The total amount of VILI generated in the model in Fig. 10.3 is some function of $\Omega_{voltrau}$ and $\Omega_{atelectrau}$. What this function should be is currently unknown, but the mathematically simplest possibilities are either a weighed sum of $\Omega_{voltrau}$ and $\Omega_{atelectrau}$ or alternatively their product. The first option assumes that volutrauma and atelectrauma are independent contributors to VILI, but this is contradicted by evidence that these two injury mechanisms must be present together to produce VILI, at least in an initially normal lung [9]. The second option, namely

$$\Omega_{VILI} = \Omega_{atelectrau} \Omega_{voltrau} \tag{10.13}$$

is consistent with this evidence because VILI is zero if either $\Omega_{atelectrau}$ or $\Omega_{voltrau}$ is zero. Indeed, Eq. 10.13 predicts the occurrence of VILI in a way that bears some resemblance to experimental data obtained in mice [12], when the model in Fig. 10.3 is extended to account for the well-known observation that lung units close during expiration at lower pressures than they open at during inspiration [15]. If this pressure difference is ΔP, and Eq. 10.5 describes the fraction of open lung during inspiration, then the fraction during expiration is given by

$$z = \alpha + \frac{(\gamma - \alpha)}{1 + e^{\frac{P_{1/2} - \Delta P - P_A}{\tau}}}. \tag{10.14}$$

The model in Fig. 10.3 is, of course, an extremely simplified representation of a complex phenomenon that takes place in a regionally heterogeneous fashion throughout the lung. Nevertheless, it suggests that it may be possible to model VILI development in a given lung in order to predict injury before it occurs. Such a model could be interrogated numerically for minimally injurious ventilation strategies that could then be implemented to reduce the changes of VILI that actually occur, thereby increasing the chances that a patient with ARDS will survive. This is an idea that has yet to be developed to the point of clinical utility, but there is incontrovertible proof that some ventilation strategies are safer than others [8, 17], so it seems certain that an optimal strategy for a given lung exists. The challenge is to find a way to identify it.

10.4 Dynamics of Recruitment and Derecruitment

Like all models, the one shown in Fig. 10.3 makes simplifying assumptions. This is acceptable so long as these assumptions do not impair the ability of the model to make useful predictions. The model in Fig. 10.3 fails in this regard in terms of its capacity to account for the dynamic nature of recruitment and derecruitment. These phenomena do not manifest instantaneously as soon as P_A changes. Rather, the closing and opening of lung units require some amount of time to occur, even when P_A is held constant . If these dynamic effects are not taken into account, it is impossible to accurately predict how much intra-tidal recruitment, and thus how much atelectrauma will be produced by a given regimen of mechanical ventilation.

The dynamic nature of recruitment and derecruitment is readily demonstrated experimentally by tracking the changes in lung elastance that occur following a so-called *recruitment maneuver*, which consists of a deep lung inflation that serves to open as much of the lung as possible. If, after the recruitment maneuver, the lung is ventilated with a normal tidal volume at a normal breathing frequency, the lung units that were recruited by the deep inflation will gradually derecruit over time. As this process continues, the amount of parenchymal tissue in communication with the airway opening (i.e., the fraction of open lung) gradually decreases, causing the overall stiffness of the lung to increase correspondingly. This is illustrated in Fig. 10.4 which shows respiratory system elastance (E_{rs}) versus time following recruitment maneuvers in mice during mechanical ventilation at different lung volumes controlled by adjusting PEEP [18]. Normal mice challenged with a non-injurious intratracheal instillation of saline show modest transient increases in E_{rs} over the 180 s immediately following deep lung inflation. Some of this increase may reflect a small amount of derecruitment, but transients of this nature are also expected as a result of stress recovery in the viscoelastic tissues of the lung and chest wall (see Chap. 5). In mice with injured lungs, however, the transient increases in E_{rs} are greatly accentuated to an extent that depends on the nature and degree of the injury as illustrated by the dramatic effects shown in Fig. 10.4 [18, 19]. Imaging with micro-computed tomography [18] and in vivo microscopy [20] have shown that these additional increases are indeed due to ongoing derecruitment.

Accounting for the time dependence of recruitment and derecruitment requires an extension to the model shown in Fig. 10.3. Figure 10.5 demonstrates a simple way in which this can be done by adding a dashpot with resistance $R_{R/D}$ in parallel with the spring $E_{R/D}$. Now, whenever P_A changes, the horizontal dimension of the alveolar

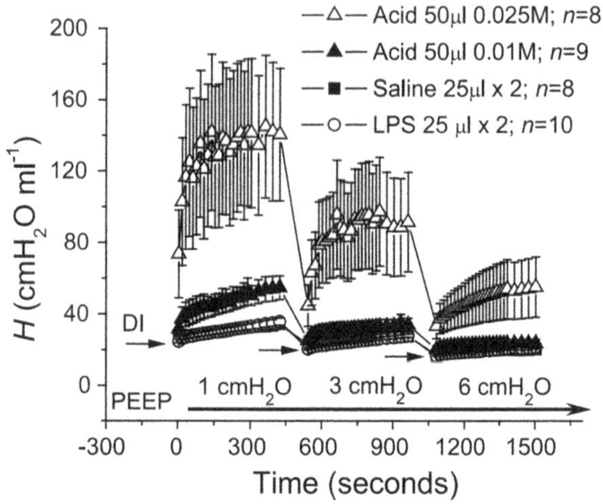

Fig. 10.4 Mean (± standard error) of respiratory system stiffness (H) in mice following recruitment maneuvers at levels of positive end-expiratory pressure (PEEP) of 1, 3 and 6 cmH_2O. The different symbols correspond to mice with normal lungs (saline) and with lungs injured by lipopolysaccharide (LPS) and hydrochloric acid at two different concentrations. (Reproduced with permission from Ref. [18])

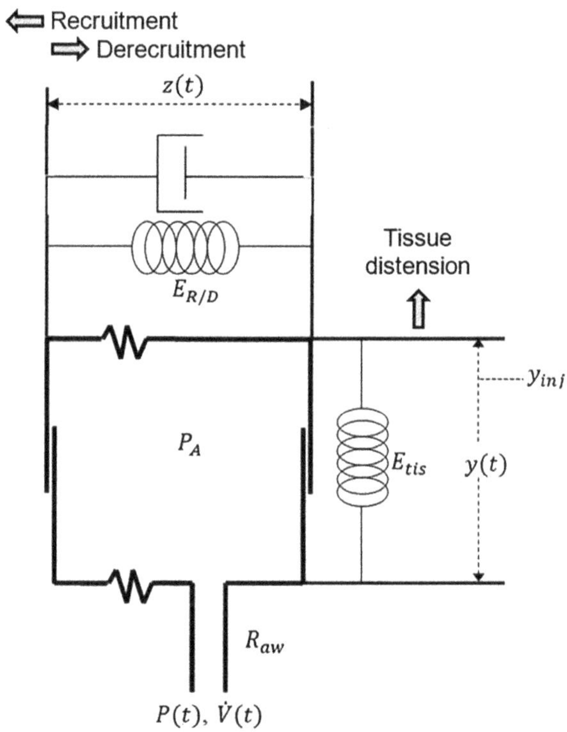

Fig. 10.5 An extension of the model in Fig. 10.3 with the addition of a dashpot in parallel with spring $E_{R/D}$. The resistance, $R_{R/D}$, of the dashpot prevents the spring from immediately assuming its steady-state length determined by P_A that corresponds to a given fraction of open lung. Instead, when P_A changes, the spring-and-dashpot pair cause recruitment or derecruitment to take place transiently at a rate determined by the ratio $R_{R/D}/E_{R/D}$

compartment cannot respond immediately because the dashpot takes a finite amount of time to change its length. Thus, for example, while a step change in P_A will result in an instantaneous change in y (assuming the tissue is purely elastic), it will cause z to change at a rate determined by the ratio $R_{R/D}/E_{R/D}$. This model has been used to explore the effects of expiratory duration on derecruitment during a form of mechanical ventilation known as airway pressure release ventilation (APRV). APRV is applied by driving inspiration and expiration with fixed pressures P_{Hi} and P_{Low}, respectively, where $P_{Hi} > P_{Low}$. The durations of inspiration and expiration are T_{Hi} and T_{Low}, respectively, and are chosen such that $T_{Hi} >> T_{Low}$. The protective effects of APRV appear to arise from the fact that if T_{Low} is short enough there is insufficient time for derecruitment to occur during expiration, which avoids injurious recruitment during the subsequent inspiration [16].

The model in Fig. 10.5 can be further extended in some obvious ways to make it more realistic. For example, the assumption of a purely elastic parenchyma encapsulated in Eq. 10.2 can be replaced by a nonlinear Kelvin body or some other appropriate structure to account for the viscoelastic behavior of the lung (Sect. 5.2). Also, the dashpot resistance $R_{R/D}$ can be assigned different values depending on whether the dashpot is lengthening or shortening to account for the observation that recruitment and derecruitment typically take place at different rates [21, 22]. These extensions do not, however, attend to the fact that the lung is not a single uniformly ventilated compartment, particularly in disease when regional heterogeneities in behavior can become pronounced.

The following model has been proposed as a means of representing the regionally distributed nature of recruitment and derecruitment dynamics throughout the lung [21]. At the micro-scale level, a lung unit is assumed to be either open or closed, and to transition between these two states depending on whether the pressure, P, applied to it is above a critical opening pressure P_o or below a critical closing pressure P_c, where $P_o > P_c$. The transition between the two states does not happen instantaneously, however, but rather occurs with a latency determined by a variable $x(t)$ associated with the unit. This variable is considered to follow a "virtual trajectory" between the limits 0 and 1, and to move toward either limit at a rate proportional to how much P is either below P_c or above P_o, while remaining unchanged when P lies between P_c and P_o. Furthermore, if a unit is open and its associated value of x reaches a value of 0, the unit immediately closely, while a closed unit immediately opens if its value of x reaches 1. These conditions are expressed mathematically as

$$\begin{aligned} \frac{dx(t)}{dt} &= s_o[P(t) - P_o]; P(t) > P_o, x < 1 \\ &= 0, P_c \le P \le P_0 \\ &= s_c[P(t) - P_c]; P(t) < P_c, x > 0 \end{aligned} \tag{10.15}$$

where s_o and s_c are constants of proportionality that control the rates of closing and opening, respectively. The virtual trajectory does not correspond to any physical process in particular but rather represents, in an empirical sense, all those processes that might be involved in the opening and closing of a lung unit. These processes include, for example, the movement of a liquid plug along an occluded airway to the point where it breaks apart (a recruitment event) or the accumulation of fluid at a point along the airway where it bulges into the lumen and eventually forms an occluding plug (a derecruitment event).

Each individual unit in an entire lung obeys Eq. 10.15 with its own values of P_c, P_o, s_c and s_o. To avoid an explosion in the number of model parameters, these 4 parameters are assumed to be drawn from appropriate probability distributions. There is experimental evidence to support the idea that P_c and P_o conform to Gaussian distributions [15], in which case it suffices merely to specify their means (μ_c and μ_0) and variances (σ_c^2 and σ_0^2). That is, the probability distributions, p, of P_c and P_o are given by

$$p(P_c) = \frac{1}{\sqrt{2\pi\sigma_c^2}} e^{\frac{-(P_c-\mu_c)^2}{2\sigma_c^2}}, \quad p(P_o) = \frac{1}{\sqrt{2\pi\sigma_o^2}} e^{\frac{-(P_o-\mu_0)^2}{2\sigma_0^2}}. \tag{10.16}$$

These distributions imply that recruitment and derecruitment events occur predominately over a limited range of airway pressures and become increasingly rare outside these ranges. This is compatible with frequent observations of a knee in the pressure-volume relationship of the lung during inflation from residual volume, which is thought to correspond to a relatively sudden increase in the fraction of open lung [21]. In fact, closed lung units may open in avalanches that are so sudden they can cause the pressure-volume relationship to curve back on itself, giving the brief appearance of a negative elastance [14].

How s_c and s_o should be distributed is less clear, but they must always be greater than zero and there is no limit in principle to how large they could become. This suggests a distribution with a long tail, so one possibility is that they conform to a power law thus:

$$p(s_c) = \frac{S_c}{\text{unif}[0,1]}, \quad p(s_o) = \frac{S_o}{\text{unif}[0,1]} \tag{10.17}$$

where unif[0, 1] is the uniform probability distribution on the interval [0,1], and S_c and S_c are constants. Both s_c and s_o computed in this manner are distributed according to s^{-2}.

The entire virtual trajectory model is now specified by only 6 parameters—μ_c, μ_0, σ_c^2, σ_0^2 and S_c and S_c—presenting the possibility of fitting the model to experimental data. Figure 10.6 shows fits of the model to the data shown in Fig. 10.4 under the constraints that $\sigma_c^2 = \sigma_o^2$ and $\mu_o = \mu_c + \Delta P$, where ΔP is a constant [22]. Interestingly, these fits indicate that the lung injury data shown in Fig. 10.4 can be explained by the statistical distributions for P_c and P_0 moving to progressively higher pressures while leaving the remaining parameters relatively unaffected. This is what would be expected to happen with progressive increases in surface tension at the air-liquid interface due to impairment of pulmonary surfactant function [23–25]. Impaired surfactant function also increases the intrinsic elastance of the lung tissues independent of the degree of derecruitment [26].

The above approach to describing the dynamics of recruitment and derecruitment has been incorporated into a number of multi-scale computational models of the lung [19, 27, 28]. Nevertheless, it is quite computationally expensive to solve this model

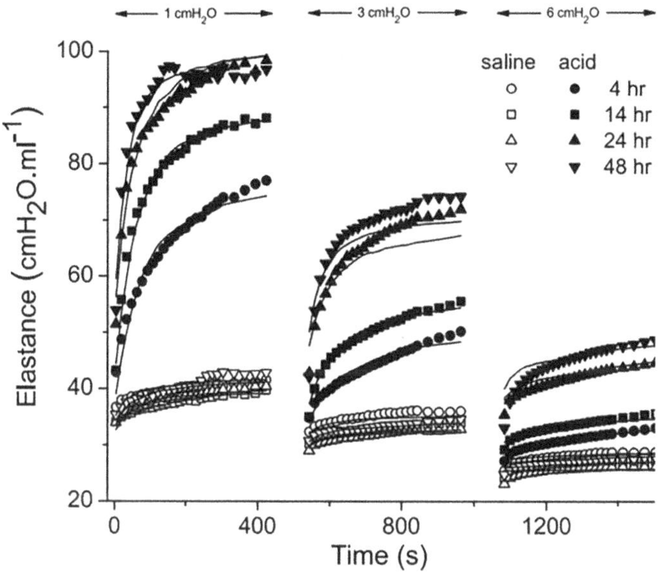

Fig. 10.6 Fits of the distributed virtual trajectory model to the elastance versus time data (lines) shown in Fig. 10.4. (Reproduced with permission from Ref. [22])

numerically for a large number of independent alveolar units that collectively represent the statistical distributions of the opening and closing pressures and velocities. However, an analytic form of the model can be constructed for the special case of a derecruited lung being suddenly subjected to a step increase in pressure from P_1 to P_2 [29]. Because of the time dependence of the recruitment events that follow the step change, the lung will recruit asymptotically to the steady-state level of open fraction consistent with P_1. From Eq. 10.15,

$$\frac{dx(t)}{dt} = s_o(P_2 - P_1) \tag{10.18}$$

for those units for which $P_2 > P_1$. The variables P_o and s_o are drawn from the probability distributions $p(P_o)$ (Eq. 10.16) and $p(s_o)$ (Eq. 10.17), respectively. If $F(P)$ is the steady-state fraction of open lung at pressure P then

$$F(P_2) = F(P_1) + \int_{P_0}^{P_1} p(P_o)dP_o. \tag{10.19}$$

The time taken to open a lung unit depends on the velocity of x along the virtual trajectory for that unit. Assuming that x must traverse the entire trajectory from 0 to 1, this gives the opening time as

$$t_o = \frac{1}{s_o(P_2 - P_1)}. \tag{10.20}$$

For an initially closed lung unit to become open at time t after the step up in pressure, the two conditions $P_2 > P_1$ and $s_o > 1/t(P_2 - P_1)$ must be met. If the values of s_o and P_0 for a given unit are independent, then

$$F(P_2, t) = F(P_1) + \int_{P_1}^{P_2} \left[\int_{1/t(P_2 - P_1)}^{\infty} p(s_o) ds_o \right] p(P_o) dP_o. \tag{10.21}$$

This equation cannot be solved analytically if $p(s_o)$ is a power-law function of s as in Eq. 10.17, but a solution is possible if the power law is replaced with a decaying exponential function of s. This yields time profiles for open lung fraction that fit to experimental data well [29].

10.5 Synergy Between Volutrauma and Atelectrauma

While volutrauma and atelectrauma are apparently caused by different biophysical mechanisms, it has long been known that they interact in a synergistic fashion to cause greater VILI than either mechanism alone [30]. This is exemplified by the data shown in Fig. 10.7, which shows how VILI develops in initially normal mice subjected to four different regimens of mechanical ventilation [9]. The readouts shown in Fig. 10.7 are 3-min derecruitability tests similar to those in Fig. 10.4, this time administered every 15 min for 4 h. The magnitude of the rise in lung elastance over 3 min following a recruitment maneuver is a sensitive index of lung health. The small rises in elastance that remain stable over the 4 h experiment in mice ventilated with a modest tidal volume at 3 cmH_2O PEEP (Fig. 10.7a) show that this non-injurious regimen did not cause VILI. Interestingly, employing a very large tidal volume (close to inspiratory capacity) in these animals also did not cause VILI (Fig. 10.7b) even though one might expect this to pose a risk for volutrauma. Similarly, ventilating with a modest tidal volume at zero PEEP to maximize the chances of cyclic end-tidal derecruitment did not lead to atelectrauma (Fig. 10.7c). However, imposing a high tidal volume with zero PEEP at the same time caused a progressive rise in lung derecruitability with time (Fig. 10.7d), demonstrating that over-distension and inspiratory recruitment are required simultaneously to cause VILI.

Several possible explanations have been proposed for this synergy between over-distension and recruitment. One is that regions of alveolar parenchyma at the interface with atelectatic regions of the lung experience excessive levels of strain compared to the rest of the lung tissue. Such stress concentrators [7] result in damage

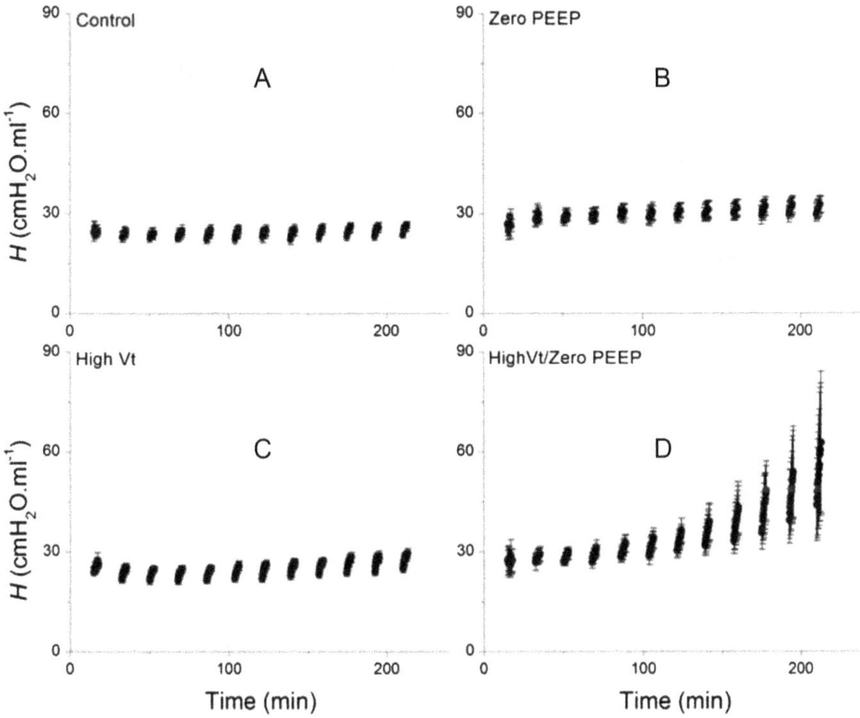

Fig. 10.7 Derecruitability tests performed every 15 min in mice mechanical ventilated for 4 h with (**a**) a modest tidal volume (V_T) at a PEEP of 3 cmH_2O, (**b**) a modest V_T at zero PEEP, (**c**) a V_T equal to total inspiratory capacity at a PEEP of 3 cmH_2O, and D) a V_T equal to total inspiratory capacity at zero PEEP

developing at these interface locations, which causes the atelectatic regions to grow (Fig. 10.8). CT imaging of injured lungs supports the existence of this mechanism [31], which is essentially a form of volutrauma at the micro level.

Another explanation for volutrauma-atelectrauma synergy is suggested by the way that perforations appear in the blood-gas barrier in the lungs of mice with VILI. The appearance of intravenously injected dextran molecules of different sizes in the lung airspaces indicates that holes form in the blood-gas barrier during injurious mechanical ventilation, and that these holes have a power-law distribution of sizes [32]. In other words, there are many small holes and few very large ones. This is yet another example of the ubiquitousness of power laws in nature [33], at least some of which can be explained by the rich-get-richer mechanism [34]. This mechanism can be applied to the generation of VILI as follows [35]. Suppose that the injurious micro-stresses generated by recruitment cause small perforations to appear in the plasma membrane of airway epithelial cells. There is evidence to suggest that such perforations, if not overly extensive, will normally reanneal quite quickly without lasting consequences [36]. If, however, recruitment occurs on a background of tissue

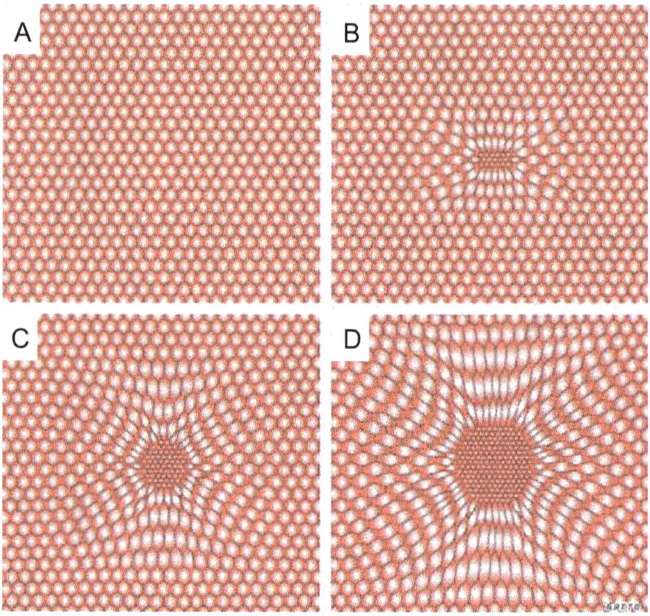

Fig. 10.8 Artist rendering of the growth of consolidated lung due to damage of high-stress alveoli at the border between atelectasis and normal lung. (Reproduced with permission from Ref. [7])

over-distension, the perforations may be prevented from reannealing and will thus grow. Furthermore, if bigger holes are bigger because they are more readily expanded by tissue stretch compared to smaller holes, then the rich (i.e., big) holes will become richer.

This process can be modeled as follows [35]. Let the number of injurious events acting on the blood-gas barrier per unit time be λ. Suppose that a fraction $0 < \beta < 1$ of these events cause existing holes in the barrier to each increase in area by a fixed amount A_0. The remaining fraction $1 - \beta$ of events cause new holes to be created, each with initial area A_0. The total hole area at time t is therefore $A_0\lambda t$, and the fraction of this total area due to expansion of existing holes is $A_0\beta\lambda$t (assuming hole creation begins at $t = 0$). The rich-get-richer mechanism in this model is invoked by assuming that larger holes are more likely to expand than smaller ones. The probability that a given hole of area A_i becomes expanded by the next expansion event is thus $A_i/(A_0\lambda t)$. The expected rate of change of area for this particular hole is then

$$\frac{dA_i(t)}{dt} = A_0\beta\lambda\frac{A_i(t)}{A_0\lambda t} = \frac{\beta A_i(t)}{t}. \tag{10.22}$$

Each hole was created at some point in time $0 < t_i < t$, at which point it had an initial area of $A_i(t_i) = A_0$. Using this as an initial condition, Eq. 12.22 can be solved by rearranging and integrating as follows:

$$\int_{A_0}^{A_i} \frac{1}{\beta A(t)} dA(t) = \int_{t_i}^{t} \frac{1}{t} dt. \tag{10.23}$$

Evaluating the integrals on both sides and solving for t_i gives

$$t_i = \left(\frac{A_0}{A_i(t)}\right)^{\frac{1}{\beta}} t. \tag{10.24}$$

Since holes are created at the constant rate of $\lambda(1 - \beta)$, the probability that a randomly selected hole was generated prior to time t_i is t_i/t. If the hole has an area less than some value A, its expected time of creation occurred, by Eq. 10.24, at some point after $\left(\frac{A_0}{A}\right)^{\frac{1}{\beta}} t$, the probability of which is $1 - \left(\frac{A_0}{A}\right)^{\frac{1}{\beta}}$. In other words, the cumulative distribution is

$$F[A_i(t) < A] = \alpha(t)\left[1 - \left(\frac{A_0}{A}\right)^{\frac{1}{\beta}}\right] \tag{10.25}$$

where $\alpha(t)$ is a time-dependent normalization factor.

When the total number of holes is large, we can consider A to be a continuous variable taking values $\geq A_0$. The probability density function for A over this range can then be calculated as the derivative with respect to A of the cumulative probability function given by Eq. 10.25, which is

$$p(A, t) = \frac{\alpha(t)}{\beta} A_0^{\frac{1}{\beta}} A^{-1-\frac{1}{\beta}} = \alpha_1(t) A^{-1-\frac{1}{\beta}}. \tag{10.26}$$

The factor $\alpha_1(t) = \frac{\alpha(t)}{\beta} A_0^{\frac{1}{\beta}}$ ensures the area under the probability distribution is unity. That is,

$$\int_{A_0}^{A_{max}} \alpha_1(t) A^{-k} dA = 1 \tag{10.27}$$

where $k = 1 + \frac{1}{\beta}$, and A_{max} is the largest possible hole that can exist at time t. Being the largest hole, this also means it was created at the earliest time point, t_0. Substituting t_0 into Eq. 10.24 and rearranging then gives

$$A_{max} = A_0 \frac{t^\beta}{t_0^\beta}. \tag{10.28}$$

If t_0 is assumed to be a very small relative to t, the integral in Eq. 10.27 reduces to

$$\alpha_1(t)\frac{A_0^{-k+1}}{-k+1}\left(t_0^{\beta(k-1)}t^{\beta(1-k)}-1\right)=1. \tag{10.29}$$

Since $k=1+\frac{1}{\beta}$, its substitution into Eq. 10.29 gives

$$\alpha_1(t)\beta A_0^{-1/\beta}\frac{t-t_0}{t}=1. \tag{10.30}$$

Equation 10.26 can then be rewritten as

$$p(A,t)=\frac{A_0^{1/\beta}}{\beta}\frac{t}{t-t_0}A^{-k} \tag{10.31}$$

which converges in the limit of large t to

$$p(A)=\frac{A_0^{\frac{1}{\beta}}}{\beta}A^{-k}. \tag{10.32}$$

When $\beta=0.5$, Eq. 10.32 leads to $k=3$ [35], which is approximately the power-law exponent found experimentally [32].

The above simple model of progressive creation and size-dependent expansion of holes in the blood-gas barrier illustrates how a rich-get-richer mechanism can potentially explain the synergy between volutrauma and atelectrauma in the generation of VILI. The key signature of this mechanism is power-law behavior as exemplified in this case by a power law distribution of hole areas. Interestingly, power laws also characterize the size distributions of areas of tissue damage in the injured lung, which may thus also be explicable on the basis of a rich-get-richer type mechanism [37].

10.6 Mechanical Failure of the Blood-Gas Barrier

The most consequential effect of both volutrauma and atelectrauma, as well as their synergistic combination, is physical failure of the blood-gas barrier, or exacerbation of existing failure if the lung is already injured. These events have a major deleterious effect on lung mechanics and operate in a positive feedback loop, taking the lung toward eventual catastrophic failure. The models considered above make a purely empirical link between the injurious impact of mechanical ventilation and the development of perforations in the blood-gas barrier. A potential way to make this link mechanistic for volutrauma is illustrated in Fig. 10.9, which represents the alveolar epithelium as a 2D patchwork of cells stitched together by focal adhesions along the cell borders [38]. Simulations with this model support the existence of a stretch threshold. When this threshold is exceeded, damage is initiated,

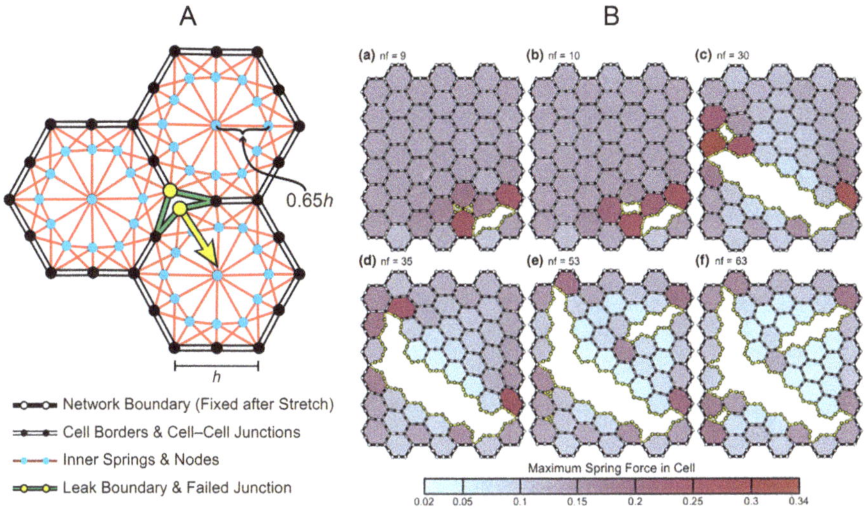

Fig. 10.9 (**a**) The junctions on the cell borders can separate to form leaks between the cells if the resultant force from the attached springs in one cell exceeds the threshold value for that node. (**b**) Leak progression in a 45-cell network as the network is stretched biaxially. The panels (a–f) show how the leak forms as stretch force increases. Only one node fails at a time; *nf* indicates how many nodes have failed in each panel. The shading represents the value of the maximum spring force within each cell. Spring stiffness $k = 1$, force threshold $Ft = 0.2$, and cell edge length $h = 1$. (Reproduced with permission from Ref. [38])

corresponding to an injurious tidal volume above which volutrauma begins to occur. The model also predicts that although the area of the leak in the epithelium increases with increasing strain, it does so at a rate somewhat lower than the rate of increase in strain itself due to the release in strain caused by the appearance of tears between the cells [38].

The biophysics of atelectrauma has been investigated experimentally at the micro level by repeatedly passing air bubbles over cell monolayers cultured under a layer of fluid. Measuring the electrical impedance across the cell monolayer provides a means of assessing barrier integrity continuously so that its degradation can be tracked. It has been shown that once air bubble passage is initiated, impedance starts to decrease only after an initial period of stability has elapsed [39]. This can potentially be explained by a multi-hit hypothesis in which multiple degrading events are required before functional deficit begins to manifest. More specifically, the damaging effects of bubble passage must be repeated numerous times before the damage that occurs with each passage aggregates to the point that it begins to manifest as an impairment in monolayer barrier function.

To model this process, suppose that the integrity of a layer of N cells is normally maintained by n focal adhesions between each cell and its neighbors (similar to Fig. 10.9), where the attachments each bear equal stress and contribute equally to the integrity of the monolayer. Further assume that a certain minimum number of adhesions around any given cell must be broken before that cell starts to allow transepithelial leak to occur around its border [40]. Let each passage of a bubble over the monolayer have a small but finite chance of breaking each inter-cellular adhesion. When all attachments are intact, $n = n_{max}$ for each cell, and the resistance, R, of the entire monolayer is at its maximum value R_{max}. When all the attachments to every cell have been disrupted, $n = 0$ for each cell, and $R = R_{min}$. The N cells in the monolayer act in parallel to restrict the passage of electric current between the apical and basal surfaces, so each cell in the monolayer has a maximum resistance of $r = R_{max}N$ (achieved when $n = n_{max}$). The minimum cellular resistance is $r = R_{min}N$ when $n = 0$.

At each time-step in the model, a single bubble passes across the monolayer, which creates a biomechanical stress on each cell that has probability, p_{break}, of breaking one of its attachments. As the number of broken attachments for a given cell, $n_{broken} = (n_{max} - n)$, increases, p_{break} for that cell increases according to

$$p_{break} = \rho\left[\frac{1}{n_{max}} + \frac{n_{broken}}{n_{max}}\right] \tag{10.33}$$

where $0 \leq n_{broken} < n_{max}$ and $0 < \rho < 1$ is a constant. This means that cells with fully intact attachments ($n_{broken} = 0$) have only a very small probability of $\frac{\rho}{n_{max}}$ of experiencing an attachment break due to bubble passage. p_{break} increases as attachments continue to break because this causes the remaining attachments to bear all the stress of subsequent bubble passages. This also makes the model conform to the rich-get-richer mechanism described above because cells in the monolayer that have a large number of broken attachments are more likely to experience more breaks than are cells with fewer broken attachments.

The multi-hit mechanism requires that a critical fraction $0 < \gamma < 1$ of the n_{max} adhesions to any given cell must be broken before that cell starts to fail in its barrier function, as reflected in a drop in r below $R_{max}N$. A simple way to represent the progressive degradation in barrier function once failure begins (i.e., $n_{broken} > \gamma n_{max}$) is to assume a linear relationship between additional breaks and decreases in r according to

$$\begin{aligned} r &= R_{max}N;\, n_{broken} \leq \gamma n_{max} \\ &= R_{min}N + (R_{max} - R_{min})N\left(\frac{n}{n_{max} - \gamma n_{max}}\right);\, n_{broken} > \gamma n_{max} \\ &= R_{min}N;\, n = 0. \end{aligned} \tag{10.34}$$

Figure 10.10 shows experimental measurements of R versus number of bubble passages together with the best fit predictions of the multi-hit model [40]. The model recapitulates the key features of the data, specifically an initial plateau followed by a sigmoidal descent in predicted R.

An injured airway epithelium is also capable of spontaneous repair if injurious mechanical ventilation is ceased, resulting in the reversal of the degraded barrier function caused by VILI. This can be readily incorporated into the framework of the

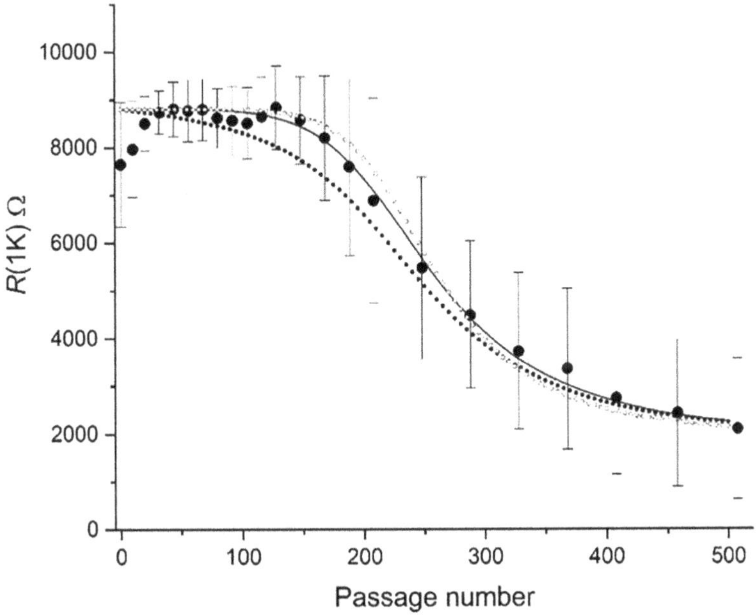

Fig. 10.10 Time evolution of the decay in monolayer resistance R (mean ± SD) measured at 1 kHz (large symbols with SDs) as a function of the number of bubble passages over the monolayer. Predictions by the stochastic model of barrier repair are shown by the solid line. The closed circles show the prediction when γ is forced to be zero. The open circles show the prediction with alternate values for n_{max} and ρ. (Reproduced with permission from Ref. [40])

multi-hit model by assigning first-order kinetics to the repair process. That is, if there is a fixed probability that a broken attachment will become repaired per unit time, then the probability p_{attach} that the number of attachments to a cell will increase by 1 per unit time is proportional to n_{break} according to.

$$p_{attach} = \gamma n_{break} \tag{10.35}$$

where γ is a constant. This gives rise to an exponential rise in R, asymptoting toward R_{max} [40].

10.7 VILI and Lung Mechanics

Injurious mechanical ventilation exacerbates the leak across the blood-gas barrier in the injured lung, accelerating the accumulation of plasma-derived material in the airspaces where it displaces the air and interferes with the function of pulmonary surfactant. These events have a major deleterious effect on lung mechanics and operate in a positive feedback loop, taking the lung toward eventual catastrophic

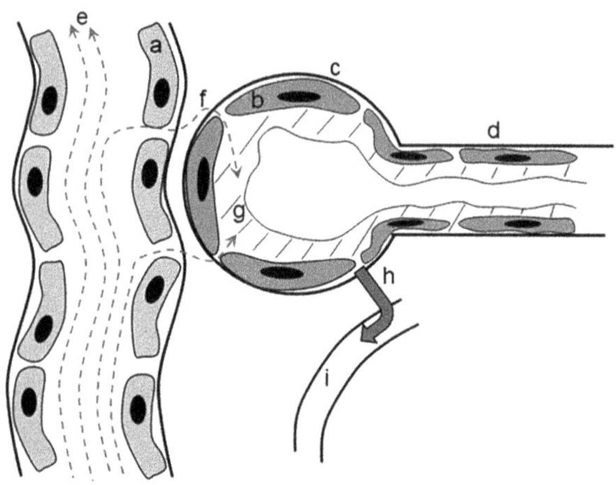

Fig. 10.11 Schematic of the events taking place during VILI development and resolution that are represented in the model. Endothelial cells (**a**) line a pulmonary capillary in close proximity to the epithelial cells (**b**) that line an alveolus (**c**) served by a peripheral bronchiole (**d**). Injurious mechanical ventilation causes impairment of the endothelial-epithelial barrier so that plasma fluid and proteins flowing through the capillary (**e**) leak through gaps in the blood-gas barrier (**f**). This material accumulates in the alveolus as V_{excess} (hatched area **g**) where it causes an elevation in E_{rs}. At the same time, clearance (**h**) of V_{excess} from the alveolus into the lymphatic system (**i**) leads to eventual resolution of VILI. (Reproduced with permission from Ref. [40])

failure. Nevertheless, not all patients who develop ARDS go on to die. The majority (roughly 70%), in fact, survive due to the natural reparative mechanisms that clear exudates in the lung and repair the blood-gas barrier as outlined in Fig. 10.11.

VILI is accentuated by tissue overdistension in a dose-dependent manner in the presence of repetitive recruitment. This effect can be incorporate into the multi-hit model of atelectrauma development above by having p_{break} increase with V_T through an increase in the parameter ρ in Eq. 10.33 [40]. To account for the fact that VILI does not occur when V_T is low, assume that $\rho = 0$ when V_T is less than a critical value, V_{Tcrit}, and that p_{break} increases asymptotically toward 1 as V_T rises above V_{Tcrit}. These assumptions are incorporated into the model by having ρ in Eq. 10.33 conform to a nonlinear expression constrained to lie between 0 and 1 thus:

$$\begin{aligned} &\rho = 0 \;\; if \;\; V_T < V_{Tcrit}, \\ &\rho = \frac{(V_T - V_{Tcrit})}{(V_T - V_{Tcrit}) + D} \;\; if \;\; V_T \geq V_{Tcrit} \end{aligned} \tag{10.36}$$

where D is a constant that defines the value of V_T at which $\rho = 0.5$.

These barrier failure events are translated into an effect on the whole lung via their effects on the rate at which plasma fluid and proteins, labeled as (e) in Fig. 10.11, flow into and accumulate within the airspaces. Assuming that the flow of liquid into the alveolar space, $\dot{V}_{in}$, of this material is inversely proportional to the blood-gas barrier resistance R, and that the drainage of fluid from the airspaces is inversely proportional to R_{max}, the net flow of material into the alveolar space is

$$\dot{V}_{net} = \dot{V}_{in} - \dot{V}_{out} = B\left(\frac{1}{R} - \frac{1}{R_{max}}\right) \tag{10.37}$$

where B is a constant. VILI leads to $R < R_{max}$, resulting in the accumulation of an excess volume, V_{excess}, of plasma-derived material in the airspaces.

The clearance of V_{excess} from the airspaces, labeled as (h) in Fig. 10.11, is assumed to occur at a rate, $\dot{V}_{clear}$, given by

$$\dot{V}_{clear} = k_{clear} V_{excess} \tag{10.38}$$

where k_{clear} is a clearance rate constant. (In reality, $\dot{V}_{clear}$ is a complex function of alveolar interstitial geometry and fluid pressure distributions [41]). This means that V_{excess} accrues over time according to

$$V_{excess} = \int_0^t \left[\dot{V}_{net}(t) - \dot{V}_{clear}(t)\right] dt. \tag{10.39}$$

Plasma-derived material in the airspaces impairs pulmonary function both as a result of fluid displacing air and proteins inactivating pulmonary surfactant. This causes a rise in respiratory system elastance, E_{RS}, which is a key physiologic marker of VILI. Assume the relationship to be

$$E_{RS} = E_{RS,baseline} + qV_{excess} \tag{10.40}$$

where q is a constant and $E_{RS,\ baseline}$ is the elastance of the uninjured respiratory system. Figure 10.12 shows predictions of E_{RS} made by this model compared to measurements made in injuriously ventilated mice as they developed VILI over a 4-hour period. Despite the simplicity of the model, it recapitulates the initial (~1 hour) period of no change in lung function followed by a tidal volume dependent acceleration in E_{RS} [19].

10.8 Interactions Between Surfactant Release and Mechanical Ventilation

A novel form of mechanical ventilation for the injured lung that has received increasing attention in recent years is *variable ventilation* in which tidal volume is varied randomly from breath to breath [42, 43]. One of the rationales for variable ventilation is that mimicking the natural tidal volume variability of spontaneous

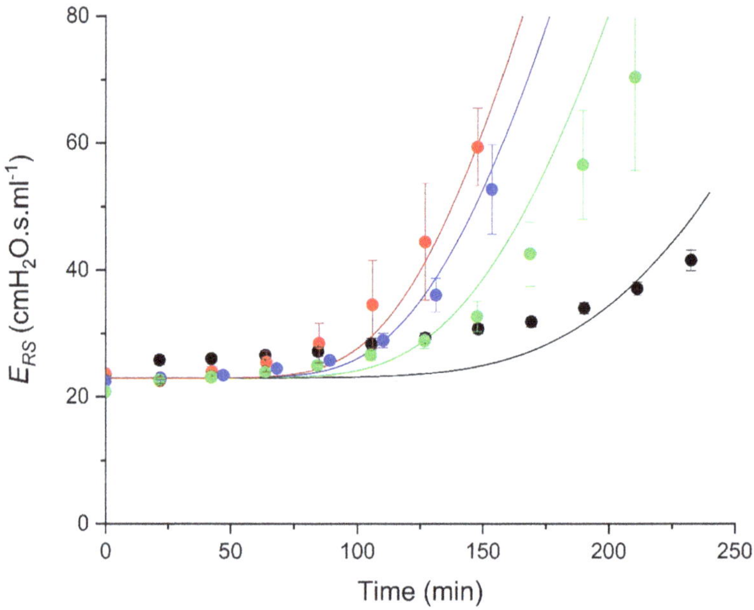

Fig. 10.12 Respiratory system elastance versus time in four groups of mice ventilated for 4 h at a PEEP of zero with tidal volumes of 1.0 (black), 1.1 (green), 1.2 (blue) and 1.3 (red) ml. The predictions of E_{RS} by the computational model are shown as correspondingly colored solid lines (see text for parameter details). (Reproduced with permission from Ref. [40])

breathing ought to be beneficial during mechanical ventilation because the lung has evolved to operate in this way. Significant experimental evidence bears this out. For example, variable ventilation with a given mean tidal volume has been shown to improve gas exchange compared to conventional ventilation with the same fixed tidal volume. Variable ventilation has also been shown to be superior to conventional mechanical ventilation in pre-clinical models of acute lung injury in terms of inducing endogenous surfactant secretion [44, 45], improving alveolar recruitment [46, 47], lowering cytokine levels [48], and reducing histologic evidence of VILI [49]. A small clinical trial also showed that variable ventilation improved respiratory compliance in human subjects with mild acute lung injury [50]. Several mechanisms have been proposed to explain these various benefits, including variations along the nonlinear pressure-volume curve of the injured lung resulting in reduced mean tissue stress [51], and time-dependent closure and opening of lung units that helps avoid focused atelectrauma [47].

There is also convincing evidence that variable ventilation improves surfactant release from Type-II alveolar cells [44, 45], resulting in improvements in tissue compliance with commensurate reductions in damaging tissue stress. For example, single large stretches have been shown to be a potent stimulus for surfactant release [52], while continual large-amplitude cyclic stretching of epithelial cells in culture

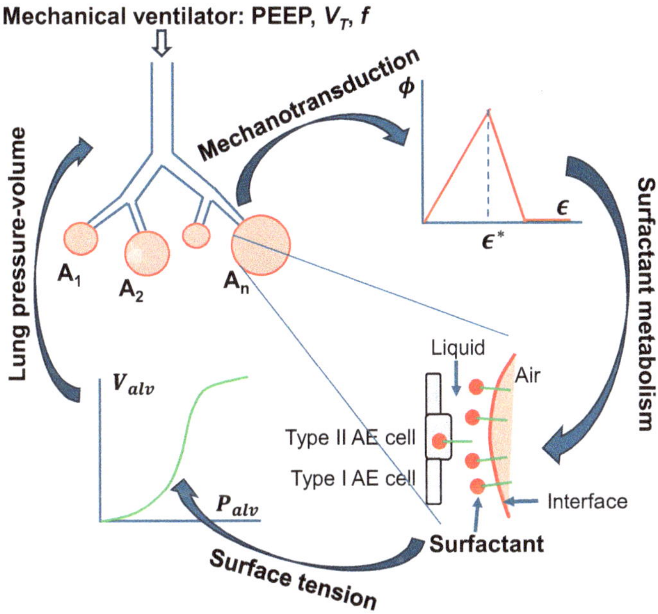

Fig. 10.13 A feedback model of surfactant release during mechanical ventilation. During quasi-static inflation, different units of alveolar parenchyma receive regional tidal volumes that determine alveolar surface areas (A_1, A_2, …) and local tissue strains (ϵ) between end-expiration and end-inspiration. The local rate of surfactant secretion (ϕ) by Type II alveolar epithelial (AE) cells is determined via mechanotransduction from the degree of stretch of each alveolar unit increasing up to a critical strain ϵ^* beyond which strain inhibits surfactant production. Surfactant is also removed from the interface at a rate determined by various processes such as degradation and uptake. The sigmoidal pressure-volume ($P_{alv} - V_{alv}$) curve of each alveolar unit is a function of its interfacial surfactant concentration, which is modulated by the strain excursions of previous recent breaths. Stretching units beyond ϵ^* results in a negative feedback loop that keeps surface tension within certain limits in the normal lung. Assuming negligible contributions from airway phenomena, P_{alv} is nearly the same throughout the lung and the sum of the individual $P_{alv} - V_{alv}$ curves gives the lung $P - V$ curve, which in turn determines P_{alv} for the next cycle for a given PEEP and V_T

down-regulates surfactant release [44]. Computational modeling is a useful tool for studying the interactions between PEEP and tidal volume during mechanical ventilation, surfactant release and lung physiology in a systematic way.

The coupling between mechanical ventilation and lung pathophysiology can be represented by a bifurcating airway tree terminating in a parallel arrangement of acinar units (Fig. 10.13), where the units have independent surfactant production and secretion dynamics and alveolar pressure-volume relationships. Within an acinar unit, surfactant secreted by Type-II alveolar epithelial cells accumulates onto the air-liquid interface to reduce surface tension and increase local alveolar compliance. Superimposed on a given PEEP, the model is driven by a prescribed tidal volume that it apportions between the different lung units in proportion to their relative compliances [53]. The epithelial cells within each unit secrete surfactant at a rate proportional to the magnitude of the strain (ϵ) between end-expiration and end-inspiration that the unit experiences. Highly compliant units thus stretch more, and thus receive more surfactant, compared to poorly compliant units. This should result in a runaway phenomenon overinflating the softer units and collapsing the stiffer ones. However, the peak in surfactant production ϕ versus ϵ stabilizes the model because the relationship between these variables becomes inverted beyond ϵ^* (the critical strain beyond which strain inhibits surfactant production). This means that when $\epsilon > \epsilon^*$ for a unit that starts off with a relatively favorable compliance, the unit receives a larger regional V_T and ϵ so that ϕ is reduced and the unit becomes stiffer. This negative feedback mechanism stabilizes units and reduces the heterogeneity of regional ventilation.

The situation is different when injury is introduced because certain units can be trapped in the region $\epsilon < \epsilon^*$, leading to permanent collapse. To investigate this possibility, the model was used to investigate the relationship between tidal volume, PEEP and lung injury defined by collapse representing atelectrauma or overdistension representing volutrauma [48, 53] (Fig. 10.14) during variable versus conventional ventilation. The top left panel of Fig. 10.14 shows simulations comparing the results of variable and conventional ventilation in terms of heterogeneous surfactant release and heterogeneous acinar stiffness. For a disruption of normal homogeneous surfactant release under conventional ventilation (top left panel), alveolar collapse occurs at low-PEEP/low-tidal volume settings, while over-

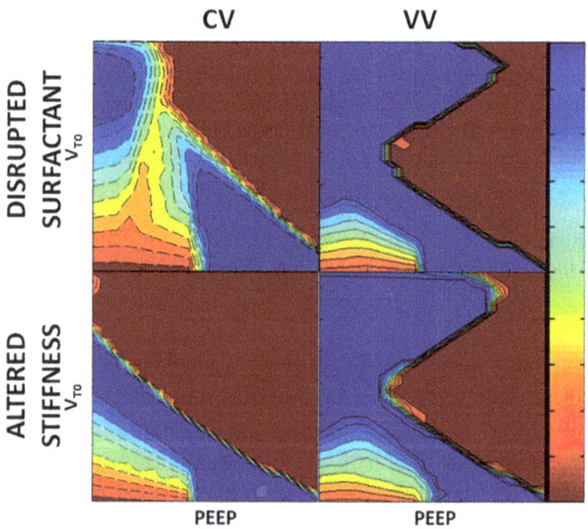

Fig. 10.14 Comparison of best PEEP-V_T settings for conventional ventilation (CV) versus variable ventilation (VV) for two different pathologies. Contour plots describe the fraction of the lung that is neither collapsed nor over-distended when ventilating with a given PEEP and V_T. For example, the top left panel shows the PEEP and tidal volume settings that result in the best outcomes (number of units recruited but not overdistended) in blue, and injury due to collapse or overdistention in red. Top left: Heterogeneous surfactant release under CV. Top right: Heterogeneous surfactant release under VV. Bottom left: Heterogeneous alveolar stiffness under CV. Bottom right: Heterogeneous alveolar stiffness under VV. (Reproduced with permission from Ref. [48, 53])

distention occurs for high-PEEP/high-tidal volume settings, resulting in two relatively small disjoint regions (blue) whereby alveolar injury is prevented. When variable ventilation is applied (top right panel), one contiguous region of PEEP and tidal volume allows for full recruitment without over-distention. Overall, the range of possible non-injurious combinations of PEEP and tidal volume during variable ventilation is much greater than during conventional ventilation. For the heterogeneous acinar stiffness case, the size of the region over which no injury occurs is significantly larger during variable ventilation. Despite these interesting results, however, the model has several limitations including its quasi-static nature and the absence of branching airways, gravity, and tissue viscoelastic behavior. Nevertheless, this simple model demonstrates the importance of patient-ventilator coupling through surfactant mechanotransduction, alveolar pressure-volume behavior and the stretch pattern that a given ventilation mode delivers.

10.9 Conclusions

VILI is a common and unfortunate consequence of providing mechanical ventilation to a patient who is unable to breathe unassisted. VILI is a particular risk in ARDS patients whose lungs are already injured. Ventilating the lungs in a manner that minimizes the risk of incurring VILI is thus a major medical need, but determining how to do this optimally is a complex and very patient-specific problem. The mathematical/computational models presented in this chapter show how the micro-level biophysical processes that govern the generation of volutrauma and atelectrauma can be related to their macro-level manifestations as measurable derangements in epithelial barrier function and lung mechanics. These models thus provide a means for defining how mechanical ventilation leads to VILI in quantitative terms. Such models will eventually provide a platform for the search for personalized ventilation strategies that are minimally injurious.

10.10 Problems

1. Simulate the model shown in Fig. 10.5. Use Eq. 10.12 to calculate the amount of atelectrauma ($\Omega_{atelectrau}$) resulting from 1 hour of sinusoidal breathing at a rate of 12 breaths/minute with a tidal volume of 0.5 L. Recalculate $\Omega_{atelectrau}$ when a recruitment maneuver (deep inspiration of 2 L) is applied every 10 min. What do your findings suggest about the benefits of recruitment maneuvers?
2. Solve Eq. 10.21 for $p(s_o) = \frac{1}{\tau}e^{-\frac{s_o}{\tau}}$ and for $p(s_o) = \frac{1}{a}; 0 < s_o < a$, where τ and a are constants. Do the two $F(P_2, t)$ profiles obtained look equally plausible to represent the time-course of lung recruitment at fixed pressure, or does one seem more likely than the other?

3. Simulate the monolayer barrier model given by Eqs. 10.33 and 10.34. Repeat the simulation after replacing Eq. 10.34 with an equation that assumes r decreases by a fixed amount (i.e., independently of n) with each bubble passage once $n_{broken} \leq \gamma n_{max}$. Repeat the simulation assuming r decreases according to the square of n. What do your results suggest about the influence of the rich-get-richer mechanism on the progress of VILI?
4. Simulate the complete model of the effects of VILI on lung mechanics (Eqs. 10.33, 10.34, 10.35, 10.36, 10.37, 10.38, 10.39 and 10.40). Explore the effects of (a) imposing a maximum rate of fluid clearance from the airspaces (i.e., $\dot{V}_{clear}$ reaches a plateau), and (b) giving E_{RS} a quadratic dependence on V_{excess} in Eq. 10.40.

References

1. Matthay MA. Resolution of pulmonary edema. Thirty years of progress. Am J Respir Crit Care Med. 2014;189:1301–8.
2. Ware LB, Matthay MA. The acute respiratory distress syndrome. N Engl J Med. 2000;342: 1334–49.
3. Fernando SM, Ferreyro BL, Urner M, Munshi L, Fan E. Diagnosis and management of acute respiratory distress syndrome. Can Med Assoc J. 2021;25:E761–8.
4. Bos LDJ, Ware LB. Acute respiratory distress syndrome: causes, pathophysiology, and phenotypes. Lancet. 2022;400:1145–56.
5. Slutsky AS. Lung injury caused by mechanical ventilation. Chest. 1999;116:9S–15S.
6. Hamlington KL, Smith BJ, Dunn CM, Charlebois CM, Roy GS, Bates JHT. Linking lung function to structural damage of alveolar epithelium in ventilator-induced lung injury. Respir Physiol Neurobiol. 2018;255:22–9.
7. Gaver DP 3rd, Nieman GF, Gatto LA, Cereda M, Habashi NM, Bates JHT. The POOR get POORer: a hypothesis for the pathogenesis of ventilator-induced lung injury. Am J Respir Crit Care Med. 2020;202:1081–7.
8. Acute Respiratory Distress Syndrome Network, Brower RG, Matthay MA, Morris A, Schoenfeld D, Thompson BT, Wheeler A. Ventilation with lower tidal volumes as compared with traditional tidal volumes for acute lung injury and the acute respiratory distress syndrome. N Engl J Med. 2000;342:1301–8.
9. Seah AS, Grant KA, Aliyeva M, Allen GB, Bates JH. Quantifying the roles of tidal volume and PEEP in the pathogenesis of ventilator-induced lung injury. Ann Biomed Eng. 2011;39:1505–16.
10. Bilek AM, Dee KC, Gaver DP. Mechanisms of surface-tension induced epithelial damage in a model of pulmonary airway repoening. J Appl Physiol. 2003;94:770–83.
11. Kay SS, Bilek AM, Dee KC, Gaver DP. Pressure gradient, not exposure duration, determines the extent of epithelial cell damage in a model of pulmonary airway reopening. J Appl Physiol. 2004;97:269–76.
12. Hamlington KL, Smith BJ, Allen GB, Bates JH. Predicting ventilator-induced lung injury using a lung injury cost function. J Appl Physiol. 1985;2016(121):106–14.
13. Salazar E, Knowles JH. An analysis of pressure-volume characteristics of the lungs. J Appl Physiol. 1964;19:97–104.
14. Alencar AM, Arold SP, Buldyrev SV, Majumdar A, Stamenovic D, Stanley HE, Suki B. Physiology: dynamic instabilities in the inflating lung. Nature. 2002;417:809–11.

15. Crotti S, Mascheroni D, Caironi P, Pelosi P, Ronzoni G, Mondino M, Marini JJ, Gattinoni L. Recruitment and derecruitment during acute respiratory failure: a clinical study. Am J Respir Crit Care Med. 2001;164:131–40.
16. Bates JHT, Gaver DP, Habashi NM, Nieman GF. Atelectrauma versus volutrauma: a tale of two time-constants. Crit Care Explor. 2020;2:e0299.
17. Allen GB, Suratt BT, Rinaldi L, Petty JM, Bates JH. Choosing the frequency of deep inflation in mice: balancing recruitment against ventilator-induced lung injury. Am J Physiol Lung Cell Mol Physiol. 2006;291:L710–7.
18. Allen GB, Leclair T, Cloutier M, Thompson-Figueroa J, Bates JH. The response to recruitment worsens with progression of lung injury and fibrin accumulation in a mouse model of acid aspiration. Am J Physiol Lung Cell Mol Physiol. 2007;292:L1580–9.
19. Smith BJ, Grant KA, Bates JH. Linking the development of ventilator-induced injury to mechanical function in the lung. Ann Biomed Eng. 2013;41:527–36.
20. Allen GB, Pavone LA, DiRocco JD, Bates JH, Nieman GF. Pulmonary impedance and alveolar instability during injurious ventilation in rats. J Appl Physiol (1985). 2005;99:723–30.
21. Bates JH, Irvin CG. Time dependence of recruitment and derecruitment in the lung: a theoretical model. J Appl Physiol (1985). 2002;93:705–13.
22. Massa CB, Allen GB, Bates JH. Modeling the dynamics of recruitment and derecruitment in mice with acute lung injury. J Appl Physiol (1985). 2008;105:1813–21.
23. Perun ML, Gaver DP 3rd. An experimental model investigation of the opening of a collapsed untethered pulmonary airway. J Biomech Eng. 1995;117:245–53.
24. Perun ML, Gaver DP 3rd. Interaction between airway lining fluid forces and parenchymal tethering during pulmonary airway reopening. J Appl Physiol (1985). 1995;79:1717–28.
25. Naureckas ET, Dawson CA, Gerber BS, Gaver DP 3rd, Gerber HL, Linehan JH, Solway J, Samsel RW. Airway reopening pressure in isolated rat lungs. J Appl Physiol (1985). 1994;76:1372–7.
26. Takahashi A, Bartolak-Suki E, Majumdar A, Suki B. Changes in respiratory elastance after deep inspirations reflect surface film functionality in mice with acute lung injury. J Appl Physiol (1985). 2015;119:258–65.
27. Geitner CM, Becher T, Frerichs I, Weiler N, Bates JHT, Wall WA. An approach to study recruitment/derecruitment dynamics in a patient-specific computational model of an injured human lung. Int J Numer Methods Biomed Eng. 2023;39:e3745.
28. Smith BJ, Bates JH. Variable ventilation as a diagnostic tool for the injured lung. IEEE Trans Biomed Eng. 2015;62:2106–13.
29. Albert SP, DiRocco J, Allen GB, Bates JH, Lafollette R, Kubiak BD, Fischer J, Maroney S, Nieman GF. The role of time and pressure on alveolar recruitment. J Appl Physiol (1985). 2009;106:757–65.
30. Webb HH, Tierney DF. Experimental pulmonary edema due to intermittent positive pressure ventilation with high inflation pressures. Protection by positive end-expiratory pressure. Am Rev Respir Dis. 1974;110:556–65.
31. Cereda M, Xin Y, Hamedani H, Bellani G, Kadlecek S, Clapp J, Guerra L, Meeder N, Rajaei J, Tustison NJ, et al. Tidal changes on CT and progression of ARDS. Thorax. 2017;72:981–9.
32. Hamlington KL, Bates JHT, Roy GS, Julianelle AJ, Charlebois C, Suki B, Smith BJ. Alveolar leak develops by a rich-get-richer process in ventilator-induced lung injury. PLoS One. 2018;13:e0193934.
33. Newman ME. Power laws, Pareto distributions and Zipf's law. Contemp Phys. 2005;46:323–51.
34. Barabasi A-L, Albert R. Emergence of scaling in random networks. Science. 1999;286:509–12.
35. Mori V, Smith BJ, Suki B, Bates JHT. Linking physiological biomarkers of ventilator-induced lung injury to a rich-get-richer mechanism of injury progression. Ann Biomed Eng. 2019;47:638–45.
36. Gajic O, Lee J, Doerr CH, Berrios JC, Myers JL, Hubmayr RD. Ventilator-induced cell wounding and repair in the intact lung. Am J Respir Crit Care Med. 2003;167:1057–63.

37. Gottman DC, Smith BJ. A scale-free model of acute and ventilator-induced lung injury: a network theory approach inspired by seismology. Front Netw Physiol. 2024;4:1392701.
38. Hamlington KL, Ma B, Smith BJ, Bates JH. Modeling the progression of epithelial leak caused by overdistension. Cell Mol Bioeng. 2016;9:151–61.
39. Yamaguchi E, Yao J, Aymond A, Chrisey D, Nieman GF, Bates JHT, Gaver DP. Electric cell-substrate impedance sensing (ECIS) as a platform for evaluating barrier-function susceptibility and damage from pulmonary atelectrauma. Biosensors. 2022;12:390.
40. Bates JHT, Nieman GF, Kollisch-Singule M, D.P. G: ventilator-induced lung injury as a dynamic balance between epithelial cell damage and recovery. Ann Biomed Eng. 2023;51: 1052–62.
41. Grotberg JB, Romano F. Computational pulmonary edema: a microvascular model of alveolar capillary and interstitial flow. APL Bioeng. 2023;7:036101.
42. Mutch WA, Eschun GM, Kowalski SE, Graham MR, Girling LG, Lefevre GR. Biologically variable ventilation prevents deterioration of gas exchange during prolonged anaesthesia. Br J Anaesth. 2000;84:197–203.
43. Mutch WA, Harms S, Lefevre GR, Graham MR, Girling LG, Kowalski SE. Biologically variable ventilation increases arterial oxygenation over that seen with positive end-expiratory pressure alone in a porcine model of acute respiratory distress syndrome. Crit Care Med. 2000;28:2457–64.
44. Arold SP, Bartolak-Suki E, Suki B. Variable stretch pattern enhances surfactant secretion in alveolar type II cells in culture. Am J Physiol Lung Cell Mol Physiol. 2009;296:L574–81.
45. Arold SP, Suki B, Alencar AM, Lutchen KR, Ingenito EP. Variable ventilation induces endogenous surfactant release in normal Guinea pigs. Am J Physiol Lung Cell Mol Physiol. 2003;285:L370–5.
46. Arold SP, Mora R, Lutchen KR, Ingenito EP, Suki B. Variable tidal volume ventilation improves lung mechanics and gas exchange in a rodent model of acute lung injury. Am J Respir Crit Care Med. 2002;165:366–71.
47. Bellardine CL, Hoffman AM, Tsai L, Ingenito EP, Arold SP, Lutchen KR, Suki B. Comparison of variable and conventional ventilation in a sheep saline lavage lung injury model. Crit Care Med. 2006;34:439–45.
48. Thammanomai A, Hueser LE, Majumdar A, Bartolak-Suki E, Suki B. Design of a new variable-ventilation method optimized for lung recruitment in mice. J Appl Physiol (1985). 2008;104: 1329–40.
49. Spieth PM, Carvalho AR, Pelosi P, Hoehn C, Meissner C, Kasper M, Hubler M, von Neindorff M, Dassow C, Barrenschee M, et al. Variable tidal volumes improve lung protective ventilation strategies in experimental lung injury. Am J Respir Crit Care Med. 2009;179:684–93.
50. Bou Jawde S, Walkey AJ, Majumdar A, O'Connor GT, Smith BJ, Bates JHT, Lutchen KR, Suki B. Tracking respiratory mechanics around natural breathing rates via variable ventilation. Sci Rep. 2020;10:6722.
51. Suki B, Alencar AM, Sujeer MK, Lutchen KR, Collins JJ, Andrade JS Jr, Ingenito EP, Zapperi S, Stanley HE. Life-support system benefits from noise. Nature. 1998;393:127–8.
52. Wirtz HR, Dobbs LG. Calcium mobilization and exocytosis after one mechanical stretch of lung epithelial cells. Science. 1990;250:1266–9.
53. Amin SD, Majumdar A, Alkana P, Walkey AJ, O'Connor GT, Suki B. Modeling the effects of stretch-dependent surfactant secretion on lung recruitment during variable ventilation. J Biomed Sci Eng. 2013;6:61–70.

Chapter 11
Lung Defenses

Abstract The blood-gas barrier of the lung is topologically on the outside of the body and thus is at risk of attack from environmental insults. The lung has evolved elaborate defense mechanisms to deal with this threat, the final line of defense being the immune system. Models of immunity in the lung based on systems of ordinary differential equations can be used to explore the effectiveness of pharmaceutical therapies. Models based on systems of partial differential equations can be used to explore how viruses and bacteria affect the integrity of the pulmonary epithelial barrier. Agent-based models of the interactions between the cells of the immune system and the tissue on which they operate can recapitulate the transient nature of allergic inflammation in the lung. An agent-based model has been used to test the plausibility of the idea that allergic inflammation progresses with a predetermined time course much like a muscle twitch or a nerve action potential, albeit over a much longer duration. A corollary of this idea is that there should be a refractory period to the initiation of allergic inflammation, something for which there is experimental evidence. This suggests that such an "inflammatory twitch" might fail to resolve in asthma.

Keywords Blood-gas barrier · Immune system · Differential equations · Agent-based model · Allergic inflammation · Refractory period

11.1 Introduction

Although the lung is an internal organ, its airway and alveolar surfaces are essentially just a continuation of the skin, similar to the gastrointestinal and urogenital tracts. From a topological perspective, this means that much of the lung is on the outside surface of the body. This places the lung in direct contact with the environment, making it vulnerable to being breached by environmental threats. One potential strategy Nature might use to deal with this would be to endow the exposed surface of the lung with the physical robustness needed to repel such threats. Indeed, most of the exposed surface comprises the blood-gas barrier, which must be resilient enough to keep the capillary blood and alveolar air separated. Unlike the skin, however, the lung cannot protect itself with layers of dead epithelial cells because

B. Suki, J. H. T. Bates, *Mathematical Modeling of the Healthy and Diseased Lung*,
https://doi.org/10.1007/978-3-031-95322-4_11

its primary function of gas exchange relies on an extremely thin blood-gas barrier with an enormous surface area. The two missions of this physical interface between air and blood—to act as a barrier and to facilitate the diffusive exchange of oxygen and carbon dioxide—are thus inherently in conflict. To meet the challenges posed by this conflict, the lung has evolved an elaborate set of defense mechanisms to protect the blood-gas barrier against breach, and to deal with the consequences should breach occur [1].

One way that breach of the blood-gas barrier can occur is as a result of acute lung injury (see Chap. 10). This is a highly consequential event because it allows fluid and proteins from the blood plasma to enter the alveolar airspace where they interfere with gas exchange by displacing alveolar air and increasing the distance that gases must diffuse to cross between air and blood. They also interfere with the surface tension lowering properties of pulmonary surfactant, which further stresses the barrier and can worsen the breach. Recovery relies on the lung's innate reparative mechanisms and lymphatic clearance of the plasma-derived material that has accumulated in the airspaces (Sect. 10.7).

The blood-gas barrier can also be threatened by small particles such as dust, viruses, bacteria and fungal spores inhaled from the environment. The first line of defense against these invaders is removal by coughing and sneezing instigated by sensitive receptors in the epithelial lining of the airways. Most invaders that fail to be forcibly expelled are trapped in the sticky mucus layer that coats the airway surface. This mucus is continually being produced by goblet cells in the airway lining and is continually swept up to the tracheal opening by beating cilia protruding from cells in the airway epithelium. Most particles that are not removed by this so-called *mucociliary escalator* are usually phagocytized by roaming *alveolar macrophages*. If the particles are inert and ultimately fatal to the macrophages, their continued presence can give rise to chronic diseases such as silicosis in the case of silica dust [2], or asbestosis in the case of asbestos particles [3].

Foreign particles can also elicit a variety of other responses from the immune system. Examples include direct release of bactericidal agents from the pulmonary epithelium as well as recruitment of *neutrophils*, *eosinophils* and macrophages from the blood into the pulmonary interstitium and airspaces [1]. These immune cells may then enter into various activated states that collectively orchestrate the *inflammatory response* and, eventually, its resolution. Mathematical and computational models provide both quantitative and qualitative insight into these complex dynamic events.

11.2 Ordinary Differential Equation Models of the Immune Response

When the pulmonary epithelium becomes damaged by, for example, injurious mechanical ventilation, products from the damaged cells can rapidly elicit an immune response that involves cells of the *innate immune system* (the component of the immune system that responses rapidly and nonspecifically to anything that is perceived as a threat). Principal among these are neutrophils and macrophages, some of which are already resident in the lung tissue while others are recruited from the circulation. Neutrophils can operate in several roles including releasing cytotoxic agents that kill invading pathogens, phagocytozing debris, and eventually undergoing apoptosis when they have done their jobs. Macrophages exist initially in an agnostic state denoted M0 but can differentiate into proinflammatory and anti-inflammatory phenotypes denoted M1 and M2, respectively. It is also possible for M1 macrophages to convert to M2, and vice versa, and between them they contribute to the repair of damaged epithelial cells and removal of dead cells. A model of these various players and their interactions has been developed to explain the response to ventilator-induced lung injury [4].

Expressing this model in mathematical form is readily achieved by assuming that each of the various cell types comprise a distinct compartment. That is, each cell type exists as a set of identical members distributed evenly throughout some space. For example, activated neutrophils are assumed to be uniformly mixed in the circulation so that their concentration everywhere in the blood is the same. Similarly, healthy, damaged and dead epithelial cells are assumed to have, at any point in time, uniform concentrations across the entire pulmonary epithelium. These concentrations change with time in various ways. However, the only spatial information in this model is that cells are either in the blood or in the lung tissue. The activities of a given cell type are also the same everywhere throughout each space. The model thus makes no further spatial refinement within each of the two spaces. These assumptions allow the model to be specified as a set of coupled ordinary differential equations in time, t, as follows.

Let $E_h(t)$, $E_d(t)$ and $E_e(t)$ designate, respectively, the fractions of epithelial cells that are healthy, damaged and dead, such that their sum is always 1. The rates of change of these three quantities can be expressed as sums of various contributing processes. One of these processes is the rate, κ_{prolif}, at which epithelial cells proliferate to produce new healthy cells. Assuming that healthy and damaged cells are equally capable of proliferating, and that they all have the same average doubling time, the total rate at which new epithelial cells are produced is proportional to the total number of healthy and damaged cells. New epithelial cells will only be produced, however, if there is space for them to be accommodated in the epithelium, and space only becomes available when established epithelial cells die and are removed. Therefore, the total production rate of new cells is proportional to the number of dead cells that need to be replaced. This means that κ_{prolif} is proportional to the product of $[E_h(t) + E_d(t)]$ and $E_e(t)$. The proportionality factor is comprised by a baseline rate constant b_p plus an additional component $k_{ep}p_{med}(t)$ due to the level of pro-inflammatory mediators, $p_{med}(t)$, secreted by activated neutrophils and M1 macrophages in the tissue, where k_{ep} is another rate constant. That is,

$$\kappa_{prolif}(t) = \left(b_p + k_{ep}p_{med}(t)\right)[E_h(t) + E_d(t)]E_e(t). \tag{11.1}$$

Another source of healthy cells is repair of damaged cells. The rate, κ_{repair}, at which repair takes place has a baseline rate proportional to the number of damaged cells with constant of proportionality b_r plus an additional rate due to the actions of repair mediators produced by M2 macrophages. This second rate is assumed to result from a given initial concentration, C_R, of mediators interacting with a fixed concentration, C_S, of substrate receptors. This results in a product with concentration C_P that determines the rate of repair. The rate at which mediators and receptors combine is proportional to the concentrations of both species, which are $C_R - C_P$ and $C_S - C_P$, respectively, since the reaction started with $C_P = 0$. However, the mediators are much more abundant than the receptors (i.e., $C_R >> C_S$) so $C_S - C_P \approx C_S$. At the same time, the product dissociates into mediators and receptors at a rate proportional to C_P. The rates of combination and dissociation are equal at steady state, in which case $(C_R - C_P)C_S = k_{er}C_P$ where k_{er} is a constant. This expression rearranges to $C_P = C_R C_S/(k_{er} + C_R)$, which is known as the Michaelis-Menton equation [5]. Accordingly,

$$\kappa_{repair}(t) = b_r E_d(t) + E_d(t)\frac{k_{er}C_R(t)}{x_{er} + C_R(t)} \tag{11.2}$$

where k_{er} is the maximum rate due to repair mediators and x_{er} is the mediator concentration at half maximum rate.

Healthy cells also lose numbers at a rate κ_{damage} as a result of damage caused by the concentrations of activated neutrophils, $N(t)$, and M1 macrophages, $M_1(t)$. This rate is proportional to the number of healthy cells and is assumed to conform to a second-order Hill function (a higher-order form of the Michaelis-Menten equation, see section 8.3) with maximum rate and half-maximum concentration of k_{mne}, respectively. Thus,

$$\kappa_{damage}(t) = E_h(t)\frac{k_{mne}[N(t) + M_1(t)]^2}{x_{mne}{}^2 + [N(t) + M_1(t)]^2}. \tag{11.3}$$

Finally, healthy cells can be damaged by mechanical ventilation at a rate, κ_{vent}, proportional to the number of healthy cells with rate constant s_d thus:

$$\kappa_{vent}(t) = s_d E_h(t). \tag{11.4}$$

Putting this all together gives

$$\frac{dE_h(t)}{dt} = \kappa_{prolif}(t) + \kappa_{repair}(t) - \kappa_{damage}(t) - \kappa_{vent}(t). \tag{11.5}$$

The term $\kappa_{repair}(t)$ detracts in equal measure from the pool of damaged cells, while the terms $\kappa_{damage}(t)$ and $\kappa_{vent}(t)$ add to the pool. At the same time, damaged cells are reduced due to phagocytosis by neutrophils at a rate

$$\kappa_{Nphag}(t) = k_{en}N(t)E_d(t) \tag{11.6}$$

and phagocytosis by M1 macrophages at a rate

$$\kappa_{M1phag}(t) = k_{em1}M_1(t)E_d(t)\frac{1}{1+\left(\frac{a(t)}{a_\infty}\right)^2} \tag{11.7}$$

where k_{en} and k_{em1} are rate constants. The quotient in Eq. 11.7 represents another Hill-type function governing the influence of the concentration of anti-inflammatory mediators, a, and a_∞ is another constant.

A final contributor to the rate of change of damaged cells is death, $\kappa_{death}(t)$, given by

$$\kappa_{death}(t) = b_d E_d(t). \tag{11.8}$$

The total rate of change of damaged cells is thus

$$\frac{dE_d(t)}{dt} = -\kappa_{repair}(t) + \kappa_{damage}(t) + \kappa_{vent}(t) - \kappa_{Nphag}(t) - \kappa_{M1phag}(t) - \kappa_{death}(t). \tag{11.9}$$

Similar reasoning leads to

$$\frac{dE_e(t)}{dt} = -\kappa_{prolif}(t) + \kappa_{Nphag}(t) + \kappa_{M1phag}(t) + \kappa_{death}(t). \tag{11.10}$$

Equations 11.5, 11.9 and 11.10 together describe the dynamics of the three compartments of epithelial cells. These equations are coupled because they share common κ terms. This means that rather than varying independently of each other, changes in either E_h, E_d or E_e affect at least one of the other two quantities (being fractions, they always add to 1).

Corresponding equations can be developed to describe the rates of change of the various classes of neutrophils and macrophages as well as the various mediators they produce or are influenced by. The result is a set of 18 coupled first-order nonlinear differential equations containing a total of 68 parameters [4]. This daunting model is nevertheless still an extremely simplified representation of reality. For example, the numerous mediators that act to up- or down-regulate the immune response have been lumped together into only two groups labeled pro-inflammatory and anti-inflammatory. The values of many of the 68 parameters in the model are as yet unknown, which makes it very difficult to generate meaningful model predictions that apply to any given set of circumstances. Nevertheless, a model such as this one can be used to determine which parameters are likely to have the strongest effects on outcomes, which can inform subsequent experimentation. For example, Fig. 11.1 shows how one of the most influential parameters, k_{mne} (Eq. 11.3), affects the fate of the pulmonary epithelium when it and the other parameters are randomly varied over

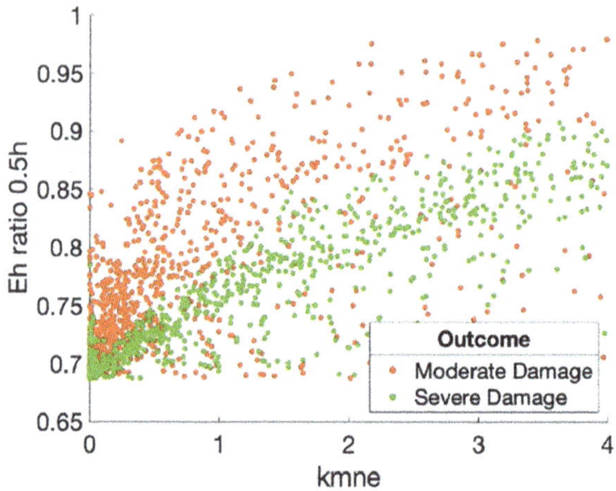

Fig. 11.1 Monte-Carlo predictions of the collateral damage to epithelial cells caused by neutrophils and macrophages on epithelial health, governed by the value of the parameter k_{mne}. The measure of damage (vertical axis) is the ratio of the number of healthy epithelial cells after 200 hours of mechanical ventilation to the initial number of healthy cells. The red and green dots indicate the cases that had moderate and severe injury, respectively, at 200 h. (Adapted with permission from [6])

their biologically possible ranges in a Monte-Carlo simulation [4]. The outcomes are divided into groups that correspond to moderate versus severe injury after the simulation of 200 h of mechanical ventilation. The plot shows how an early indication of outcome (the fraction by which healthy epithelial cells are reduced at 0.5 h after the start of mechanical ventilation) correlates with the value of k_{mne} in the two groups. These simulations demonstrate the potential for therapies that target k_{mne} as a means of mitigating ventilator-induced lung injury.

Similar models can be proposed to describe the immune response to viruses and bacteria, which, when severe enough, can also induce an acute form of lung injury. An example model structure is shown in Fig. 11.2 for pneumococcal infection [6]. The structure of this model is similar in many respects to that described by Eqs. 11.1, 11.2, 11.3, 11.4, 11.5, 11.6, 11.7, 11.8, 11.9, and 11.10, although this time the immune trigger is provided by a bacterium instead of fragments from damaged epithelial cells. Also, the cytokines and chemokines that serve as the links in the cellular network are included in more detail instead of simply being classified as pro- or anti-inflammatory.

The equation describing pneumococcal (P) kinetics in this model is

$$\begin{aligned}\frac{dP(t)}{dt} &= P_0 + k_p P(t)\left(1 - \frac{P(t)}{P_{max}}\right) - [k_{P_{AM}} M_0(t) + k_{P_N} N(t) + k_{P_{IM}} M_I(t)] \\ &\quad \frac{P(t)}{x + P(t)} - k_{CFUBp} P(t) - A(t)P(t)\end{aligned} \tag{11.11}$$

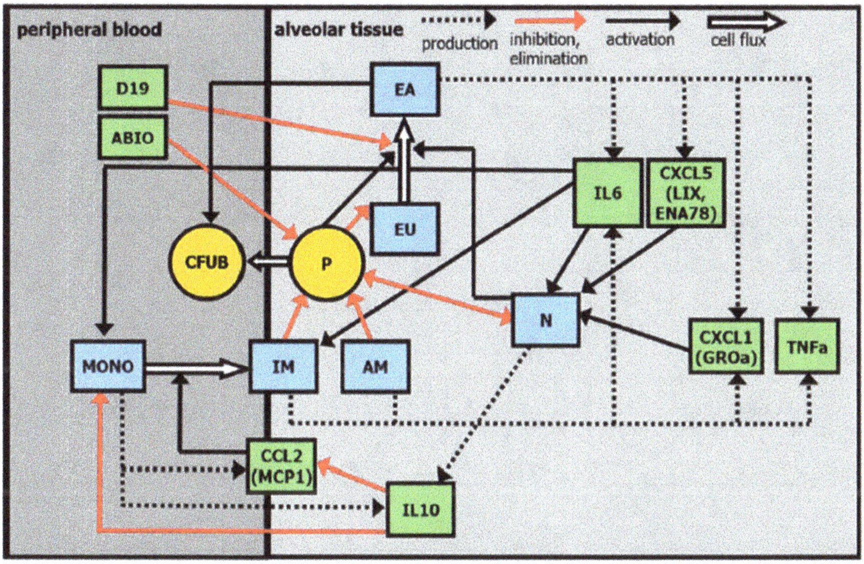

Fig. 11.2 A model of the pulmonary immune response to pneumococcal infection. The model compartments include bacteria in lungs (P) and in peripheral blood (CFUB), uninfected epithelial/endothelial cells (EU), infected epithelial/endothelial cells (EA), alveolar macrophages (AM), inflammatory macrophages (IM), neutrophils (N), and monocytes in peripheral blood (MONO). IL-6, IL-10, CCL2, CXCL1, CXCL5 and TNFα comprise the cytokine/chemokine network. ABIO and D19 represent antibiotic and NOX-D19 treatments, respectively. (Adapted with permission from [6])

where P_0 is bacterial proliferation rate, k_p is the initial growth rate constant, and P_{max} is the bacterial carrying capacity of the tissue. The constants $k_{p_{AM}}$, k_{p_N} and $k_{p_{IM}}$ govern the rates of bacterial killing by macrophages (M_0), neutrophils (N) and inflammatory macrophages (M_I), respectively, all of which operate according to Michaelis-Menton kinetics with half-maximum concentration x. Corresponding equations define the dynamics of the various cell and cytokine compartments in the model. Antibiotics were injected into the model in a series of pulses that pass through two transitional compartments in order to cause a delay in their effects.

The model was fit to time-series measurements of cells and cytokines collected from mice infected with pneumococci and then treated with ampicillin or NOX-D19. Figure 11.3 shows model fits to various outcome measures versus time when the animals were treated with ampicillin at 48 h after bacterial inoculation, showing that the model captures the overall dynamics of the infection. The model was then used to examine the long-term effectiveness of treatment with ampicillin and/or NOX-D19 therapy. The results indicate that treatment is only possible when ampicillin is included in the regimen.

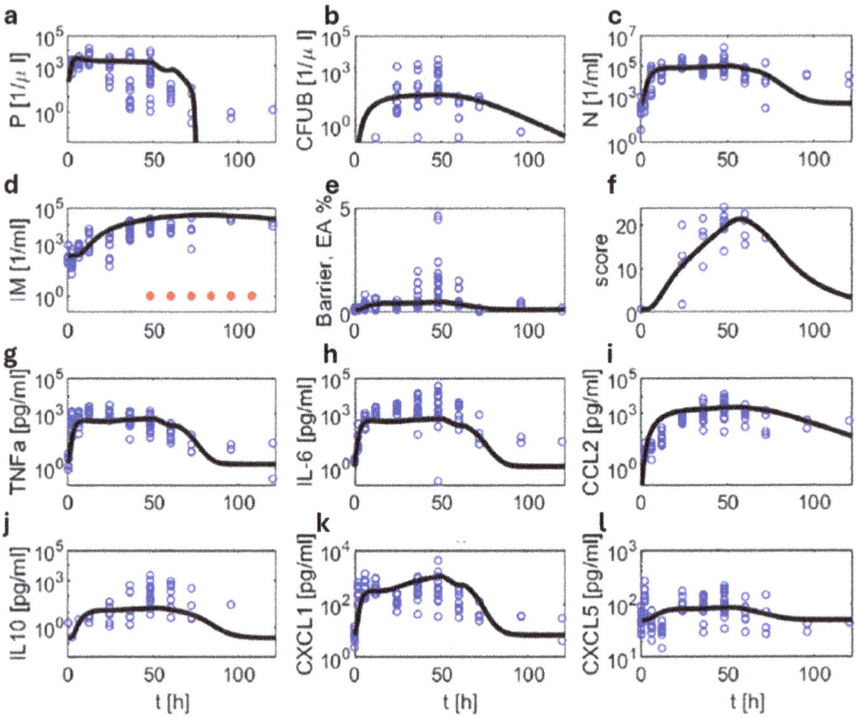

Fig. 11.3 Time course of response to infection with streptococcus pneumoniae with ampicillin given at 48 h. The panels show pneumococcal population in (**a**) BAL and (**b**) blood, (**c**) neutrophils in BAL, (**d**) inflammatory macrophages in BAL, (**e**) epithelial barrier function, (**f**) histologic score, (**g**) TNF-α in BALF, (**h**) IL-6 in BALF, (**i**) CCL2 in BALF, (**j**) IL-10 in BALF, (**k**) CXCL1 in BALF and (**l**) CXCL5 in BALF. Black curves—simulation results. Blue circles—experimental data. Red dots show the times of antibiotic treatment. (Adapted with permission from [6])

11.3 Partial Differential Equation Models of the Immune Response

A compartment model of a biologic system described by a system of coupled ordinary differential equations accounts for how the system evolves in continuous time. On the other hand, the spatial resolution of such a model is typically crude, being limited to the small number of distinct compartments it is comprised of. The models depicted above, for example, contain only two compartments—lung tissue and blood—and make no predictions as to precisely where events take place within these two spaces.

It is frequently necessary to account for events that evolve continuously in both time and space. This requires the use of partial differential equations. A pulmonary example concerns the way that the SARS-CoV-2 virus spreads after attaching to angiotensin converting enzyme 2 (ACE2) receptors on the epithelium of the upper airway tract. This attachment attracts the attention of the innate immune system, which responds by mounting an inflammatory response that grows in intensity and extent, leading in some cases to a cytokine storm, severe illness, and even death [7]. T cell involvement is a key component of the innate immune response to SARS-CoV-2 infection [8]. This leads to a mathematical model of viral dynamics [9] over the airway epithelium that begins by considering the infection of T cells and how these cells spread. In this model, a patch of upper airway epithelium is represented as a 2-dimensional surface. The rate of change of viral load $V(x, y, t)$ on the surface is determined by the diffusivity, D_V, of the virus over the surface, the rate constant, k, governing virus production by inflamed T cells, and the rate constant, c, governing viral clearance by the immune system. These factors are accounted for in the equation

$$\frac{\partial V}{\partial t} = \nabla \cdot (D_V \nabla V) + kI - cV \tag{11.12}$$

where I is the concentration of infected T cells. The symbol ∇, known as del or nabla, is the vector operator $(\boldsymbol{i}\partial/\partial x + \boldsymbol{j}\partial/\partial y)$ that calculates the gradient of a multi-dimensional quantity, in this case $V(x, y, t)$, and $\boldsymbol{i}$ and $\boldsymbol{j}$ are unit vectors in the directions of the x and y axes. The gradient of $V(x, y, t)$ is itself a vector, $(\boldsymbol{i}G_x+\boldsymbol{j}G_y)$, so the first term in Eq. 11.12 is the divergence operator $\nabla \cdot ()$ acting through the dot product on the vector $D_V(\boldsymbol{i}G_x+\mathbf{j}G_y)$. If the diffusion coefficient D_V is constant across the tissue, the first term in Eq. 11.12 is simply $D_V\nabla^2 V(x, y, t)$.

The T cell concentration, T, on the epithelial surface also changes continually because T cells diffuse with diffusivity D_T, die with rate constant d_T, and are infected by virus with rate constant k_T. In addition, T cells are increased by the cytokine IL-6 (concentration L_6) and by M1 macrophages (concentration M_1) and are decreased by the cytokine IL-10 (concentration L_{10}) and by M2 macrophages (concentration M_2). The rate constants governing these last four reactions are k_{L6T}, k_{L10T}, k_{M1T} and k_{M2T}, respectively. This model also assumes [9] that uninfected T cells are attracted by infected T cells but are repulsed by virus particles with respective strength coefficients $\alpha_{attract}$ and $\alpha_{repulse}$, and that the tissue has a maximum carrying capacity for T cells given by T_{max}. Putting all these factors together gives the following expression for the kinetics of T [9]:

$$\begin{aligned}\frac{\partial T}{\partial t} &= \nabla \cdot \{D_T \nabla T - T[\alpha_{attract}(1 - \frac{T}{T_{max}})\nabla I - \alpha_{repulse}(1 - \frac{T}{T_{max}})\nabla V]\} \\ &- d_T T - k_T VT + k_{L6T}L_6 + k_{L10T}L_{10} - k_{M1T}M_1 - k_{M2T}M_2\end{aligned} \tag{11.13}$$

where the explicit dependencies of V, T and I on x, y and t have been dropped to save space, and L_{10} is the concentration of IL-10.

Infected T cells are assumed to diffuse at the same rate as uninfected T cells, and to be produced by interactions between uninfected T cells and virus particles with rate constant k_{infect}. T cells are assumed die with rate constant d_I, and to be increased by their interactions with IL-6 with rate constant k_{L6I}. These various processes give [9]

$$\frac{\partial I}{\partial t} = \nabla \cdot (D_T \nabla I) + k_{infect} VT - d_I I + k_{L6I} I L_6. \tag{11.14}$$

The assumed dynamics of M1 macrophages include diffusion with diffusivity D_{M1}, preferential movement toward infected T cells proportional to $\alpha_{attract}$, increases due to IL-6 proportional to k_{L6M1}, and decreases due to IL-10 proportional to k_{NM1}. These effects are collectively expressed as

$$\frac{\partial M_1}{\partial t} = \nabla \cdot \left[D_{M1} \nabla M_1 - \alpha_{attract} M_1 \left(1 - \frac{M_1}{M_{1max}} \right) \nabla I \right] + k_{L6M1} L_6 - k_{L10M1} L_{10} \tag{11.15}$$

where M_{1max} is the carrying capacity of the tissue for M1 macrophages.

Following the same nomenclature, consideration of the factors that affect the kinetics of M2 macrophages, IL-6 and IL-10 give

$$\frac{\partial M_2}{\partial t} = \nabla \cdot (D_{M2} \nabla M_2) + k_{L10M2} L_{10} - k_{L6M2} L_6, \tag{11.16}$$

$$\frac{\partial L_6}{\partial t} = \nabla \cdot (D_{L6} \nabla L_6) + k_{M1L6} M_1 - k_{L10L6} L_{10}, \tag{11.17}$$

and

$$\frac{\partial L_{10}}{\partial t} = \nabla \cdot (D_{L10} \nabla L_{10}) + k_{M2L10} M_2 - k_{L6L10} L_6. \tag{11.18}$$

Note that Eqs. 11.13, 11.14, 11.15, 11.16, 11.17, and 11.18 are written as if the diffusivity factors D_T, D_{M1}, D_{M2}, D_{L6} and D_{L10} are functions of x and y. If they are constants, they can be placed in front of ∇ in each equation.

The partial differential equations 11.12 and 11.13, 11.14, 11.15, 11.16, 11.17, and 11.18 are coupled in the same sense as the ordinary differential equations 11.5, 11.9 and 11.10. Both sets of equations must be integrated numerically starting from specified initial conditions while the partial differential equations also require boundary conditions. Figure 11.4 shows model predictions of how $V(x, y, t)$ and $T(x, y, t)$ develop over time on a patch of model tissue [9].

11.4 Agent-Based Models of the Immune Response

Immune system dynamics can also be represented by agent-based models (ABMs). In fact, ABMs are particularly well suited to this task because individual cells act essentially as autonomous agents, while the environment on which these agents live and operate corresponds to the tissues that the immune system is designed to protect.

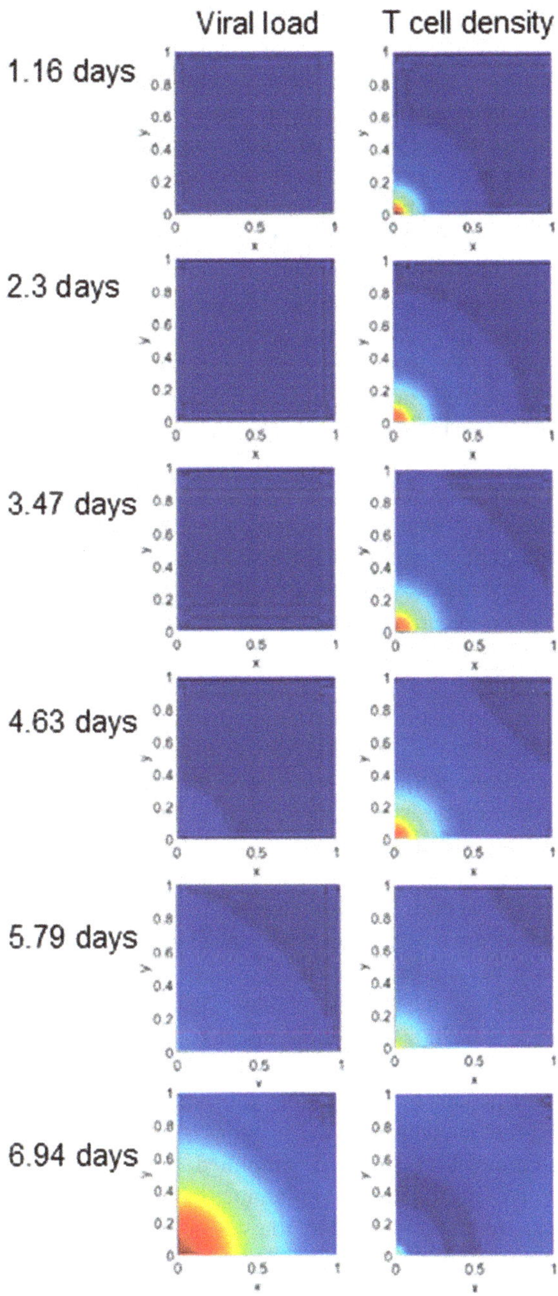

Fig. 11.4 Predictions from Eqs. 11.12 and 11.14, 11.15, 11.16, 11.17, and 11.18 of viral load and T cell density on a patch of tissue (arbitrary dimensions) for a given set of parameter values and initial conditions. The model shows how these two quantities develop over time at different rates. (Adapted with permission from [9])

It is also relatively straightforward to imbue the ABM environment with complex structures that would be challenging to represent in systems of partial differential equations, especially if these structures include discontinuous boundaries. On the other hand, the computational cost of running ABMs increases rapidly with the number of pixels (voxels) comprising the environment, the number of agents, and the complexity of rule sets defining agent behavior, so compromises may have to be made concerning model scale. Nevertheless, ABMs provide a powerful tool for simulating events that take place in both time and space.

An interesting and important question that has been addressed using an ABM is how the immune inflammatory response in the lung is controlled. The inflammatory response has evolved as a necessary defense mechanism to project against foreign invaders (bacteria, viruses, allergens, etc.), so it is vital that this response be mounted when needed. Inflammation is driven by a series of mechanisms that are initiated when a sentinel system is triggered [1, 10]. There are a variety of such sentinels that include the toll-like receptors and other pattern recognition receptors found on immune cells such as macrophages and dendritic cells. These receptors recognize molecular signatures on common microbes that are not found in the host, allowing them to be immediately identified as potential threats. The innate immune system is then engaged to respond accordingly. Another sentinel mechanism is provided by immune memory whereby prior exposure to a pathogen is encoded in the surface receptors of memory B and T cells so that subsequent exposure allows them to immediately orchestrate an appropriate response.

Equally important as mounting an inflammatory response, however, is closing it down when the threat has been dealt with. Failure to do this can lead to a persistent and unresolving inflammatory state, something that seems to characterize most chronic diseases of modern society. But this raises an interesting question. How does the immune inflammatory response turn itself off when it is no longer needed? It is tempting to view the immune system as an army doing battle against an invading enemy, but this military analogy can be misleading. The actions of an army of human soldiers are always under the control of a commander who surveys the battle scene and gives the order to attack or retreat as appropriate. There is no such sentient controller in the immune system. Thus, while it is easy to see how the command to attack is facilitated by the sentinel systems described above, it is far less clear how the order to cease and desist arises. The resolution of inflammation is brought about by the actions of T regulatory cells and certain anti-inflammatory cytokines [1, 11], but what is controlling their actions?

Figure 11.5 shows a simplified schema of the interactions proposed to be involved in the allergic inflammatory response in the lungs [12]. Allergens (granules) are first processed by the sentinel mast and antigen presenting cells that detect the invader and instigate the inflammatory response. The response is then amplified through the involvement of T helper cells and pro-inflammatory cells such as eosinophils, resulting in collateral damage to the lung tissue. T regulatory cells and anti-inflammatory cells such as M2 macrophages then bring about the resolution of the inflammatory response so that the tissue can be repaired. To keep the model as simple as possible, dendritic cells, B cells, and macrophages are represented as a single antigen presenting cell type. Similarly, neutrophils, eosinophils, and other

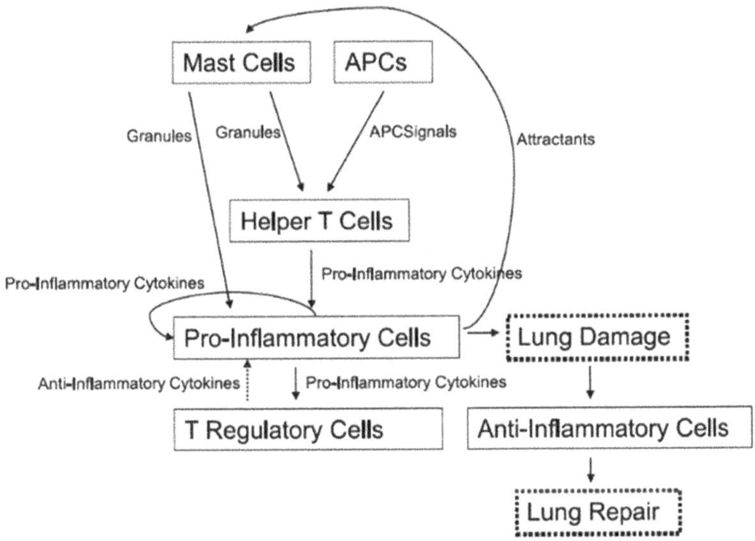

Fig. 11.5 Schema of the principal players and their interactions involved in allergic inflammation in the lung. Mast cells and antigen presenting cells (APCs) are activated upon encountering particles (corresponding to antigen). This initially leads to an inflammatory reaction that is eventually resolved by anti-inflammatory mechanisms. (Adapted with permission from [12])

cells capable of causing tissue damage are represented as a single proinflammatory cell type, while fibroblasts and other cells involved in tissue repair are represented by a single anti-inflammatory cell type. It is not obvious from this schema, however, how the actions of these various actors are coordinated to produce an inflammatory response that waxes and wanes with an appropriate time course.

A possible control strategy for the inflammatory response is suggested by the way in which allergic inflammation in the lung varies over time when allergy to a foreign protein is induced in experimental animals [13, 14]. Sensitivity to the protein ovalbumin is readily induced in mice, for example, via intraperitoneal injection. This primes the sentinel cells of the immune system so that inhalational exposure to an aerosol of ovalbumin a week or two later elicits an inflammatory reaction in the lung. The reaction is characterized by excess secretion of airway mucus and hyperresponsiveness to airway challenge by methacholine. Curiously, however, continued challenge of the animal with ovalbumin does not cause progressive increases in the level of inflammation. Rather, after approximately a month, the inflammation resolves. This phenomenon is known as immune tolerization [15], something that is exploited therapeutically to treat allergies to certain substances. But it raises the question as to why our bodies would have evolved a defense mechanism that is initially concerned about a threat, only to apparently lose interest in it after some seemingly arbitrary period of time. One possibility is that there is no alternative. While it is clear how to turn on a defensive mechanism when a threat appears, it is conceptually much more problematic to detect that the threat is no longer present. Active downregulation of the defensive machinery in the perceived absence of a threat may thus not be possible in a non-sentient control system.

It turns out that Nature has found a way to get around this conundrum in other biologic control systems such as those of nerves and skeletal muscles. The active response to stimulation in both these systems occurs as a sequence of self-limited events called action potentials or muscle twitches, respectively. The advantage of this kind of control system is that the response only needs to be repeatedly turned on so long as the stimulus is present. When the stimulus is no longer present, the response dissipates. In other words, there is no need to actively turn it off. This gets around the problem of how the absence of something (i.e., the need for inflammation) can be detected automatically.

This possibility was investigated using an ABM with an environment consisting of a capillary space and an alveolar tissue space, each divided into an array of patches. The damage status of each alveolar patch can take a value between 100 (perfect health) and 0 (complete damage). The agents in the model represent the various cell types shown in Fig. 11.5. The inflammatory response is initiated by the sudden presence of particles (representing antigen) at random locations in the alveolar space. The capillary space provides an unlimited reservoir of cells for the alveolar space as well as acting as a sink for cells in the alveolar space. The model also includes an endothelial barrier between the capillary and alveolar spaces. Pores in the barrier allow restricted cell movement between the two spaces until the inflammatory reaction reaches a certain level at which point the barrier is removed (simulating the effects of vasodilation and endothelial leak).

The agents move randomly from patch to patch in the environment with equal probability in all directions unless a patch contains a signal to which an agent is responsive, in which case the probability of movement to that patch is increased in proportion to the strength of the signal. Granules and pro- and anti-inflammatory cytokines diffuse randomly from patch to patch. Cytokine removal is simulated by decreasing the amounts on a patch by a set fraction at each time step. The agents are also given exponentially distributed lifetimes appropriate for the particular cell types they represent.

The behaviors of the agents are governed by sets of rules designed to capture the essential elements of actual biological behavior. These rules define how agents move, how they respond to signals in the environment, and what actions they take in various circumstances including how much damage or repair they cause on their current alveolar tissue patch. Figure 11.6 shows example model states at baseline (prior to instigation of inflammation) and at two points during progressively increasing tissue damage.

When this ABM is initiated by the random deposition of particles in the alveolar tissue space, its subsequent behavior initially exhibits a decrease in tissue health as pro-inflammatory agents move in to deal with the particles. The reaction subsequently becomes tempered by the arrival of anti-inflammatory agents that eventually undergo apoptosis, allowing the pro-inflammatory agents to gain the upper hand again. The result is a series of damped oscillations reflecting what is essentially a predator-prey exchange between pro-inflammatory and anti-inflammatory forces.

Fig. 11.6 Three states of the ABM of allergic inflammation in the lung shown a capillary space (red) surrounding an alveolar tissue space (grey) enclosing an airspace (white). The endothelial barrier is represented by the blue dashed border. The blue outlines are barriers between these respective spaces. The leaky barrier between the blood and alveolus can disappear during an allergic reaction (as seen in the middle panel), representing vasodilation. Darker gray areas indicate damaged lung tissue. The other shapes dispersed throughout the blood and alveolar spaces represent different cell types. (Adapted with permission from [12])

Mapping the first period of these oscillations onto experimental data reflecting of the time-course of the allergen-induced airway remodeling in mice [16] yields the profile of what could be called the "inflammatory twitch", a self-limited inflammatory event lasting about a month [12].

The conceptual utility of the inflammatory twitch hypothesis is that it suggests a possible pathophysiologic explanation for allergic asthma, a condition involving chronic airway inflammation. Any normal mouse, and presumably also any normal human, can be induced to have a transient airway inflammation by sensitization and challenge with a foreign antigen, but the effect wears off with continued antigen exposure as discussed above. Allergic asthma, in contrast, seems to be characterized by a perpetual inflammatory state [17]. Could it be that this is due to failure of the inflammatory twitch to resolve in asthma? The plausibility of this idea can be investigated using the ABM by observing what happens to the predicted duration and amplitude of the inflammatory twitch when its various components are manipulated. In other words, the ABM can serve as a virtual laboratory in which experiments are performed. This mirrors laboratory experiments in which, for example, relevant cell types and/or cytokines are knocked in or out. Figure 11.7 shows some examples of such in silico experimentation that suggest the most potent mechanism for amplifying and prolonging the inflammatory twitch may be failure of pro-inflammatory cells to undergo apoptosis in a timely manner. In fact, there is experimental evidence that this may be prime mechanism involved in [18, 19]. This is a clear example of when failure to turn off the inflammatory response leads to severe collateral tissue damage.

A corollary of the inflammatory twitch hypothesis is that instigation of allergic inflammation should be accompanied by a refractory period during which reinitiation of the twitch is not possible. This follows from the fact that both neuronal

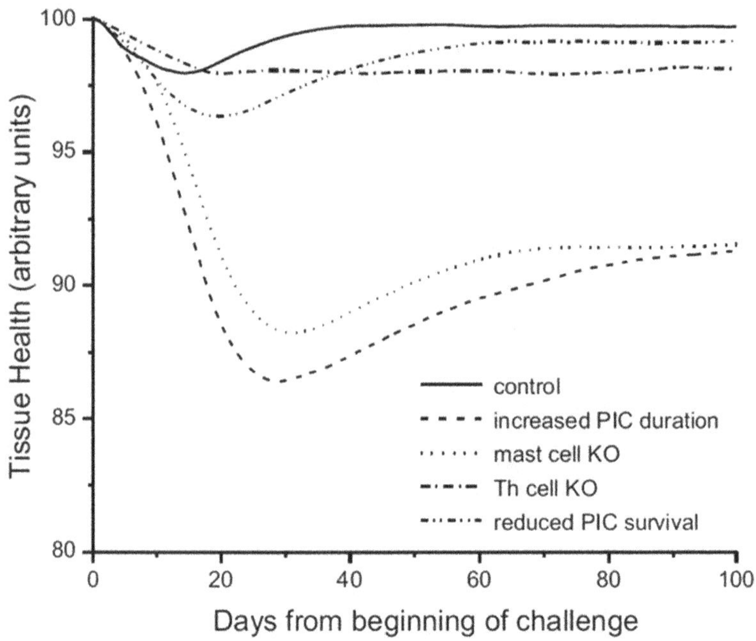

Fig. 11.7 The normal inflammatory twitch (solid line) shows a transient decrease in lung tissue health that resolves back to normal after about a month. The duration and amplitude of the twitch is modulated by the various interventions shown in the figure. The most dramatic of these was caused by a fivefold increase in the lifetime of the pro-inflammatory cells (PIC). (Adapted with permission from [21])

action potentials and muscle twitches, once initiated, are followed by refractory periods. Refractoriness is a necessary feature of systems that automatically resolve back to baseline with predetermined time profiles. In fact, laboratory experiments in which mice were allergically inflamed with ovalbumin for various durations and then rechallenged after various rest periods demonstrate clear evidence of a refractory period [20]. This does not prove the inflammatory twitch hypothesis, of course—absolute proof is never possible with a computational model that can only ever represent reality to a limited degree. Nevertheless, it does support the plausibility of the hypothesis, again demonstrating how computational modeling can lead to biological insight.

11.5 Conclusions

Because of its primary function as a gas exchanging organ, the lung communicates directly with the environment and is therefore inherently vulnerable to attack from environmental pathogens. Elaborate mechanisms have evolved to protect the lung from assault either by removing invaders before they can do any damage or by neutralizing them and repairing their damage after the fact. The inflammatory response, a critical component of this defense mechanism, is highly dynamic and involves orchestrated interactions between numerous cell types. Computational models are invaluable for understanding the behaviors of such complex systems. Coupled systems of ordinary differential equations can be used to model the immune response with high temporal resolution but are limited spatially to a relatively small number of well-mixed compartments such as the pulmonary capillary blood and the lung tissue. Much higher spatial resolution can be achieved with coupled systems of partial differential equations. Both high temporal and high spatial resolution can also be achieved with agent-based models, but often at the expense of considerably greater computational cost. Agent-based models, however, have the important additional benefit of intuitive accessibility for modelers and non-modelers alike.

11.6 Problems

1. Simulate the system of Eqs. 11.1, 11.2, 11.3, 11.4, 11.5, 11.6, 11.7, 11.8, 11.9, and 11.10 and explore its dynamics by setting the initial values of all variables and the values of all parameters to be random numbers from a uniform distribution on the interval [0, 1]. Repeat with different randomizations to get a feel for how the system responds. Now repeat the simulations but replace all Michaelis-Menten expressions in the model equations (i.e., expressions like $\frac{k_{er}C_R(t)}{x_{er}+C_R(t)}$) with just the term in the numerator (i.e., $k_{er}C_R(t)$). How do the dynamics of the system of equations change now that saturation-type kinetics are no longer represented?
2. Think about how the agent-based model in Sect. 11.4 could be extended to be more realistic. For example, how would you add more granularity to the pro-inflammatory and anti-inflammatory mechanisms beyond merely ascribing each to a single generic cell type?

References

1. Hewitt RJ, Lloyd CM. Regulation of immune responses by the airway epithelial cell landscape. Nat Rev Immunol. 2021;21:347–62.
2. Adamcakova J, Mokra D. New insights into pathomechanisms and treatment possibilities for lung silicosis. Int J Mol Sci. 2021;22
3. Warheit DB, Hill LH, George G, Brody AR. Time course of chemotactic factor generation and the corresponding macrophage response to asbestos inhalation. Am Rev Respir Dis. 1986;134: 128–33.
4. Minucci S, Heise RL, Valentine MS, Kamga Gninzeko FJ, Reynolds AM. Mathematical modeling of ventilator-induced lung inflammation. J Theor Biol. 2021;526:110738.
5. Chrisman MA, Goldcamp MJ, Rhodes AN, Riffle J. Exploring Michaelis–Menten kinetics and the inhibition of catalysis in a synthetic mimic of catechol oxidase: an experiment for the inorganic chemistry or biochemistry laboratory. J Chem Educ. 2023;100:893–9.

6. Schirm S, Ahnert P, Berger S, Nouailles G, Wienhold SM, Muller-Redetzky H, Suttorp N, Loeffler M, Witzenrath M, Scholz M. A biomathematical model of immune response and barrier function in mice with pneumococcal lung infection. PLoS One. 2020;15:e0243147.
7. Rossi F, Tortora C, Argenziano M, Di Paola A, Punzo F. Cannabinoid receptor type 2: a possible target in SARS-CoV-2 (CoV-19) infection? Int J Mol Sci. 2020;21
8. de Candia P, Prattichizzo F, Garavelli S, Matarese G. T Cells: warriors of SARS-CoV-2 infection. Trends Immunol. 2021;42:18–30.
9. Amoddeo A. A mathematical model and numerical simulation for SARS-CoV-2 dynamics. Sci Rep. 2023;13:4575.
10. Barnes PJ. Pathophysiology of allergic inflammation. Immunol Rev. 2011;242:31–50.
11. Haczku A, Cao Y, Vass G, Kierstein S, Nath P, Atochina-Vasserman EN, Scanlon ST, Li L, Griswold DE, Chung KF, et al. IL-4 and IL-13 form a negative feedback circuit with surfactant protein-D in the allergic airway response. J Immunol. 2006;176:3557–65.
12. Pothen JJ, Poynter ME, Bates JH. The inflammatory twitch as a general strategy for controlling the host response. J Immunol. 2013;190:3510–6.
13. Cojocaru A, Irvin CG, Haverkamp HC, Bates JH. Computational assessment of airway wall stiffness in vivo in allergically inflamed mouse models of asthma. J Appl Physiol. 1985;2008 (104):1601–10.
14. Riesenfeld E, Allen GB, Bates JH, Poynter ME, Wu M, Aimiand S, Lundblad LK. The temporal evolution of airways hyperresponsiveness and inflammation. J Allergy Ther. 2012;1:1–7.
15. Natarajan P, Singh A, McNamara JT, Secor ER Jr, Guernsey LA, Thrall RS, Schramm CM. Regulatory B cells from hilar lymph nodes of tolerant mice in a murine model of allergic airway disease are CD5+, express TGF-beta, and co-localize with CD4+Foxp3+ T cells. Mucosal Immunol. 2012;5:691–701.
16. Tanaka H, Masuda T, Tokuoka S, Takahashi Y, Komai M, Nagao K, Nagai H. Time course study on the development of allergen-induced airway remodeling in mice: the effect of allergen avoidance on established airway remodeling. Inflamm Res. 2002;51:307–16.
17. Locksley RM. Asthma and allergic inflammation. Cell. 2010;140:777–83.
18. Oh J, Malter JS. Pin1-FADD interactions regulate Fas-mediated apoptosis in activated eosinophils. J Immunol. 2013;190:4937–45.
19. Reddy AT, Lakshmi SP, Dornadula S, Pinni S, Rampa DR, Reddy RC. The nitrated fatty acid 10-nitro-oleate attenuates allergic airway disease. J Immunol. 2013;191:2053–63.
20. Pothen JJ, Poynter ME, Lundblad LK, Bates JH. Dissecting the inflammatory twitch in allergically inflamed mice. Am J Physiol Lung Cell Mol Physiol. 2016;310:L1003–9.
21. Pothen JJ, Poynter ME, Bates JH. A computational model of unresolved allergic inflammation in chronic asthma. Am J Physiol Lung Cell Mol Physiol. 2015;308:L384–90.

Chapter 12
The Aging Lung

Abstract Because of its overarching impact on society, aging has always been a topic of great scientific interest. The human body is capable of self-healing following injury and infection, but the healing is never quite perfect, which leaves microscale alterations to molecular and cellular structures. The slow but gradual accumulation of this microscale damage eventually percolates to the macroscale where it affects the functional performance of tissues and organs. Lung aging contributes significantly to the aging of the organism, and to its susceptibility to disease. While research on aging has mostly focused on the experimental testing of aging hypotheses, modeling has played a major role in unraveling the mysteries of aging. For example, scaling concepts applied to lung aging demonstrate how microscale aging affects macroscale function. This sets the scene for an examination of how aging can be imposed on network models of the parenchyma upon which agent-based models of cellular metabolism operate. Since a living system is a vast network of dynamic interactions, general network approaches are applicable to the question of aging. Increasingly accurate model predictions of lung aging based on its current state and environmental factors may lead to better treatments for age-related diseases.

Keywords Collagen fibers · Elastic fibers · Scaling · Stiffening · Network interactions · Interdependent networks · Network collapse

12.1 Introduction

The course of a normal human lifespan begins at birth, and for the next two or three decades is marked by physical and mental development toward a functional peak in early adulthood. Physical function and mental acuity remain relatively unchanged for several decades after that, barring accidents and personal neglect. Eventually, however, and despite the best of habits, there start to appear signs of physical and/or mental deterioration that accelerate until death. It is customary to refer to this later stage of life as aging because of its obvious and unwanted manifestations, but aging is actually a process that starts as soon as we are born.

B. Suki, J. H. T. Bates, *Mathematical Modeling of the Healthy and Diseased Lung*,
https://doi.org/10.1007/978-3-031-95322-4_12

Aging has major consequences for society. By counteracting diseases and creating safer living conditions, modern medicine and public health measures have increased human life expectancy significantly over the past century. As a consequence, the world's population is rapidly growing older. According to the United Nations, the number of people in the world aged 65 years or older will rise from 761 million in 2021 to 1.6 billion in 2050 [1], potentially exceeding 20% of the world's population. The trend is even stronger for people above 80 years of age. This has concerning implications for life in the twenty-first century because, for example, the size of the workforce compared to the total population is shrinking while the combined healthcare costs for those near the end of life are rising.

Even though modern medicine is allowing people to live longer than ever before, it has so far done little to affect the rate of human aging. Indeed, the mechanisms driving aging are far from completely understood, although it is clear that its deleterious effects are somehow due to the accumulation of damage sustained by the body as it interacts with the world. The human body has a remarkable ability to self-heal following injury and infection, but the healing is never quite perfect, which leaves microscale alterations to molecular and cellular structures. For most of life, the slow but gradual accumulation of this microscale damage is largely inconsequential, but eventually the structural alterations caused by aging percolate to the macroscopic scale where they affect the functional performance of tissues and organs, including the lung.

Lung aging contributes significantly to the aging of the entire body, and to its susceptibility to various diseases [2]. A key driver of aging is cellular oxidative stress resulting from an excess of damaging free radicals that fail to be properly controlled by antioxidants. This can be instigated in the lung by inhalation of noxious particles [3]. Lung aging in turn potentiates the development of various lung diseases such as chronic obstructive pulmonary disease (COPD) [4]. And this process is not unidirectional. COPD, particularly its emphysema sub-type (see Chap. 9), accelerates lung aging [5, 6]. Aging affects most cell types in the lung, including epithelial and endothelial cells, fibroblasts and macrophages, by causing a slow deterioration in the homeostatic maintenance of the extracellular matrix (ECM) [7, 8]. Another critical feature of aging is cellular senescence [9], a state of permanent arrest of the cell cycle that contributes to dysregulated homeostasis by slowing injury repair. These impaired cellular processes lead to gradual remodeling of the airways and parenchyma accompanied by progressive declines in lung function [10].

Because of its overarching impact on society, aging has become a topic of great scientific interest. Research on aging to date, however, has focused on the experimental testing of existing hypotheses such as the free radical theory of aging [11]. Hopefully, previous chapters in this book have convinced the reader that mathematical and computational modeling have a major role to play in unraveling the mysteries of health and disease as they relate to the lung. The present chapter expands this concept to the investigation of lung aging by first introducing the notion of scaling, which is crucial when dealing with different length and/or time scales. This sets the scene for an examination of how aging can be imposed on network models of the parenchyma upon which agent-based models (ABMs) of cellular metabolism operate. Aging, like most chronic diseases and indeed life itself, is a complex emergent phenomenon, so modeling is sure to be an indispensable tool as we seek to understand it, and perhaps eventually even control its progression.

12.2 A Scaling Approach to Lung Tissue Elasticity

Scaling analysis is often used in physics and engineering to reveal hidden relationships between variables over multiple length or time scales. Our understanding of how scaling affects physics and biology started in 1638 when Galileo published his life's work called "Discourses and Mathematical Demonstrations Relating to Two New Sciences". One of these sciences was the strength of materials. Galileo showed, by comparing bones from small and large animals, that one cannot simply scale up the bones of a small animal to create bones suitable for a large animal because the result would not be able to support the increased body weight. He recognized that a bone's cross-section, which is what determines the weight it can support, does not scale isometrically with volume or weight to the two-thirds power. Instead, large animals need thicker bones than smaller animals relative to their body weight.

Scaling also applies to lung elasticity in terms of how the mechanical properties of tiny collagen and elastic fibers determine tissue elasticity at the level of the alveolus, the parenchyma, and the whole organ. It is reasonable to suppose, for example, that organ level tissue stiffness Y is a function of fiber stiffness Y_f. Is it then possible to express this function as $Y = cY_f^{\alpha}$? If so, what is the best value of the scaling exponent α? If $\alpha = 1$, for example, this scaling relation suggests that any aging-related increases in Y_f would be reflected in a proportional increase in Y, which seems unlikely given what is known about animal bones. Another important question is what determines the proportionality constant c in this relationship. Answering questions such as these will enable us to understand the determinants of tissue elasticity over five orders of magnitude of scale from approximately 1 μm at the fiber level to 10 cm at the whole organ level.

The age-related structural changes affecting parenchymal tissue stiffness can be represented by extending published models that describe fiber elasticity and the interactions between fibers and proteoglycans [12] and the structural contributions of elastin and collagen to the septal wall and the alveolar duct [13] (also see Chap. 4). Consider a generic fiber in the parenchyma having cross sectional area A_f and unstressed length l_f. In general, the stress-strain ($\sigma_f - \epsilon_f$) relationship of the fiber is nonlinear, so it is convenient to write it in the form

$$\sigma_f = Y_f \epsilon_f e^{b_f \epsilon_f} \tag{12.1}$$

where the subscript f denotes fiber, Y_f is the linear fiber modulus in the limit of small strains, and b_f is a dimensionless constant that accounts for the exponential nonlinearity. Expanding the right-hand side of Eq. 12.1 as a power series produces

$$\sigma_f = Y_f \epsilon_f [1 + b_f \epsilon_f + \frac{1}{2} b_f^2 \epsilon_f^2 + \ldots]. \tag{12.2}$$

If the strains are not too large ($\epsilon_f < 0.1$), then Eq. 12.2 can be replaced by

$$\sigma_f = Y_f \epsilon_f + b_f Y_f \epsilon_f^2. \tag{12.3}$$

The elastic energy (E_f) stored in the fiber when it is strained from 0 to ϵ is determined by integrating the force in the fiber with respect to its length and then normalizing the force and extension to obtain stress and strain. This gives

$$E_f = A_f l_f Y_f \left[\frac{1}{2} \epsilon_f^2 + \frac{b_f l_f}{3} \epsilon_f^3 \right] \tag{12.4}$$

where the values of parameters in Eq. 12.4 vary spatially throughout the lung.

As the lung inflates, the fibers in the alveolar septal wall tissue become stretched, and both the fibers and the septal walls themselves attempt to align with the direction of strain. The alignment of the fibers, however, is opposed by the presence of proteoglycan molecules that act like compressive springs connected between adjacent fibers [13] (see Fig. 4.9). If the angle between two such fibers changes from θ_0 to θ_f following a small stretch of the tissue, the elastic energy stored in the compression of the intervening proteoglycans is $\frac{1}{2} r_b (\theta_f - \theta_0)^2$, where r_b is the compressive, or bond-bending, spring constant. The total elastic energy E_t in a tissue strip of volume V is then, summing over all M fibers in the tissue,

$$E_t = \sum_{f=1}^{M} V_f Y_f \left[\frac{1}{2} \epsilon_f^2 + \frac{b_f l_f}{3} \epsilon_f^3 \right] + \frac{1}{2} r_b \delta_f^2 \tag{12.5}$$

where $\delta_f = \theta_f - \theta_0$ and $V_f = A_f l_f$ is the volume of a single fiber. E_t is thus a function of the constitutive properties of the fibers, the proteoglycan molecules, and the size of the network. The strain energy density of the tissue is then

$$e_t = \frac{E_t}{V} = \sum_{f=1}^{M} \frac{1}{2} v_f Y_f \epsilon_f^2 + \frac{1}{2} v_f Z_f \epsilon_f^3 + \frac{1}{2} q_f \delta_f^2 \tag{12.6}$$

where $v_f = V_f/V$ is the fiber volume fraction, $Z_f = \frac{2}{3} b_f l_f Y_f$ is a nonlinear elastic constant, and $q_f = r_b/V$ is the proteoglycan bond-bending constant per unit volume.

Equation 12.6 can be expressed as two nested summations. The outer summation accounts for all the n_w septal walls in the tissue at length scales above 100 μm, while the inner summation accounts for all the n_f fibers within an individual wall at scales below 100 μm, thus:

$$e_t = \sum_{w=1}^{n_w} \sum_{f=1}^{n_f} \frac{1}{2} v_f Y \epsilon_f^2 + \frac{1}{2} v_f Z_f \epsilon_f^3 + \frac{1}{2} q_f \delta_f^2. \tag{12.7}$$

Equation 12.7 expresses the multi-scale nature of parenchymal tissue mechanics. To make this equation more amenable to calculation and interpretation, the individual wall expressions are replaced with an effective average expression. Denoting averages in the form $\frac{1}{n_f}\sum_f(\cdot) = <\cdot>_w$, Eq. 12.7 becomes

$$e_t = n_f v_f Y_f \frac{1}{2} \sum_{w=1}^{n_w} \left[<\epsilon_f^2>_w + \frac{Z_{f,w}}{Y_{f,w}} <\epsilon_f^3>_w + \frac{q_{f,w}}{v_f Y_f} <\delta_f^2>_w \right]. \tag{12.8}$$

Finally, averaging across all the septal walls gives the total energy density as

$$e_t = n_w n_f v_f Y_f S(Y_f, Z_f, v_f, q_f, \epsilon_f, \delta_f). \tag{12.9}$$

The dimensionless function S depends on the joint distribution of the linear and nonlinear fiber stiffnesses, the properties of the proteoglycans, the collagen and elastin volume fractions, and the strain and angular orientation of the fibers. The scaling relationship in Eq. 12.9 demonstrates that e_t depends linearly on n_w and n_f.

Equation 12.8 can also be partitioned into separate contributions from collagen and elastin (denoted by subscripts c and e, respectively) expressed in terms of stress (computed as the derivative of the energy density with respect to macroscopic strain) and the incremental modulus Y of the tissue strip (computed as the second derivative of the energy density) thus:

$$\sigma = n_w n_c v_c Y_c \frac{\partial S_c}{\partial \epsilon} + n_w n_e v_e Y_e \frac{\partial S_e}{\partial \epsilon} \tag{12.10}$$

$$Y = n_w n_c v_c Y_c \frac{\partial^2 S_c}{\partial \epsilon^2} + n_w n_e v_e Y_e \frac{\partial^2 S_e}{\partial \epsilon^2} \tag{12.11}$$

where n_c and n_e represent the effective number of collagen and elastic fibers, respectively, in a single average wall. Partitioning of the energy can be done for any structural elements such as septal walls, alveolar ducts, small airways, or macroscopic segments of the lung such as the apex and base.

Unfortunately, analytic solutions to Eqs. 12.10 and 12.11 cannot be obtained, in general, because of the nonlinearity of the function S and the network connectivity. Even numerical solutions may be difficult to obtain because of the complexity of the problem. Nevertheless, there is something useful to be learned from the above derivation. First, for small displacements, Eq. 12.8 can be simplified by neglecting the second and third terms under the summation and introducing effective wall-averaged strains that can be linked to the macroscopic strain ϵ so that

$$e_t = n_f v_f Y_f \frac{1}{2} \sum\nolimits_w \left[< \epsilon_f^2 >_w \right] \approx \frac{1}{2} n_w n_f v_f Y_f \varphi_f^2 \epsilon^2 \tag{12.12}$$

where $0 < \varphi_f \leq 1$ expresses the fact that an average fiber in an average wall experiences less than the macroscopic strain because of septal wall and fiber alignment as well as the straightening of wavy fibers. The stress and stiffness of the strip can then be written in a similar fashion to Eqs. 12.10 and 12.11 as

$$\sigma = n_w n_c v_c \varphi_c^2 Y_c \epsilon + n_w n_e v_e \varphi_e^2 Y_e \epsilon \tag{12.13}$$

$$Y = n_w n_c v_c \varphi_c^2 Y_c + n_w n_e v_e \varphi_e^2 Y_e \tag{12.14}$$

where φ_c^2 and φ_e^2 are straightening factors for collagen and elastin, respectively.

Equations 12.13 and 12.14 state that the macroscopic stress and the total incremental modulus of the tissue strip can be modeled by two parallel springs corresponding to collagen and elastin, respectively, with their moduli weighted by the product of certain morphometric parameters. These weighting factors, $n_w n_c v_c \varphi_c^2$ and $n_w n_e v_e \varphi_e^2$, represent scaling relations that provide simple and important insight into the mechanical function of the parenchyma. For example, when advanced glycation end products enhance the cross-link density in the aging lung [14], both the fiber diameter [15], and hence its volume fraction as well as the fiber modulus of collagen [16], are expected to rise. The corresponding stress and modulus in turn scale synergistically with the product $v_c Y_c$.

The multiscale relations expressed in Eqs. 12.10, 12.11, 12.12, 12.13 and 12.14 also suggest structure-function interpretations for the function S. Proteoglycan molecules lodged between fibers make the network elasticity nonlinear through q_f even if the fibers themselves are linearly elastic (i.e. $Z_f = 0$). Hence, macroscopic nonlinearity always contains nonlinear contributions from the proteoglycans in the form of the factor $q_f/v_f Y_f$ that describes the strength of proteoglycan compressibility relative to fiber stiffness. Since the proteoglycan content of tissue decreases with age, as has been demonstrated in cartilage [17], a corresponding drop in $q_f/v_f Y_f$ would facilitate fiber alignment during stretching, resulting in straighter fibers at lower strains. This is supported by histologic data showing that the radius of curvature of collagen fibers decreases from 0.052 μm^{-1} at age 20 years to 0.034 μm^{-1} at age 80 years [15]. In addition, certain glycosaminoglycans in the lung gradually decline during cell senescence [18, 19], although others may not change significantly [20].

Equations 12.13 and 12.14 also indicate that stress and incremental modulus scale with the average value of the fiber modulus, the number of septal walls in the tissue, the fiber density per wall, and the volume fraction of fibers. The products $n_w n_c v_c$ and $n_w n_e v_e$ are the absolute fiber volume fractions. Aging has been found to decrease n_w as a result of airspace enlargement, while n_f appears to decrease and v_f increases [15]. Biochemical tissue analysis has shown that these volume fractions decrease by a factor of 2 for collagen between the ages of 20 and 80 years, but show little change for elastin [21].

Arguments based on scaling thus suggest that collagen fiber modulus in the lung increases with age. The derivation of the model equations (Eqs. 12.10, 12.11, 12.12, 12.13 and 12.14) involve several assumptions, of course, including that Y_f can be taken out of the inner summation in Eq. 12.7, that the effects of the non-fiber matrix on stiffness can be ignored (although this would be expected to increase septal wall thickness, which also increases with age [22]), that surface tension and mechanical interactions between collagen and elastin can be neglected, and that incremental stress-strain analysis (which is reasonably appropriate for normal breathing) can be applied. For larger deformations such as those that occur during deep inspiration, the second Piola-Kirchhoff stress and the Green strain tensors reviewed in Chap. 4 should be used in the derivations. Nevertheless, the model predictions are in accord with the known age-related increases in lung tissue stiffness [12]. In fact, collagen stiffness increases almost everywhere in the body during aging [23]. This compensates for the progressive decrease that occurs in collagen weight per unit volume of tissue. Figure 12.1a shows uniaxial stress-strain curves measured in sub-pleural tissue strips from a 5-month old (young adult) and a 21-month old (aged adult) mouse together with fits provided by a theoretical model that predicts these age-related changes to be due to a mild (25%) increase in septal wall modulus from 12.4 to 15.5 kPa [16].

The tissue strips used to create the curves in Fig. 12.1a were estimated to contain about 1000 alveoli. These curves are very different from those shown in Fig. 12.1b, which were obtained from micro-strips of lung tissue containing only 1–2 alveoli, one from a 29-year-old healthy subject and the other from a 80-year-old patient with obstructive syndrome (Fig. 12.1b) [24, 26]. The curves from the human micro-strips are much more nonlinear and reach stress magnitudes that are substantially larger than those from the macroscopic mouse strips. These differences may be due to the

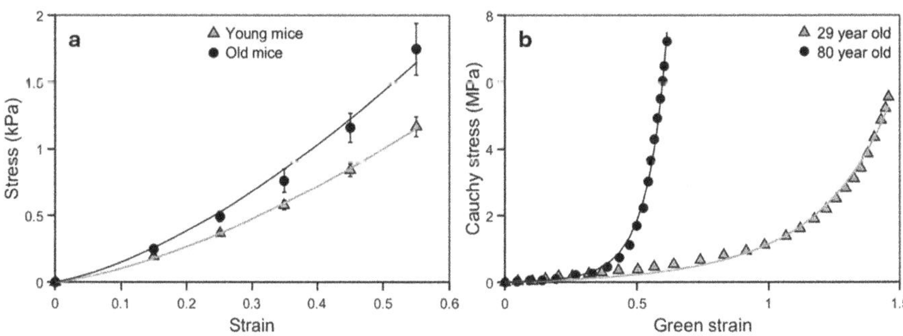

Fig. 12.1 Stress-strain curves of lung parenchymal strips and their respective model fits from young and aged subjects. (**a**) Comparison of mean and standard deviation of normal (5 months) and aged (21 months) mouse stress-strain curves (symbols) and the corresponding model fits (solid lines). (**b**) Cauchy stress as a function of Green strain calculated from the force-extension data published in Ref. [24] corresponding to lung tissue from a normal 29-year-old human subject and an 80-year-old patient with obstructive lung disease. Fits (solid lines) were obtained using Fung's hyperelastic strain energy formulation [25]. (Data in panel **a** was reproduced with permission from Ref. [16])

fact that, in the macroscopic strips, alveolar walls with their collagen fibers are free to realign in the direction of stretch whereas the wall and thus the fibers in the micro-strips are already largely aligned and so have little option but to elongate. Furthermore, fitting the Green strain vs second Piola-Kirchhoff stress curves (see Chap. 4) with Fung's hyperelastic strain energy model [25], provides a good description of the data. The stiffness around zero strain is 0.5 kPa for the healthy young subject and is 5 kPa from the old strip. Atomic force microscopy (AFM) indicates that lung tissue has stiffness values of only ~1 kPa at 10 years of age and ~2 kPa at 60 years of age [27]. Macroscopic indentation tests together with pressure-volume curves have also provided estimates of the shear and the bulk moduli of isolated human lungs [28]. The Young's modulus has been reported to increase from 7.5 kPa at 10 years of age to 10.7 kPa at 100 years of age [12]. Thus, different measurement techniques and conditions provide widely varying stiffness estimates for the lung parenchyma, even though all approaches show that tissue stiffness increases with age.

12.3 Agent-Based Modeling of Aging

One of the most remarkable and distinctive attributes of the healthy human body is its ability to self-repair after sustaining injury. It is therefore not surprising that this ability wanes with age. Indeed, aberrant repair is thought to be behind many of the chronic diseases associated with old age, a prime example being idiopathic pulmonary fibrosis (Chap. 8). This can be modeled by adapting the computational model of self-repair introduced in Chap. 6 [29]. In this model, metabolically active agents modify the local stiffness of a tissue substrate in an attempt to keep it at homeostatic equilibrium. The tissue substrate is represented by a network of springs, while the agents move randomly between adjacent spring junctions. The equations that describe how agents and springs interact (Eqs. 6.17 and 6.18) are

$$\frac{da}{dt} = \frac{k}{1+k} - a \tag{12.15}$$

and

$$\frac{dk}{dt} = (a-q)(-Dk) + P \tag{12.16}$$

where a is the level of activity of an agent, and k is the elastic constant of the spring it has just moved across. The parameters q, D and P represent, respectively, the threshold at which the agent behavior switches between digesting and rebuilding tissue, the rate of tissue digestion, and the rate of tissue production. Equations 12.15 and 12.16 have a fixed point (a^*, k^*) defined by setting their left-hand sides to zero and solving for a and k (Sect. 6.3). The model thus displays a form of homeostasis following injury (simulated by a perturbation in tissue properties) by always returning toward the fixed point.

To explore the possible consequences of aging on self-healing using this model with baseline parameter values $q = 0.2$, $D = 0.32$ and $P = 0.1$, one can ask how these values might change with age. There are several possibilities. For example, aging attenuates the turnover of collagen [30, 31] and the activity of lysyl oxidase, a key extracellular enzyme responsible for cross-linking of collagen and elastin [32]. Also, the cell surface receptor β1 integrin [33] is downregulated in aged skin [34], causing cells that are normally attached to the ECM [35] to have reduced speed and directional persistency of movement over the tissue, possibly due to reduced cytoskeletal tension [36]. These aged-related changes can be incorporated into the model by changing its parameter values to $q = 0.4$, $D = 0.08$ and $P = 0.0427$, which moves the fixed point (a^*, k^*) of the system from (0.50559, 1.022612) to (0.6667, 2), corresponding to increased fibrosis.

The model described above also self-heals a localized fibrotic injury in such a way that a halo is first created around the injury site and then decays away over time. Figure 12.2 demonstrates that the dynamics of this halo are significantly altered with the age-adjusted parameters. In particular, the halo in the aged model is wider and peaks somewhat later, with complete repair of the injury taking 10 times longer compared to the baseline (non-aged) case. Figure 12.2 also shows that the behavior of the model is altered somewhat when the random walk of the agents is preferentially biased in the direction of higher substrate stiffness, corresponding to either altered mechanotransduction or altered tissue stiffness.

There are many possible ways to extend the model expressed in Eqs. 12.15 and 12.16. For example, one can use a higher-order sigmoidal Hill function in either or both equations to describe the feedback and/or crosstalk between a and k. However, the model still lacks the ability to account for damage. This can be taken into account by adjusting Eqs. 12.15 and 12.16 as follows:

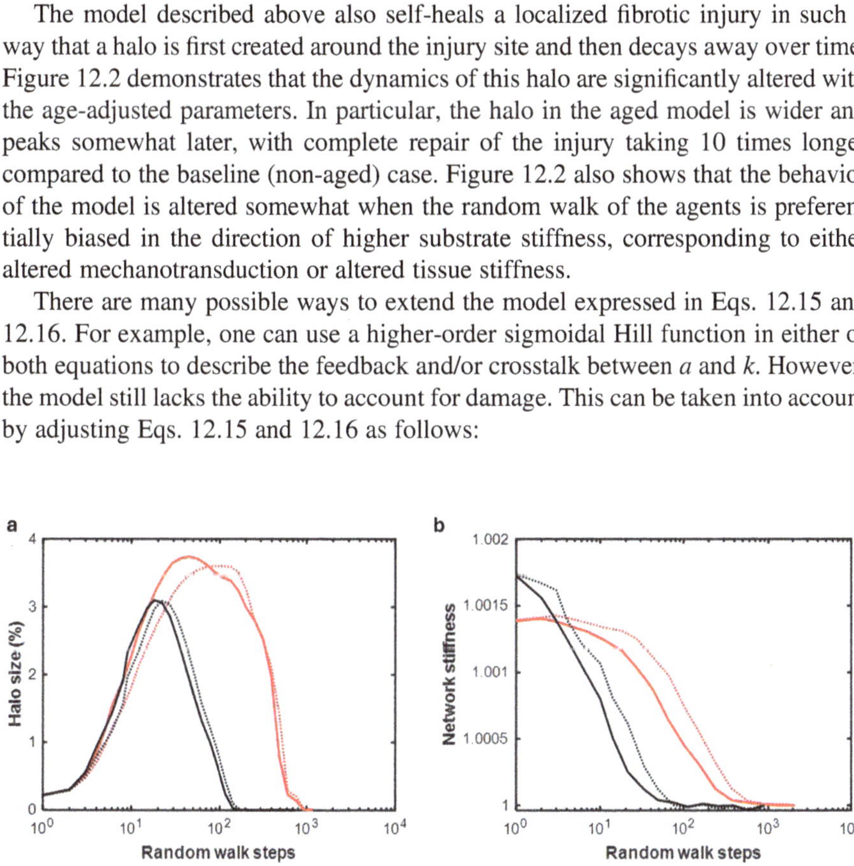

Fig. 12.2 Evolution of network properties in the baseline and the aging models following an initial injury of setting a single spring constant to a value twice the fixed point. Time courses of halo size relative to the size of the network (**a**) and normalized stiffness (**b**) for the stiffness-biased (solid lines) and pure random (dotted lines) migration for the baseline (black) and aging (red) models. The network stiffness increases only by 0.15% above the equilibrium value because less than 4% of the springs are affected by the halo. (Reproduced with permission from Ref. [29])

$$\frac{da}{dt} = \frac{k}{1+k} - a + x \tag{12.17}$$

$$\frac{dk}{dt} = (a-q)(-Dk) + P \tag{12.18}$$

and

$$\frac{dx}{dt} = F\frac{xa}{1+a^2} - Cx \tag{12.19}$$

where x represents the extent of damage, F is the strength of the crosstalk between damage and activity, and C accounts for the spontaneous decay of the damage. Using the parameter values $q = 0.4$, $D = 0.08$ and $P = 0.0427$, and setting $F = 0.1$ and $C = 0.05$, the new fixed point is $(a^*, k^*, x^*) = (0.6667, 2, 0)$, showing that homeostasis corresponds to the complete repair of damage because $x^* = 0$. Figure 12.3 compares how damage to the baseline model (representing healthy youthfulness) and to the aged model affect agent activity (Fig. 12.3a) and substrate stiffness (Fig. 12.3b). When the initial conditions for the aged model were set to the fixed point of the baseline model (0.50559, 1.022612,0), its subsequent transition to the aged fixed point (0.6667, 2,0) mimics the slow drift that occurs with aging (red

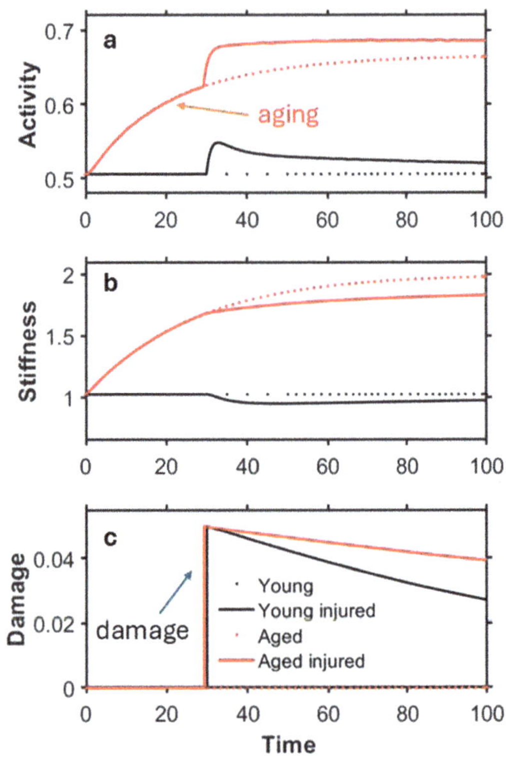

Fig. 12.3 Response to an injury in the baseline (young) model (black curves) and during simulated aging (red curves). Panels (**a**), (**b**) and (**c**) correspond to agent activity, spring stiffness and damage magnitude, respectively. Initial values at time 0 are equal to the fixed-point values of the baseline model. Solid lines show the effects of a sudden damage at time 30 mimicking a wound that lowers the stiffness. The black dotted lines do not change since they correspond to the baseline model at the fixed-point. The red dotted lines represent normal aging

dotted lines in Fig. 12.3). Of course, the fixed point itself is never fixed in time because age continues to increase, as reflected in the progressive loss in the abilities of epithelial and immune cells to maintain normal homeostasis [37].

Consider what happens when an injury has been imposed on both the young and the aged models at time step 30 (Fig. 12.3c). Agent activity (Fig. 12.3a) immediately increases in both models and then decays towards the respective fixed points. In agreement with the halo dynamics shown in Fig. 12.2, activity approaches its fixed point visibly faster in the young model. Substrate stiffness also drops immediately following injury, and its return to the original model trajectory is again significantly slower in the aged model. Importantly, the value of C is the same in both the young and aged models, so delayed repair is a consequence of how activity interferes with repair through the first term on the right-hand side of Eq. 12.19. Moreover, for positive damage, activity always increases above the fixed point of the aged system. This can be interpreted as injury accelerating the aging process. Indeed, it is now believed that elderly people are not only more susceptible to COPD, but the disease itself accelerates aging [5, 38]. This view is supported by clinical trial data showing that COPD patients between 56 and 65 years of age showed comorbidities and signs of aging similar to those of healthy controls 15 to 20 years older [39].

The above model could be further extended by having the Eqs. 12.17, 12.18 and 12.19 govern agent activity on a parenchymal spring network. This would allow the study of how tissue damage affects the spatial dynamics of the halo in Fig. 12.2. The agents themselves could also be specifically tailored to represent cells involved in the pathogenesis of COPD (see Chap. 9) or pulmonary fibrosis (see Chap. 8), perhaps with additional detail about cellular mechanisms incorporated through the use of differential equations describing cytokine signaling (see Fig. 8.2). We leave these extensions as an exercise for enthusiastic readers.

12.4 Stochastic Network Modeling of Aging

Throughout this book, we have considered various network models that describe specific anatomic or biologic systems. These include the physics-based networks comprising the airway tree and the elastic lung parenchyma (Figs. 2.2 and 2.3), or the active metabolic network defining cell-protein interactions in pulmonary fibrosis (Fig. 8.2). A network can also be viewed more abstractly, however, as a graph with nodes and links, without reference to any particular physical, chemical or biological entity. This facilitates the study of general network properties related to topology and complexity. One such property is the connectivity of a network, defined by the degree distribution that is represented by the histogram of the number of links each node has with other nodes [40]. The complexity of the network is associated with the second moment of this distribution [41] in addition to the sheer number of nodes it contains [42]. Since a living system is essentially a vast network of dynamically interacting components, it is natural to take a network approach to the question of aging.

An early network model of aging focuses on how the kinetics of protein turnover are affected by defective mitochondria, aberrant proteins, oxygen free radicals, antioxidants and proteolytic scavengers [43]. The premise of this model is the "disposable soma theory", which posits that the gradual accumulation of random damage in somatic cells contributes to the advancing decrepitude that comes with increasing age.

Large-scale network mapping now shows that evolution has limited the capacity of organisms to repair themselves indefinitely. This leads to progressive loss of normal cell function with age, thereby regulating life expectancy [44]. Key predictions of the model included elevated levels of inactive protein, dysfunction and loss of mitochondria, and reduced ATP production [43], all of which seem likely of genes and proteins, making it possible to study the topology of such networks in detail. This has been applied to protein-protein and gene regulatory networks in yeast, revealing some of the ways that cellular pathways influence chronological aging [45]. Not surprisingly, the degree distribution of these networks follows power laws of the form

$$p(K) \sim K^{-\gamma} \tag{12.20}$$

where K is the node degree (number of links) connecting to a node, and the exponent γ has a value of roughly 1.5, signifying complex structure and behavior [46]. However, an organ such as the lung is much more than a collection of gene and protein networks. For example, breathing imposes dynamic mechanical forces on virtually all cells in the lung, and most of these cells have critical functions mediated by mechanotransduction [47, 48]. The cells and tissues in the lung thus comprise yet another complex network that operates at the level of scale of the organ itself. Modeling the effects of aging on the lung therefore requires a more general approach.

An example of such an approach is based on the idea that evolution of organisms inevitably increases complexity by causing intricate multi-scale interdependencies to develop from the intra-cellular to the inter-organ levels. It is then the breakdown of these interdependencies that gives rise to the physiologic features of aging [42]. This idea is expressed in network form by allowing nodes to represent any important functional entity in the body, including different proteins, cell types and tissues. Each node is assigned some function as well as a set of directed links that specify how the node influences the function of other nodes. For example, the normal mechanical functioning of an alveolus depends on it having an appropriate concentration of pulmonary surfactant. A node representing an alveolus in the model thus receives a link from another node representing its resident type II alveolar epithelial cells. This link expresses the rate of surfactant secretion into the alveolus. At a coarser level of resolution, a single node could represent all the alveoli, or even an entire organ. For example, a node representing the lung has a link to the node representing the heart that denotes transport of oxygenated blood. Evolution works on the topology of the network by adding or removing nodes and links. At any point in evolutionary time, the population of a species is represented by a generic network while individuals within the population are represented by perturbations on the generic network.

The aging of an individual is simulated in this model by the implementation of three rules [42]. First, each node must have at least one incoming and one outgoing link so that it receives enabling function from at least one node and provides enabling function to at least one other node. Second, a functioning node can stop functioning at any point with a fixed probability α, and a nonfunctioning node can start functioning again with probability β. The state X of a node is 1 when the node functions, and 0 when it does not. Finally, a node stops functioning when most of the nodes from which it receives input stop functioning, and it starts functioning again when the majority of its input nodes resume their normal function.

Aging proceeds in this network model as follows [42]. At any time t, the configuration of the network corresponding to a given individual in the population is given by the state vector

$$\psi(t) = [X_1(t), X_2(t), \ldots, X_N(t)] \tag{12.21}$$

where the $X_i(t)$ are either 0 or 1. The vitality of the network has a value between 0 and 1 given by

$$\phi(t) = \frac{1}{N}\sum\nolimits_{i=1}^{N} X_i(t). \tag{12.22}$$

Initially, all nodes are assumed to be functional: $X_i(0) = 1$, $(i = 1, \ldots, N)$. To start the aging process, the state of a small randomly selected fraction of the nodes is set to be non-functional (i.e., $X_j(1) = 0$; $j\in$ initially damaged nodes). At each subsequent time step, $\psi(t)$ is updated by setting $X_i(t) = 0$ for each functioning node with probability p_0, and $X_i(t) = 1$ for each nonfunctioning node with probability p_1. The state of a node thus remains unaltered with probability $1 - p_0 - p_1$. At each time step, if the majority of the nodes supplying function to a given node are non-functional then the node also becomes non-functional, and conversely. This process is continued until $\phi(t)$ reaches the critically low level of $\phi_c = 0.01$ corresponding to death. The number of time steps to reach ϕ_c corresponds to the lifetime of the individual. A different stochastic realization of the network (with the same governing parameters) corresponds to a different individual in the population, so by performing a Monte-Carlo simulation with this model it is possible to determine the distribution of lifetimes in the population. If $s(t)$ is the fraction of the population that remains alive at time t, the mortality rate of the population is then

$$m(t) = -\frac{s(t+1) - s(t)}{s(t)}. \tag{12.23}$$

The mortality behavior of this model was investigated using two different network topologies. One was the Erdős-Rényi random network, first introduced in 1958 [49], in which the probability of having a link between any two nodes is constant. The

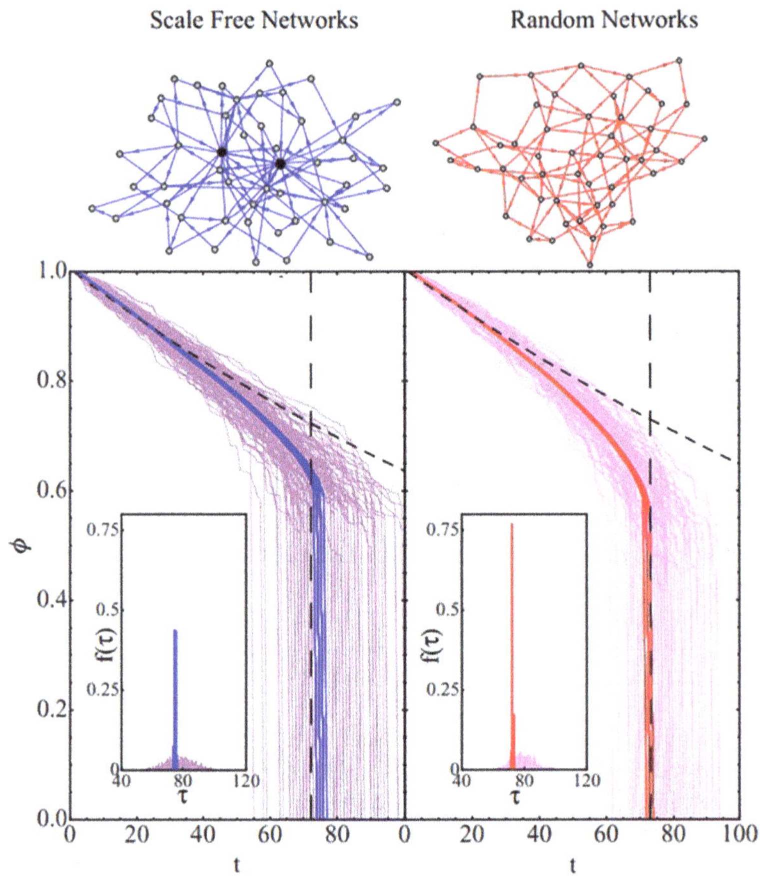

Fig. 12.4 Aging of scale-free (left column) and random (right column) networks. The graphs show the fraction of surviving nodes in 100 simulations. Example network configurations are above the graphs. The pattern of aging is similar in the two network types. The inset shows survival time distributions. Increasing the network size from 2500 (light purple and pink) 10^6 (dark blue and red) sharpens the survival time distributions. The black lines denote theoretical predictions for the initial slope. (Reproduced with permission from Ref. [42])

other was the Barabási-Albert scale-free network, introduced in 1999 [50], in which the degree distribution is given by Eq. 12.20 with $\gamma = 3$. The main difference between these two topologies is that the scale-free network has a few highly connected hub nodes and many nodes with few links. Death appears suddenly (Fig. 12.4), even when the probability that a nonfunctional node becoming functional again is equal to the probability of the reverse happening. The reason for this perhaps puzzling behavior is that although any functional node can become non-functional, not all non-functional nodes eventually become functional again. This suggests that an apparent balance between damage and repair in a living network is not enough to imbue the network with immortality. The fact that the random and scale-free networks behave so similarly (Fig. 12.4) also indicates that these findings are independent of network topology.

The mortality rates (Eq. 12.23) predicted by the model match those of C. elegans, drosophila, medflies, beetles, mice, and Himalayan goats [42]. The model also predicts the observed differences in mortality rate between a wild-type and a long-lived mutant of C. elegans. These two organisms were exposed to similar environmental conditions and thus were subject to similar initial conditions and damage processes. This implies that the only difference between the wild-type and the mutant was their rates of damage repair, which is how the model accounts for their differing mortality rates. Model sensitivity analysis also demonstrates that the larger networks succumb more rapidly to aging than smaller networks [42]. An intriguing implication of this finding is that, to the extent network size can be taken as a measure of complexity, more complex organisms age more rapidly.

The changes taking place within the network model at a given time step often involve multiple nodes. The number of nodes involved is usually relatively small, but is also highly variable, so there can be occasional avalanches that sweep through the system killing many nodes simultaneously. These give to large values of $\Delta\phi(t) = \phi(t) - \phi(t-1)$. In fact, $\Delta\phi(t)$ has a power-law distribution given by $p(\Delta\phi) \sim \Delta\phi^{-2.7}$. This suggests that aging progresses gradually for the most part, but is punctuated by sudden significant decrements in function similar to how cognitive slowing has been predicted to occur in discrete steps [51]. Interestingly, the model also predicts that immortality is possible if the repair mechanisms in the network are enhanced sufficiently.

12.5 Network Modeling of the Aging Lung

Aging affects the lung in numerous ways. At the alveolar level, aging gradually increases the average airspace size [22] together with residual volume and functional residual capacity [52]. While this airspace enlargement is probably due, in part, to increased outward recoil of a stiffening chest wall [53], age-related increases in lung compliance have also been reported [54, 55] accompanied by septal walls that are thicker [22] and have straighter and thicker collagen and elastic fibers [15]. The slow but continuous process of advanced glycation [14] increases the stiffness of collagen fibrils and fibers by adding new crosslinks [16]. This, in turn, alters cellular mechanotransduction and the effects it has on homeostasis and injury repair [56]. These changes begin at the microscopic level, but they eventually have macroscopic level consequences gas exchange and exercise capacity. Nevertheless, the precise mechanisms involved in the aging of the lung, both at the micro and macro scales, are extremely complex and not fully understood [12]. This creates yet another opportunity for computational modeling to play an explanatory role [57].

Although not explicitly about the lung, the network model described above has some clear implications for lung aging. First, however, it must be pointed out that different cellular and organ systems age at different rates [58]. The lung links all these organ systems together via its critical role in determining the body's oxygen supply and carbon dioxide waste removal capabilities [59]. In analogy with the overall network collapse illustrated in Fig. 12.4, collapse of the network that comprises the lung can occur via massive airway and alveolar closure. These events can be driven, for example, by surfactant deactivation due to drowning or acute respiratory distress syndrome (see Chap. 7). This destroys the outgoing links from the lung to all the other nodes in the organ network that depend on a steady supply of oxygen. Such a sudden avalanche-like removal of so many network links leads rapidly to death in the absence of immediate and extreme intervention. Death is thus more likely to occur, and with greater rapidity, in elderly individuals in whom the rest of the network is already somewhat compromised with a reduced complexity (i.e., number of functional nodes). Put in terms of the network model, the consequences of an insult that causes a sudden avalanche of link failures may be compensated for in a young individual who still has enough functioning nodes to restore critical links. Older individuals, on the other hand, may lack sufficient functioning nodes to survive the insult. Indeed, cellular senescence in one organ has been associated with increased risk of multi-organ failure [60].

Environmental factors also influence aging and susceptibility to disease. Here, again, the lung plays a key role as prime mediator between the environment and the internal workings of the body. Events begin when airway epithelial cells are exposed to toxic inhaled particles that can compromise the regenerative capacity of alveolar epithelial type II cells. This effectively increases the probability α that certain nodes in the network will fail. Since the aged lung is already compromised in this regard, a given amount of toxic exposure will move the network closer to the critical collapse point compared to a younger network. For example, the negative effects of aging on mechanotransduction [61] have been associated with an increased risk of ventilator-induced lung injury in the elderly [62] and an impaired response of the airway smooth muscle to mechanical stimulation [63]. Similarly, during spontaneous breathing, the cyclic stretching of the lung tissue [64] strongly influences cellular processes [65] and exhibits a natural variability that decreases significantly during sleep in young adults, but not in the elderly [66]. This signifies an age-related lack of adaptability to the circadian rhythm due again to reduced network complexity.

Although the above network model allows general inferences to be made about the effects of aging on the lung, more specific inferences require that the nodes and links of the model be assigned specific identities. The nodes of such a model might represent, for example, stiffness of the ECM, local time varying tissue strains, cytoskeletal proteins and their network organization, and the expression dynamics of integrin and nuclear factors such as YAP/TAZ, etc. The corresponding network links could then be integrin-mediated adhesion of cells to the ECM, the force generated by the cytoskeleton, and its consequent effects on local ECM strain and collagen secretion. Aging affects many of these nodes and links, such as by reducing the amount of cellular actin [67], which in turn alters mechanotransduction. The cells of the immune system also play a major role in normal repair mechanisms; hence, aging of these cells surely has a negative impact on tissue repair and defense

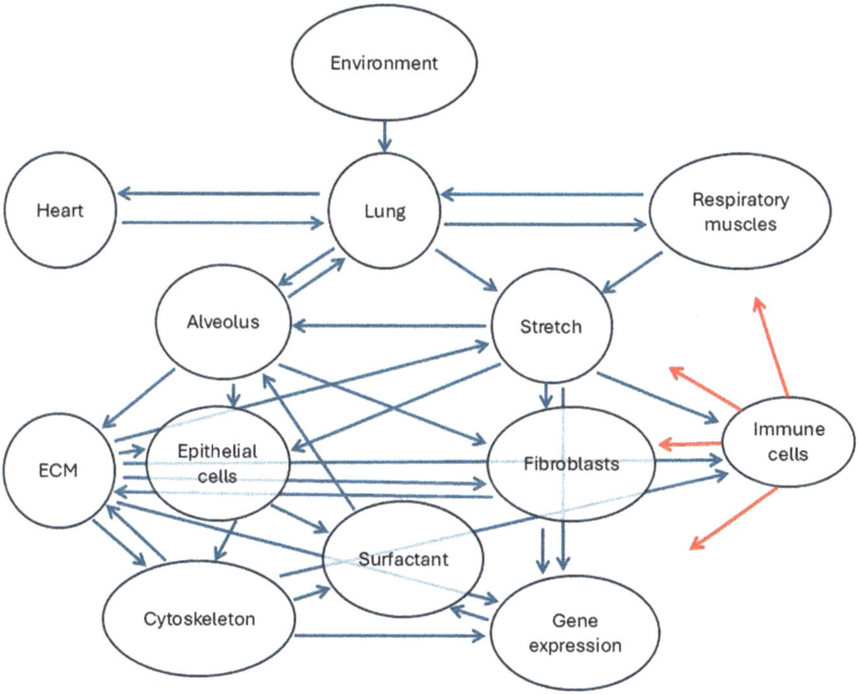

Fig. 12.5 A greatly simplified dependency interaction network for the lung. The red arrows for immune cells are shown to suggest that they can connect virtually anywhere. The effects of aging can be represented by weakening links and/or eliminating nodes

mechanisms against infections and the diseases of old age [68]. A possible arrangement of some of these processes into a dependency network, albeit highly limited in scope, is proposed in Fig. 12.5.

12.6 Conclusions

Aging has important functional and structural consequences for the lung. Even though it is inescapable, there is some hope we can stave off the ravages of aging to some degree while maximizing quality of life in the meantime. Mathematical and computational modeling can potentially help us do this. Nevertheless, the modeling of aging in general, and of the lung in particular, is still in its infancy. This reflects the enormity of the task involved in computationally recapitulating all the complex multiscale effects that aging has on an organ. As models progress through the incorporation of increasing levels of anatomic and biologic detail from the experimental laboratory, we can expect increasingly accurate predictions of the future state of the lung based on its current state and the environmental factors that impact it. This will hopefully lead to better treatments for age-related diseases, and to more informed lifestyles aimed at preventing the need for treatment in the first place. In the meantime, we hope that the reader, having aged somewhat during the reading of this book, has nevertheless been rejuvenated by the intellectual stimulation that mathematical and computational modeling provides.

12.7 Problems

1. Equations 12.13 and 12.14 describe stiffness and stress as scaling relations. Search the literature for numerical values relevant to the human lung of all parameters such as volume fractions and fiber moduli (some of these can be found in the current chapter). If you don't find values for all parameters, make educated guesses about their values based on data found in different species.
   (a) Lung tissue stiffness has been reported in the literature. In principle, knowing all the parameters on the right-hand side of Eq. 12.14 should allow estimating tissue modulus. Does such a calculation provide an estimate of lung tissue modulus in reasonable agreement with healthy normal adult values in humans? What do you have to assume about φ_c^2 and φ_e^2 for this prediction to be correct within an order of magnitude.
   (b) Research the literature further to find out how these values change with aging. Try to predict again the stress and the modulus, and adjust parameter values to match published values. In general, how do lung structural values change with aging? Speculate on the biological mechanisms that may account for such alterations.
2. Tissue aging can be modeled using the systems of differential equations described in Sect. 12.3.
   (a) Find the analytical expressions for the fixed point of the system described by Eqs. 12.15 and 12.16. Code up these equations and use a standard ODE solver such as the Runge-Kutta fourth-order algorithm to simulate aging with the model as described in Sect. 12.3.
   (b) Implement Eqs. 12.17, 12.18 and 12.19 and examine how aging influences healing of damaged tissue. Try to recreate Fig. 12.3.
   (c) Add additional realism to the model such as inflammation. Find relevant literature to understand how persistent inflammation contributes to aging. How would you add this to the model? Explore the behavior of the model by implementing these additional mechanisms.
   (d) Implement Eqs. 12.17, 12.18 and 12.19 on a hexagonal network model and study halo dynamics similar to those shown in Fig. 12.2. Does the inclusion of damage affect the halo dynamics?

References

1. United Nations: leaving no one behind in an ageing world. World Social Report 2023.
2. Schneider JL, Rowe JH, Garcia-de-Alba C, Kim CF, Sharpe AH, Haigis MC. The aging lung: physiology, disease, and immunity. Cell. 2021;184:1990–2019.
3. Wang SN, Shi YC, Lin S, He HF. Particulate matter 2.5 accelerates aging: exploring cellular senescence and age-related diseases. Ecotoxicol Environ Saf. 2024;284:116920.
4. Agusti A, Hogg JC. Update on the pathogenesis of chronic obstructive pulmonary disease. N Engl J Med. 2019;381:1248–56.
5. Ito K, Barnes PJ. COPD as a disease of accelerated lung aging. Chest. 2009;135:173–80.
6. MacNee W. Accelerated lung aging: a novel pathogenic mechanism of chronic obstructive pulmonary disease (COPD). Biochem Soc Trans. 2009;37:819–23.
7. Suki B, Parameswaran H, Alves C, Araújo AD, Bartolák-Suki E. Cellular and extracellular homeostasis in fluctuating mechanical environments. In: Zhang Y, editor. Multi-scale extracellular matrix mechanics and mechanobiology. Cham: Springer; 2020. p. 83–121.
8. Tschumperlin DJ. Matrix, mesenchyme, and mechanotransduction. Ann Am Thorac Soc. 2015;12(Suppl 1):S24–9.
9. Childs BG, Durik M, Baker DJ, van Deursen JM. Cellular senescence in aging and age-related disease: from mechanisms to therapy. Nat Med. 2015;21:1424–35.
10. Cho SJ, Stout-Delgado HW. Aging and lung disease. Annu Rev Physiol. 2020;82:433–59.
11. Sohal RS, Weindruch R. Oxidative stress, caloric restriction, and aging. Science. 1996;273:59–63.
12. Suki B, Bartolák-Suki E. Biomechanics of the aging lung parenchyma. In: Derby B, Ahktar R, editors. Mechanical properties of aging soft tissues. Springer; 2015. p. 95–133.
13. Cavalcante FS, Ito S, Brewer KK, Sakai H, Alencar AM, Almeida MP, Andrade JS Jr, Majumdar A, Ingenito EP, Suki B. Mechanical interactions between collagen and proteoglycans: implications for the stability of lung tissue. J Appl Physiol. 2005;98:672–9.
14. Bellmunt MJ, Portero M, Pamplona R, Muntaner M, Prat J. Age-related fluorescence in rat lung collagen. Lung. 1995;173:177–85.
15. Sobin SS, Fung YC, Tremer HM. Collagen and elastin fibers in human pulmonary alveolar walls. J Appl Physiol. 1985;1988(64):1659–75.
16. Bou Jawde S, Takahashi A, Bates JHT, Suki B. An analytical model for estimating Alveolar Wall elastic moduli from lung tissue uniaxial stress-strain curves. Front Physiol. 2020;11:121.
17. Lee HY, Han L, Roughley PJ, Grodzinsky AJ, Ortiz C. Age-related nanostructural and nanomechanical changes of individual human cartilage aggrecan monomers and their glycosaminoglycan side chains. J Struct Biol. 2013;181:264–73.
18. Schachtschabel DO, Wever J. Age-related decline in the synthesis of glycosaminoglycans by cultured human fibroblasts (WI-38). Mech Ageing Dev. 1978;8:257–64.
19. Konno K, Arai H, Motomiya M, Nagai H, Ito M, Sato H, Satoh K. A biochemical study on glycosaminoglycans (mucopolysaccharides) in emphysematous and in aged lungs. Am Rev Respir Dis. 1982;126:797–801.
20. Schmid K, Grundboeck-Jusco J, Kimura A, Tschopp FA, Zollinger R, Binette JP, Lewis W, Hayashi S. The distribution of the glycosaminoglycans in the anatomic components of the lung and the changes in concentration of these macromolecules during development and aging. Biochim Biophys Acta. 1982;716:178–87.
21. Andreotti L, Bussotti A, Cammelli D, Aiello E, Sampognaro S. Connective tissue in aging lung. Gerontology. 1983;29:377–87.
22. Verbeken EK, Cauberghs M, Mertens I, Clement J, Lauweryns JM, Van de Woestijne KP. The senile lung. Comparison with normal and emphysematous lungs. 1. Structural aspects. Chest. 1992;101:793–9.
23. Avery NC, Bailey AJ. Enzymic and non-enzymic cross-linking mechanisms in relation to turnover of collagen: relevance to aging and exercise. Scand J Med Sci Sports. 2005;15:231–40.

24. Sugihara T, Martin CJ, Hildebrandt J. Length-tension properties of alveolar wall in man. J Appl Physiol. 1971;30:874–8.
25. Fung YC. Elasticity of soft tissues in simple elongation. Am J Phys. 1967;213:1532–44.
26. Martin CJ, Chihara S, Chang DB. A comparative study of the mechanical properties in aging alveolar wall. Am Rev Respir Dis. 1977;115:981–8.
27. Sicard D, Haak AJ, Choi KM, Craig AR, Fredenburgh LE, Tschumperlin DJ. Aging and anatomical variations in lung tissue stiffness. Am J Physiol Lung Cell Mol Physiol. 2018;314:L946–55.
28. Lai-Fook SJ, Hyatt RE. Effects of age on elastic moduli of human lungs. J Appl Physiol. 1985;2000(89):163–8.
29. Suki B, Herrmann J, Bates JHT. An analytic model of tissue self-healing and its network implementation: application to fibrosis and aging. Front Physiol. 2020;11:583024.
30. Mays PK, McAnulty RJ, Campa JS, Laurent GJ. Age-related changes in collagen synthesis and degradation in rat tissues. Importance of degradation of newly synthesized collagen in regulating collagen production. Biochem J. 1991;276(Pt 2):307–13.
31. Pierce JA, Resnick H, Henry PH. Collagen and elastin metabolism in the lungs, skin, and bones of adult rats. J Lab Clin Med. 1967;69:485–93.
32. Poole A, Myllyla R, Wagner JC, Brown RC. Collagen biosynthesis enzymes in lung tissue and serum of rats with experimental silicosis. Br J Exp Pathol. 1985;66:567–75.
33. Gershlak JR, Black LD 3rd. Beta 1 integrin binding plays a role in the constant traction force generation in response to varying stiffness for cells grown on mature cardiac extracellular matrix. Exp Cell Res. 2015;330:311–24.
34. Giangreco A, Goldie SJ, Failla V, Saintigny G, Watt FM. Human skin aging is associated with reduced expression of the stem cell markers beta1 integrin and MCSP. J Invest Dermatol. 2010;130:604–8.
35. Hynes RO. Integrins: bidirectional, allosteric signaling machines. Cell. 2002;110:673–87.
36. Park JW, Han SB, Hah J, Lee G, Kim JK, Kim SH, Kim DH. Biological aging modulates cell migration via lamin A/C-dependent nuclear motion. Micromachines (Basel). 2020;11:801.
37. Jia H, Chang Y, Chen Y, Chen X, Zhang H, Hua X, Xu M, Sheng Y, Zhang N, Cui H, et al. A single-cell atlas of lung homeostasis reveals dynamic changes during development and aging. Commun Biol. 2024;7:427.
38. Suki B, Bates JHT, Bartolak-Suki E. Remodeling of the aged and emphysematous lungs: roles of microenvironmental cues. Compr Physiol. 2022;12:3559–74.
39. Divo MJ, Celli BR, Poblador-Plou B, Calderon-Larranaga A, de-Torres JP, Gimeno-Feliu LA, Berto J, Zulueta JJ, Casanova C, Pinto-Plata VM, et al. Chronic obstructive pulmonary disease (COPD) as a disease of early aging: evidence from the EpiChron cohort. PLoS One. 2018;13: e0193143.
40. Albert R, Barabasi AL. Statistical mechanics of complex networks. Rev Mod Phys. 2002;74: 47–97.
41. De Lucia M, Bottaccio M, Montuori M, Pietronero L. Topological approach to neural complexity. Phys Rev E Stat Nonlinear Soft Matter Phys. 2005;71:016114.
42. Vural DC, Morrison G, Mahadevan L. Aging in complex interdependency networks. Phys Rev E Stat Nonlinear Soft Matter Phys. 2014;89:022811.
43. Kirkwood TB, Kowald A. Network theory of aging. Exp Gerontol. 1997;32:395–9.
44. Kirkwood TB. Evolution of ageing. Nature. 1977;270:301–4.
45. Borklu Yucel E, Ulgen KO. A network-based approach on elucidating the multi-faceted nature of chronological aging in S. cerevisiae. PLoS One. 2011;6:e29284.
46. Suki B, Bates JH, Frey U. Complexity and emergent phenomena. In: Fredberg JJ, Sieck GC, Gerthoffer WT, editors. Comprehensive physiology, the respiratory system, respiration mechanics: organ, cell, molecule, vol. 1. Wiley-Blackwell; 2011. p. 995–1029.
47. Tschumperlin DJ, Boudreault F, Liu F. Recent advances and new opportunities in lung mechanobiology. J Biomech. 2010;43:99–107.

48. Waters CM, Roan E, Navajas D. Mechanobiology in lung epithelial cells: measurements, perturbations, and responses. Compr Physiol. 2012;2:1–29.
49. Erdős P, Rényi A. On random graphs. Publicationes Mathematicae Debrecen. 1959;6:290–7.
50. Barabasi AL, Albert R. Emergence of scaling in random networks. Science. 1999;286:509–12.
51. Myerson J, Hale S, Wagstaff D, Poon LW, Smith GA. The information-loss model: a mathematical theory of age-related cognitive slowing. Psychol Rev. 1990;97:475–87.
52. Frank NR, Mead J, Ferris BG Jr. The mechanical behavior of the lungs in healthy elderly persons. J Clin Invest. 1957;36:1680–7.
53. Mahler DA, Rosiello RA, Loke J. The aging lung. Clin Geriatr Med. 1986;2:215–25.
54. Knudson RJ, Clark DF, Kennedy TC, Knudson DE. Effect of aging alone on mechanical properties of the normal adult human lung. J Appl Physiol Respir Environ Exerc Physiol. 1977;43:1054–62.
55. Turner JM, Mead J, Wohl ME. Elasticity of human lungs in relation to age. J Appl Physiol. 1968;25:664–71.
56. Park S, Jung WH, Pittman M, Chen J, Chen Y. The effects of stiffness, fluid viscosity, and geometry of microenvironment in homeostasis, aging, and diseases: a brief review. J Biomech Eng. 2020;142:100804.
57. Kim J, Heise RL, Reynolds AM, Pidaparti RM. Aging effects on airflow dynamics and lung function in human bronchioles. PLoS One. 2017;12:e0183654.
58. Weale RA. Have human biological functions evolved in support of a life-span? Mech Ageing Dev. 1993;69:65–77.
59. Burtscher M. Exercise limitations by the oxygen delivery and utilization systems in aging and disease: coordinated adaptation and deadaptation of the lung-heart muscle axis – a mini-review. Gerontology. 2013;59:289–96.
60. Kiourtis C, Terradas-Terradas M, Gee LM, May S, Georgakopoulou A, Collins AL, O'Sullivan ED, Baird DP, Hassan M, Shaw R, et al. Hepatocellular senescence induces multi-organ senescence and dysfunction via TGFbeta. Nat Cell Biol. 2024;26:2075–83.
61. Wu M, Fannin J, Rice KM, Wang B, Blough ER. Effect of aging on cellular mechanotransduction. Ageing Res Rev. 2011;10:1–15.
62. Nin N, Lorente JA, De Paula M, Fernandez-Segoviano P, Penuelas O, Sanchez-Ferrer A, Martinez-Caro L, Esteban A. Aging increases the susceptibility to injurious mechanical ventilation. Intensive Care Med. 2008;34:923–31.
63. Fabry B, Fredberg JJ. Mechanotransduction, asthma, and airway smooth muscle. Drug Discov Today Dis Models. 2007;4:131–7.
64. Dellaca RL, Aliverti A, Lo Mauro A, Lutchen KR, Pedotti A, Suki B. Correlated variability in the breathing pattern and end-expiratory lung volumes in conscious humans. PLoS One. 2015;10:e0116317.
65. Suki B, Parameswaran H, Imsirovic J, Bartolak-Suki E. Regulatory roles of fluctuation-driven mechanotransduction in cell function. Physiology (Bethesda). 2016;31:346–58.
66. Hudgel DW, Devadatta P, Hamilton H. Pattern of breathing and upper airway mechanics during wakefulness and sleep in healthy elderly humans. J Appl Physiol. 1985;1993(74):2198–204.
67. Moshier JA, Cornell T, Majumdar AP. Expression of protease genes in the gastric mucosa during aging. Exp Gerontol. 1993;28:249–58.
68. Weyand CM, Goronzy JJ. Aging of the immune system. Mechanisms and therapeutic targets. Ann Am Thorac Soc. 2016;13(Suppl 5):S422–8.

Part III
The Future of Lung Modeling

Chapter 13
The Utility and Future of Computational Modeling

Abstract Computational modeling has become fundamental to the advancement of biology and medicine. Digital computers now make high-fidelity models of virtually unlimited complexity possible. However, as George Box cautioned, "All models are wrong, but some are useful", which implies that the ultimate usefulness of a model manifests when it is replaced by an even more useful one. Future models will become increasingly complex, exhibiting new emergent behaviors that reflect multi-scale structure and function in biologic systems. Future models, perhaps aided by machine learning, will also become increasingly personalized, facilitating patient-specific disease management. We can thus expect computational modeling to continue to advance the science of lung pathophysiology and the practice of clinical pulmonary medicine.

Keywords Model complexity · Prediction · Model usefulness · Validation · Artificial intelligence · Personalized medicine

13.1 How Far Has Modeling Come?

Scientists have always created theoretical models of natural systems in order to efficiently encapsulate understanding and make predictions. Many of these models, dating back at least as far as the Ancient Greeks, have incorporated mathematics [1]. Indeed, starting with Newton, the development of increasingly sophisticated mathematical models has marked the progress of the physical and engineering sciences for hundreds of years. More recently, computational modeling has also become fundamental to the advancement of biology [2] and medicine [3].

Throughout most of scientific history, the behavior of mathematical models could only be determined analytically, meaning it was necessary to express their predictions in terms of explicit equations. This led to many triumphs of analysis, but it was fundamentally limited by the fact that most sets of model equations cannot be solved in closed form. And even when they can, calculating the model predictions from the resulting equations is often challenging either because of the nonlinear nature of the equations or due to irregular boundary conditions. This situation changed suddenly in the middle of the twentieth century with the arrival of the

B. Suki, J. H. T. Bates, *Mathematical Modeling of the Healthy and Diseased Lung*,
https://doi.org/10.1007/978-3-031-95322-4_13

digital computer. In a matter of a few short decades, the analytical skills needed to solve mathematical equations became largely superseded by the very different skills needed to solve them numerically. Today, it is possible to calculate the predictions of models of virtually unlimited complexity using supercomputing tools. We have thus moved from the era of the purely mathematical model to that of the computational model.

One might be tempted to conclude that since computational modeling can be applied to virtually everything, it should be. Such a conclusion must be tempered by the caution of the British-born American statistician George Box, "All models are wrong, but some are useful". Usefulness means different things to different people, and often only becomes fully apparent once a model has been constructed and put to the test. Generally speaking, however, a model is useful if it has implications beyond its own existence, such as providing scientific insight or allowing predictions that matter in some definable way. Computational modeling that is not useful is, at best, a waste of the modeler's time, and at worst it is a waste of everyone's time. Accordingly, the primary purpose of the preceding chapters was to demonstrate that computational models have proven useful in the study of lung physiology and disease. Every indication is that this will continue into the future, likely in ways we cannot yet imagine.

Computational model development usually begins with an idea, often expressed verbally, about how a system is thought to work, or about cause-and-effect relationships. The idea must then be formulated in terms of mathematical or logical expressions that define how the specific components of the model interact among themselves and with the environment. As pointed out in Chap. 1, these components must collectively represent the functionally important aspects of the system being modeled. Just as important, a model must avoid inclusion of unnecessary detail that does not contribute to its usefulness, something that relies to a large extent on the experience and judgment of the modeler. Once constructed, the model must first be checked to make sure it predicts behavior that is already well known, i.e., that it performs as expected. As also pointed out in Chap. 1, this process is commonly referred to as model validation, a perhaps unfortunate term that has connotations of correctness, in contradiction to the wisdom of George Box.

Making correct predictions of known behavior, however, is no guarantee that a model will predict unknown behavior correctly. It merely means there is no immediate reason to eliminate the model from further consideration. Indeed, a model is nothing more than an organized hypothesis about how a system works, and the classic scientific method holds that the primary job of a scientist confronted with a hypothesis is to try to refute it. This means, taking George Box's idea to the limit, that the ultimate usefulness of a model manifests when it is replaced by an even more useful model. That is, regardless of how well a model performs during initial testing, there will eventually come a set of circumstances under which it fails. Usefulness thus depends on the circumstances, and progress occurs when circumstances are identified under which an accepted model is no longer useful.

Within its circumstantial domain of usefulness, a computational model can provide a number of vitally important functions. Among the most important for biology is the capacity to serve as a virtual laboratory within which numerical experiments can be performed that cannot possibly be undertaken in the experimental laboratory. For example, it is a practical impossibility to administer a smooth muscle agonist to the lungs so that only a single generation of the airway tree is affected, yet knowing how constriction of individual generations affects overall lung function is crucial to understanding airways hyperresponsiveness in asthma (Chap. 7). High-fidelity anatomically based lung models can address this question with considerable reliability. Monte-Carlo simulations using stochastic airway models can even provide the statistical reliability of such estimates. Similarly, it is currently impossible to detect the rupture of a single alveolar wall due to intense coughing in a patient with chronic obstructive pulmonary disease, yet this tiny event can be the start of a cascade involving altered mechanotransduction, gene expression, and eventual release of inflammatory mediators that cause an exacerbation of disease. Computational modeling of force balance in the parenchyma provides estimates of the conditions under which this insidious sequence of events begins (Chap. 9). Another example is the marriage of physics-based network models of tissue mechanics with agent-based models of cellular behavior that leads to an understanding of how subpleural honeycombing might emerge in idiopathic pulmonary disease (Chap. 8). *In silico* experimentation with computational models can thus provide insight into questions that are extremely difficult or even impossible to address experimentally. Perhaps even more important for ultimate scientific truth, *in silico* experimentation in the virtual laboratory can help identify the key experiments that need to be conducted in the experimental laboratory. The result can be huge savings in time and resources, with maximum positive impact on society.

13.2 Where Is Modeling Going?

Much of modeling is about prediction, but it is also worth trying to make predictions about modeling itself, even though this is much more difficult. As the great Danish physicist Niels Bohr famously said, "Prediction is very difficult, especially if it's about the future". One thing does seem certain, though. With the ever-increasing speed and memory of parallel computer networks, computational models of all biological systems, including the lung, will increase in scale and complexity. This will allow physiology and pathobiology to be described in unprecedented detail. Such descriptions will include representations of biologic process taking place over increasingly wide ranges of length and time scale. At the same time, care will have to be taken to avoid the allure of increased detail and complexity simply because it is possible. The design of a computational model should always be appropriate to the question being addressed and the data against which it will be judged, where the meaning of "appropriate" in this context is invariably somewhat subjective and should always be justified.

Current trends also portend an increasing focus on the personalization of computational models. Advances in high-resolution dynamic imaging and molecular phenotyping will allow the unique details of specific individuals to be represented in computational models. High-fidelity models of this nature might eventually represent individual physiology in sufficient detail to be truly worthy of the title "digital twin". Such models could be used, for example, to explore personalized approaches to disease treatment and prognosis, and to address questions such as why patients vary in their sensitivities to different environmental exposures, or why some patients succumb to a life-threatening condition such as acute respiratory distress syndrome (Chap. 10) while others in apparently similar condition go on to recover. Computational models are already being used to predict outcomes of medical interventions such as bronchial thermoplasty to treat asthma [4] and lung volume reduction surgery to improve quality of life in emphysema [5], and we can expect this to become increasingly important for medical treatment planning in the future. To address such issues, it will be necessary to incorporate increasing detail concerning feedback circuitry at the molecular, cellular, tissue, whole organ, and organismal levels. Despite decades of intense research, many diseases of modern society remain idiopathic and untreatable. Computational modeling is expected to play an ever more important role in revealing their ultimate causes as well as uncovering new therapeutic pathways.

With regard to therapeutic pathways, the prediction of drug delivery and uptake is another key area in which computational models of the future are poised to make important contributions. Computational fluid dynamics, for example, is already being used to simulate anatomically accurate maps of inhaled aerosol deposition along the walls of the conducting airways and in the acini [6]. As 3D medical imaging increases in resolution, we can expect these simulations to become increasingly accurate, particularly in the distal reaches of the lung where aerosol transport occurs predominately by diffusion. Inhaled medications will likely be used increasingly for non-pulmonary related diseases, the lung merely providing access to the rest of the body for medications such as vaccines [7] and insulin [8]. Understanding precisely how inhaled medications reach the vast gas-exchanging surface of the lung will be critical for optimizing systemic delivery. Once deposited on the pulmonary epithelium, delivered agents must then be absorbed into the pulmonary tissues on their way to the rest of the body. Simulating this process accurately will require being able to mimic fluid-structure interactions during breathing so that absorption rates through a cyclically changing tissue interface can be calculated. Given the multi-scale complexity of the structures involved, this in turn will necessitate the use of supercomputing.

The computational models of the future will manifest increasing degrees of emergent biologic behavior, building on phenomena such as network effects, percolation and dynamic attractors that are already recapitulated in existing models. Indeed, chronic disease, and even life itself, represent ultimate examples of emergent phenomena. An example of emergent behavior with crucial implications for medicine relates to the existence of bifurcations. Personalized management of complex disease may involve a patient who is confronting a bifurcation between recovery and deterioration. Being able to predict which is the more likely outcome with some degree of reliability is obviously of extreme clinical importance. This situation can arise, for example, in patients with acute respiratory distress syndrome (Chap. 10) where the difference between survival and death can rest on immeasurably subtle factors. Similarly, minor differences in the nature of parenchymal tissue damage can apparently determine whether an ostensibly reparative fibrotic response will resolve back to normal or lead to permanent pulmonary fibrosis (Chap. 8). Outcome prediction in situations such as these is currently extremely difficult if not impossible but is no doubt a direction in which we can expect modeling to take us.

It is both significant and prescient that the 2024 Nobel Prize in chemistry was awarded for the use of artificial intelligence to solve the long-standing and notoriously difficult problem of protein folding. Machine learning seems set to play a rapidly increasing role in all branches of science, especially those involving computational modeling. Indeed, it has already started to make its mark in pulmonary radiology [9–11], pathology [12, 13] and diagnosis [14, 15]. Machine learning has also been used recently to identify anatomical features in medical images such as segmenting the lung from the chest wall in CT images [16]. Generative artificial intelligence may eventually be used to construct realistic lung anatomies that can then be used to predict realistic function. On the one hand, artificial intelligence techniques such as deep learning, impressive though they are, might be viewed as nothing more than ultra-nonlinear regression on a massive scale. This raises the question as to whether models based on empirical relationships can ever replace those based on equations embodying underlying mechanisms. On the other hand, machine learning has led to the extraction of relationships embedded within vast data sets that would otherwise have remained hidden. Pattern recognition is suddenly no longer the unique purview of the human mind, which simply does not have the data processing and memory recall capacities to continue to compete in this domain. Even technical writing may soon move from the cradle of the human imagination to instead be nurtured within the unimaginably large knowledge repositories of large language models.

Does this portend the eventual end of human intellectual input to the scientific enterprise? It seems difficult to believe that human intuition and creativity at its most profound could ever be superseded by the workings of a machine. Could an algorithm, no matter how complex, ever have come up with Darwin's theory of evolution or Einstein's theory of general relativity *de novo*? These are deeply unsettling questions against which we can comfort ourselves with the conviction that even if such a world is on its way, it is surely a long way off. A perhaps more immediate cause for concern is what happens as the mental problem-solving skills that humans have relied on since the dawn of time, and that have reached such pinnacles of refinement during the scientific age, start to atrophy from lack of use as computers increasingly obviate their need. When it comes to computational modeling, artificial intelligence should extend our capabilities as humans, not replace them.

In this spirit, we finish by considering how the computational modeling of the future can advance the science of lung pathophysiology and the practice of clinical pulmonary medicine for the benefit of society at large. As models become increasingly accurate representations of reality that can be tailored to the individual rather than the population, they will form the foundations upon which personalized predictions of future health and disease therapies can be developed. As electronic medical records become more extensive and ubiquitous, and the security problems surrounding access to their data are overcome, these powerful future models will be informed by increasingly rich data sets that contain the information needed for precise personalized healthcare. Such models will also increasingly replace the use of human and animal experimentation in the advancement of biological science, drug discovery, and clinical trials, with commensurate savings in time, effort, money and suffering.

13.3 Conclusions

This ends our guided tour of the world of computational modeling as it applies to the lung in health and disease. Our goal was to bring together a broad range of modeling methods and techniques, and to show how they have advanced our understanding of a crucial branch of physiology and medicine. This work is far from done, so our hope is that the material covered in this book will inform and stimulate future modeling researchers to take the field forward in ways that are sure to surprise us all, for the benefit of everyone.

References

1. Strogatz SH. Infinite powers: how calculus reveals the secrets of the universe. New York: HarperCollins Publisher; 2019. p. 400.
2. Brodland GW. How computational models can help unlock biological systems. Semin Cell Dev Biol. 2015;47–48:62–73.
3. Collin CB, Gebhardt T, Golebiewski M, Karaderi T, Hillemanns M, Khan FM, Salehzadeh-Yazdi A, Kirschner M, Krobitsch S, Eu-Stands Pm C, Kuepfer L. Computational models for clinical applications in personalized medicine-guidelines and recommendations for data integration and model validation. J Pers Med. 2022;12:166.
4. Donovan GM, Noble PB, Langton D. Therapeutic response to bronchial thermoplasty: toward feasibility of patient selection based on modeling predictions. J Appl Physiol. 1985;2022(133): 1341–8.
5. Mondonedo JR, Suki B. Predicting structure-function relations and survival following surgical and Bronchoscopic lung volume reduction treatment of emphysema. PLoS Comput Biol. 2017;13:e1005282.
6. Longest PW, Bass K, Dutta R, Rani V, Thomas ML, El-Achwah A, Hindle M. Use of computational fluid dynamics deposition modeling in respiratory drug delivery. Expert Opin Drug Deliv. 2019;16:7–26.
7. Ye T, Jiao Z, Li X, He Z, Li Y, Yang F, Zhao X, Wang Y, Huang W, Qin M, et al. Inhaled SARS-CoV-2 vaccine for single-dose dry powder aerosol immunization. Nature. 2023;624: 630–8.
8. McMahon GT, Arky RA. Inhaled insulin for diabetes mellitus. N Engl J Med. 2007;356:497–502.

9. Masquelin AH, Cheney N, Kinsey CM, Bates JHT. Wavelet decomposition facilitates training on small datasets for medical image classification by deep learning. Histochem Cell Biol. 2021;155:309–17.
10. Tan H, Bates JHT, Matthew Kinsey C. Discriminating TB lung nodules from early lung cancers using deep learning. BMC Med Inform Decis Mak. 2022;22:161.
11. Singla S, Gong M, Riley C, Sciurba F, Batmanghelich K. Improving clinical disease subtyping and future events prediction through a chest CT-based deep learning approach. Med Phys. 2021;48:1168–81.
12. Diosdado J, Gilabert P, Segui S, Borrego H. LungHist700: a dataset of histological images for deep learning in pulmonary pathology. Sci Data. 2024;11:1088.
13. Viswanathan VS, Toro P, Corredor G, Mukhopadhyay S, Madabhushi A. The state of the art for artificial intelligence in lung digital pathology. J Pathol. 2022;257:413–29.
14. Dack E, Christe A, Fontanellaz M, Brigato L, Heverhagen JT, Peters AA, Huber AT, Hoppe H, Mougiakakou S, Ebner L. Artificial intelligence and interstitial lung disease: diagnosis and prognosis. Investig Radiol. 2023;58:602–9.
15. Hussain A, Marlowe S, Ali M, Uy E, Bhopalwala H, Gullapalli D, Vangara A, Haroon M, Akbar A, Piercy J. A systematic review of artificial intelligence applications in the management of lung disorders. Cureus. 2024;16:e51581.
16. Gite S, Mishra A, Kotecha K. Enhanced lung image segmentation using deep learning. Neural Comput Appl. 2022;35:1–15.

Index

B. Suki, J. H. T. Bates, *Mathematical Modeling of the Healthy and Diseased Lung*, https://doi.org/10.1007/978-3-031-95322-4

GPSR Compliance
The European Union's (EU) General Product Safety Regulation (GPSR) is a set of rules that requires consumer products to be safe and our obligations to ensure this.

If you have any concerns about our products, you can contact us on

ProductSafety@springernature.com

In case Publisher is established outside the EU, the EU authorized representative is:

Springer Nature Customer Service Center GmbH
Europaplatz 3
69115 Heidelberg, Germany

www.ingramcontent.com/pod-product-compliance
Lightning Source LLC
Chambersburg PA
CBHW070924090925
32286CB00003B/112
* 9 7 8 3 0 3 1 9 5 3 2 1 7 *